Energy, Environment, and Sustainability

Series editors

This books series publishes cutting edge monographs and professional books focused on all aspects of energy and environmental sustainability, especially as it relates to energy concerns. The Series is published in partnership with the International Society for Energy, Environment, and Sustainability. The books in these series are editor or authored by top researchers and professional across the globe. The series aims at publishing state-of-the-art research and development in areas including, but not limited to:

- Renewable Energy
- Alternative Fuels
- Engines and Locomotives
- Combustion and Propulsion
- Fossil Fuels
- Carbon Capture
- Control and Automation for Energy
- Environmental Pollution
- Waste Management
- Transportation Sustainability

More information about this series at http://www.springer.com/series/15901

Anirudh Gautam · Sudipta De
Atul Dhar · Jai Gopal Gupta
Ashok Pandey
Editors

Sustainable Energy and Transportation

Technologies and Policy

Springer

Editors
Anirudh Gautam
Ministry of Railways, Government of India
SRESTHA
Lucknow, Uttar Pradesh
India

Sudipta De
Jadavpur University
Kolkata, West Bengal
India

Atul Dhar
School of Engineering
Indian Institute of Technology Mandi
Kamand, Mandi, Himachal Pradesh
India

Jai Gopal Gupta
Department of Mechanical Engineering
Government Women Engineering College
Ajmer
Ajmer, Rajasthan
India

Ashok Pandey
CSIR-Indian Institute of Toxicology
Research
Lucknow, Uttar Pradesh
India

ISSN 2522-8366 ISSN 2522-8374 (electronic)
Energy, Environment, and Sustainability
ISBN 978-981-10-7508-7 ISBN 978-981-10-7509-4 (eBook)
https://doi.org/10.1007/978-981-10-7509-4

Library of Congress Control Number: 2017959547

Printed on acid-free paper

This Springer imprint is published by Springer Nature
The registered company is Springer Nature Singapore Pte Ltd.
The registered company address is: 152 Beach Road, #21-01/04 Gateway East, Singapore 189721, Singapore

Preface

With limited natural resources and ever-increasing energy demand, the energy sector all over the world is facing the formidable challenge of maintaining the living standard of increasing population. The human civilization observed rapid growth through industrial revolution. The key input for this growth was energy, and the source of that energy was mostly fossil fuels. With the threat of climate change, the use of this convenient form of energy also has to be compromised. Thus, finding sustainable options for future energy is an imperative need of the present. However, the use of energy in modern world is so diverse that each sector has to address this challenge of energy sustainability in a sector-specific but integrated way. Transport for mobility is a sector that consumes significant energy. Moreover, existing technologies in this sector have also several limitations regarding environmental impact. Developing sustainable energy solutions for transport is an area of significant current interest. Transformation from the existing to a sustainable option for energy conversion and use involves not only new technology developments but also suitable policy support for this transformation through transition. Energy policy thus plays a critical role for the success of implementation of sustainable energy options in real life. Several socio-economic issues have to be accessed and addressed with proper policy. In this volume, current developments of sustainable transport and energy policy have been included. The volume will be useful for postgraduate students, industry professionals as well as administrators and policy makers at different levels who are involved in implementing possible future sustainable options in the transport sector. Different policies that are either adopted or may be adopted to support sustainable transition are also included in this volume.

This work is an attempt of putting down the integrated thought process for ensuring fulfilment of energy requirements in a sustainable manner by giving consideration to various energy usage sectors and applicable technologies in those sectors. This book covers the sustainability issues in diverse fields such as smart city planning, design of transport systems in transport cities, sources of energy for mobility, thought on individual consumption for ensuring the sustainability of energy needs, technologies for emission reduction for both mobility and stationary applications. For stationary applications, it deals with case studies related to energy consumption in the manufacturing sector as well as domestic energy requirement. Along with description of technology choices, it also discusses various distribution and policy aspects related to power sectors and sources of energy such as coal and biomass.

While there are several books on smart cities or on renewable energy, or sustainability, there is hardly any comprehensive book covering topics related to technological details, planning and case studies of sustainability issues such as pollution reduction and various renewable energy options in a single book. 'Sustainability' means improving our ability to move faster and safer in such a way that it does not compromise the ability of our children to do so in future. Sustainable transport focuses mainly on conservation of resources and provision of pollution-free and safe environment. With increasing pollution in the environment, there is a need of sustainable generation and utilization of energy. Alternative renewable energy sources in the form of vegetable oil, biodiesel, biogas, producer gas and alcohols have good prospects to replace or supplement fossil fuel. Stressing the importance of alternative fuel sources for diesel engines, the present book focuses on identifying some of these arrangements on engine performance and emission characteristics.

This is a state-of-the-art book focusing on technological, managerial and policy aspects of sustainable energy production and consumption, which deals with issues such as need and planning of smart cities, alternative transport fuel options, planning of sustainable power production, pollution control technologies. Chapters focus on the changing face of transportation towards sustainability, sustainable transport solutions for the concept of smart city, role of electric vehicles in future road transport, comparative evaluation of different fuels for transports, various energy policies related to waste heat recovery, investment in clean energy and need for integration of multidimensionality and co-benefits.

We would like to place on record special thanks to all the authors for submitting their valuable work at short notice. We are also thankful to Dr. Dhiraj Patil and Dr. Shyamashree Dasgupta (IIT Mandi) for their help in the review process, which helped us in completing review of the monograph in the specified time. We hope that researchers in various fields related to sustainable energy policy such as planning of sustainable cities, waste energy recovery, efficiency improvement through planning and management of consumption, wastewater management sustainability, renewable fuel production and utilization for stationary as well as mobile applications will find this book useful in dealing with energy crises and environment degradation problems.

Lucknow, India — Anirudh Gautam
Kolkata, India — Sudipta De
Mandi, India — Atul Dhar
Ajmer, India — Jai Gopal Gupta
Mohali, India — Ashok Pandey

Contents

Editors and Contributors

About the Editors

Dr. Anirudh Gautam is from Indian Railway Service of Mechanical Engineers. He did his Mechanical and Electrical Engineering from SCRA scheme of Jamalpur and then did his Master's in Quality Management from BITS, Pilani. He has served various positions in Indian Railways. He later served in various positions at Diesel Locomotive Works including engine manufacturing, engine design and transfer of technology. He built India's first 4000-hp diesel locomotive indigenously under ToT with the EMD General Motors of the USA, for which he was honoured with national award by the Railway Minister. He has also done Master's in Engineering in Engine Design from University of Wisconsin–Madison, USA. He has carried out and successfully completed various R&D projects at RDSO at Engine Development Directorate including developing the first electronic fuel injection locomotive and the first common rail electronic fuel injection locomotive. He did his Ph.D. under the guidance of Prof. Avinash Kumar Agarwal at IIT Kanpur. He has applied for 2 patents and has published about 10 technical papers and 50 technical reports. He is presently the Executive Director, SRESTHA, Indian Railways.

Prof. Dr. Sudipta De is Professor at the Mechanical Engineering Department, Jadavpur University, India. He received his Ph.D. degree from Indian Institute of Technology (IIT), Kharagpur. He was a Guest Researcher at the Department of Energy Sciences, Lund University, Sweden, for more than 1 year. He is an Indian national expert on 'Clean and Alternative Technology' of the Ministry of Environment and Forest, Government of India. He was nominated Senior Scientist by Indian National Science Academy (INSA), New Delhi, to Germany in the field of sustainable energy under international bilateral exchange programme of the Academy. He was the selected faculty of Erasmus Mundus External Cooperation Window with specialization in sustainable energy engineering. He has visited and delivered invited lectures in many programmes/institutes including Indian National Clean Coal Technology Mission meeting in Hyderabad; Technical University of Berlin and Munich, Germany; Royal Institute of Technology (KTH) Stockholm; and University of Stavanger, Norway. He received his research funding from different institutes including UGC, DST, Government of India, DFG-Germany, Swedish Research Council, Research Council of Norway. He is/was member of several technical committees including that of Power and Energy Systems of The International Association of Science and Technology for Development (IASTED), Canada; international energy initiative 'Explore Energy' by Royal Institute of Technology, Sweden; Energy and Environment Committee of the Bengal Chamber of Commerce and Industry (oldest Chamber in India) and many others. He has published several international journal papers and three international book/invited chapters. He is also an advisory editorial board member of a book series by CRC press on 'Sustainable Energy Developments'. He has developed a Web-based learning module on 'sustainable energy' under the Indo-EU project, 'E-QUAL'. Presently, he is coordinating an Indo-Norwegian collaboration project of 3-year duration and is a member of another EU project with Finland and Germany.

Dr. Atul Dhar is working as Assistant Professor at Indian Institute of Technology Mandi since 2013. He completed M. Tech. and Ph.D. from Department of Mechanical Engineering, Indian Institute of Technology Kanpur, in 2013. He graduated in Mechanical Engineering from Harcourt Butler Technological Institute, Kanpur, in 2004. He also worked at Mahindra & Mahindra, Automotive Sector in 2006 for getting exposure to industrial working environment. He was awarded the Erasmus Mundus Fellowship of European Union for pursuing postdoctoral research at Ecole Centrale de Nantes, France, in 2013. He has been honoured with young scientist award from International Society for Energy, Environment and Sustainability in 2015. His areas of interest include reciprocating internal combustion engines, emission control technologies, alternative fuels and lubricating oil tribology. He has co-authored more than 30 international peer-reviewed journal publications.

Dr. Jai Gopal Gupta is working as Head at Department of Mechanical Engineering, Govt. Women Engineering College, Ajmer, Rajasthan, India. He completed his Ph.D. on 'Combustion, Material Compatibility and Engine Tribology Investigations of a Karanja Biodiesel Fuelled Turbo-charged Diesel Engine' from Engine Research Lab, IIT Kanpur. His areas of interest are alternative fuels, biofuels, spray characterization and engine performance characterization. He did his M.Tech. from Malviya National Institute of Technology, Jaipur, 302017, Rajasthan, India. His papers have been published in international journals such as Progress in Energy and Combustion Science, Applied Energy, Energy and Conversion Management, and also in international conferences such as ASME proceeding, SAE technical paper. He has worked with Combustion Engine and Energy Conversion (CEnEC) laboratory, Hanyang University, Seoul, South Korea, during internship under the Indo-Korean Research Internship (IKRI) Programme sponsored by Department of Science and Technology, Government of India.

Prof. Ashok Pandey, D.Phil., FBRS, FNASc, FIOBB, FISEES, FAMI is Eminent Scientist at the Center of Innovative and Applied Bioprocessing, Mohali, and former Chief Scientist and Head of Biotechnology Division at CSIR's National Institute for Interdisciplinary Science and Technology, Trivandrum. He is Adjunct Professor at MACFAST, Thiruvalla, Kerala, and Kalasalingam University, Krishnankoil, Tamil Nadu. His major research interests are in the areas of microbial, enzyme and bio-process technology, which span over various programmes, including biomass to fuels and chemicals, probiotics and nutraceuticals, industrial enzymes and solid-state fermentation. He has over 1150 publications/communications, which include 16 patents, 50+ books, 140 chapters, 423 original and review papers, with *h* index of 78 and approximately 25,000 citations (Google Scholar). He is the recipient of many national and international awards and fellowships, which include Fellow of Royal Society of Biology, UK; Academician of European Academy of Sciences and Arts, Germany; Fellow of International Society for Energy, Environment and Sustainability; Fellow of National Academy of Science, India; Fellow of the Biotech Research Society, India; Fellow of International Organization of Biotechnology and Bioengineering; Fellow of Association of Microbiologists of India; Honorary Doctorate degree from Univesite Blaise Pascal, France; Thomson Scientific India Citation Laureate Award, USA; Lupin Visiting Fellowship, Visiting Professor in the University Blaise Pascal, France; Federal University of Parana, Brazil; EPFL, Switzerland; Best Scientific Work Achievement Award, Govt of Cuba; UNESCO Professor; Raman Research Fellowship Award, CSIR; GBF,

Germany and CNRS, France Fellowship; Young Scientist Award. He was Chairman of the International Society of Food, Agriculture and Environment, Finland (Food and Health), during 2003–2004. He is Founder President of the Biotech Research Society, India (www.brsi.in); International Coordinator of International Forum on Industrial Bioprocesses, France (www.ifibiop.org); Chairman of the International Society for Energy, Environment and Sustainability (www.isees.org) and Vice-President of All India Biotech Association (www.aibaonline.com). He is Editor-in-chief of *Bioresource Technology,* Honorary Executive Advisor of *Journal of Water Sustainability* and *Journal of Energy and Environmental Sustainability,* Subject editor of *Proceedings of National Academy of Sciences (India)* and editorial board member of several international and Indian journals, and also member of several national and international committees.

Contributors

B. K. Choudhury Energy Management Department, IISWBM, Kolkata, India

Maryom Dabi Department of Mechanical Engineering, IIT Guwahati, Guwahati, India

Nandini Das Department of Economics, Jadavpur University, Kolkata, India

Shyamasree Dasgupta School of Humanities and Social Sciences, Indian Institute of Technology, Mandi, Himachal Pradesh, India

Sudipta De Department of Mechanical Engineering, Jadavpur University, Kolkata, India

Manali Deshmukh BKPS, Pune, India; Construction Management, YCMOU, Nashik, India

Atul Dhar School of Engineering, Indian Institute of Technology Mandi, Mandi, India

Madhumati Dutta Department of Humanities and Social Sciences, Indian Institute of Engineering Science and Technology, Shibpur, India

S. K. Dutta Energy Management Department, IISWBM, Kolkata, India

Anirudh Gautam RDSO, Lucknow, India

Anupa Ghosh Department of Economics, The Bhawanipur Education Society College, Kolkata, India

Duke Ghosh Science and Policy Research Unit, Global Change Research, University of Sussex, Brighton, England

Gopa Ghosh Indian Institute of Engineering Science and Technology, Shibpur, Howrah, India

Sudip Ghosh Department of Mechanical Engineering, Indian Institute of Engineering Science and Technology, Shibpur, Howrah, West Bengal, India

Jai Gopal Gupta Department of Mechanical Engineering, Government Women Engineering College, Ajmer, Rajasthan, India

Pragya Gupta Indian Institute of Engineering Science and Technology, Shibpur, India

Kuntal Jana Centre for Energy, Indian Institute of Technology Guwahati, Guwahati, India

Minu Joshi SCET, Surat, India; Theory and Practice of Sustainable Design, Cardiff University, Cardiff, UK

Punit Kumar School of Engineering, IIT Mandi, Mandi, India

Nirmal Mallick Department of Chemical Engineering, Indian Institute of Technology Guwahati, Guwahati, Assam, India

Pinakeswar Mahanta Department of Mechanical Engineering, Centre for Energy, Indian Institute of Technology Guwahati, Guwahati, India

Ashok Pandey Center of Innovative and Applied Bioprocessing (CIAB), Mohali, Punjab, India

Dibakar Rakshit Indian Institute of Technology, Delhi, India

Soundaram Ramanathan Centre for Science and Environment, Delhi, India

Joyashree Roy Department of Economics, Jadavpur University, Kolkata, India

Ujjwal K. Saha Department of Mechanical Engineering, IIT Guwahati, Guwahati, India

Pooja Sankhyayan School of Humanities and Social Sciences, Indian Institute of Technology, Mandi, Himachal Pradesh, India

Achinta Sarkar Department of Mechanical Engineering, IIT Guwahati, Guwahati, India

Piyali Sengupta Energy Management Department, IISWBM, Kolkata, India

K. N. Srivastava ABB Corporate Research, Västerås, Sweden

Ajay Vaidya SPSMBH's College of Architecture, Kolhapur, India; Urban Design, Cardiff University, Cardiff, UK

Prabu Vairakannu Department of Chemical Engineering, Indian Institute of Technology Guwahati, Guwahati, Assam, India

Ashish Verma Department of Civil Engineering and Centre for Infrastructure, Sustainable Transportation and Urban Planning (CiSTUP), Indian Institute of Science, Bangalore, India

Part I
Sustainable Energy and Transportation Technologies

Introduction to Sustainable Energy, Transportation Technologies, and Policy

Jai Gopal Gupta, Sudipta De, Anirudh Gautam, Atul Dhar and Ashok Pandey

Abstract Need of energy is continuously increasing with increasing development aspirations and world population. To meet the energy demand, world requires production of more energy from the available limited resources. Technological development is both a cause of many environmental problems as well as a key enabler for solving them. It is a matter of fact that the technologies of the past are still dominating in transport, energy production, industry, and agriculture sector, which are gradually harming our basic life supporting systems—clean water, fresh air, and fertile soil. However, in each of these sectors there are new technologies available or emerging that may essentially solve these environmental problems if used widely and wisely. Thus, new technologies have the potential to contribute in decoupling of economic growth from pressure on natural resources. To address the global challenges of energy security, climate change, and economic growth, it is a global need to develop low-carbon energy technologies such as bioenergy for heat and power, biofuels for transport, solar photovoltaic energy, solar thermal electricity, wind energy, solar heating and cooling, efficient and environment-friendly energy storage. The long-term sustainability of the global energy systems is essential to counter balance of current demographic, economic, social, and technological trends.

J. G. Gupta (✉)
Department of Mechanical Engineering, Government Women Engineering College, Ajmer 305001, Rajasthan, India
e-mail: jaigopalg@gmail.com

S. De
Department of Mechanical Engineering, Jadavpur University, Kolkata 700032, India

A. Gautam
RDSO, Lucknow, India

A. Dhar
School of Engineering, Indian Institute of Technology Mandi, Mandi 175005, India

A. Pandey
Center of Innovative and Applied Bioprocessing (CIAB), Mohali 160071, Punjab, India

A. Gautam et al. (eds.), *Sustainable Energy and Transportation*,
Energy, Environment, and Sustainability,
https://doi.org/10.1007/978-981-10-7509-4_1

Keywords Smart city · Sustainability · Renewable energy · Transport
Energy policy

During the 1990s, we have seen a substantial diffusion of renewable energy and transport technologies and further progress in industry and agriculture technology, not least biotechnology. A systematic introduction of best available technology could reduce the use of energy with 20–50%. New technologies for waste management offer a great potential; the most recent investment in this sector demonstrates utilization of more than 90% of the energy content of waste. Even more fundamental are new technologies for “up-stream” resource management in industry, offering strong synergies for productivity in production, quality in goods and services, and efficiency in the use of natural resources (Lindqvist 2002). Currently, most of the energy demand is being fulfilled by the conventional energy resources such as coal, natural gas, petroleum products, but continuous consumption of these resources leads to environmental pollution and global warming by carbon emissions. Global temperature is increasing every year because of global warming since conventional energy resources are being used for the energy supply at very high rate. This consumption trend is also depleting the available source of energy, which will be exhausted in near future (https://www.iea.org/roadmaps/).

In twenty-first century, developing and emerging economies are facing two-fold energy challenges: meeting the needs of billions of people who still lack access to basic, modern energy services while simultaneously participating in a global transition to clean, low-carbon energy systems. Globally, a large fraction of the world’s population—more than two billion people, by some estimates—still lacks access to one or several types of basic energy services, including electricity, clean cooking fuels, and adequate means of transportation. The need for a profound transformation of the world’s energy-producing and using infrastructure has been already widely recognized in the context of mounting concern about global climate change (Ahuja and Tatsutani 2009).

In order to develop an understanding of the issues associated with global carbon emissions, it is necessary to comprehend the relationship between urban growth and energy use. The energy need of cities increases with both urban growth and industrial development. Energy policies and programs must keep this spatial differentiation in mind. The dependence of cities on intensive energy consumption is a major cause of climate disruption, and there is increasing interest in the potential for many cities to facilitate a transition to sustainable energy utilization. Energy resources and their utilization are intimately related to sustainable development. Energy conservation involves efficiency improvements, formulation of appropriate pricing policies, and good housekeeping practices, and load management strategies, among other measures. For a world to attain sustainable development, much effort must be devoted not only to discovering sustainable energy resources, but also for increasing the energy efficiencies of processes utilizing these resources. Increasing the efficiency of energy-utilizing devices is therefore important, and due to increased awareness of the benefits of efficiency improvements, numerous agencies have undertaken efficiency-related initiatives. Many energy conservation and efficiency improvement programs have been and are being developed to reduce present levels of energy use (Dhakal 2009; Bi et al. 2011; Milner et al. 2012; Song et al. 2017).

Renewable energy technologies become important as environmental concerns increase, utility (electricity) costs climb, and labor costs escalate. In addition, the development and implementation of renewable energy technologies depend in part on the global economy. The attributes of renewable energy technologies, which include modularity, flexibility, and low and relatively stable operating costs, are considerably different than those for traditional, fossil-based technologies, whose attributes include large capital investments, long implementation lead times, and uncertain operating costs regarding future fuel prices. The overall benefits of renewable energy technologies are often not well understood, and consequently, these technologies are often evaluated to be not as cost effective as traditional technologies. To assess renewable technologies comprehensively, however, all of their benefits, some of which are often not considered, must be taken into consideration (Rosen 1995).

Many books have been written on the subject of sustainable energy, but very few have approached these issues specifically from a developing country viewpoint. In nations where a significant portion of the population still lacks access to basic energy services, concerns about long-term environmental sustainability often are overshadowed by more immediate concerns about energy access and affordability. For discussing the issues, the present book covers the issues of sustainable energy, transportation technologies, and energy policy. The changing face of transportation toward sustainability discusses about to eliminate redundancy of capacity and competition within public transport modes, and to achieve the highest level of efficiency and network-wide connectivity. Smart city dream in Indian scenario describes various successful models of sustainable transportation and makes recommendations for Indian cities. Role of electric vehicles in future road transport provides the insight into the battery electric vehicle technology, needed resources, and the economics, which is developing around this technology. This volume also covers need and importance of alternative fuel sources for diesel engines including engine performance and emission characteristics of these arrangements. Discussion on coal energy sustainability in India discusses the estimate of available energy resources and its exploitation through unconventional technologies such as underground coal gasification (UCG), coal bed methane (CBM) for sustainable energy production. Investments in Clean Energy in South Asia article discuss the challenges, constraints, and risks (facing investors) for increasing investments in clean energy in South Asia. General policy framework has been included with command and control, price and quantity instruments to boost renewable energy implementation.

Energy and sustainable development within a society requires a supply of energy from resources that are sustainably available at reasonable cost and can be utilized for all required tasks without causing negative societal impacts. Supplies of such energy resources as fossil fuels (coal, oil, and natural gas) and uranium are generally acknowledged to be finite; other energy sources such as sunlight, wind, and falling water are generally considered renewable and therefore sustainable over the relatively long term (Dincer and Rosen 1998). Fossil fuel restrictions as well as concerns about change of climate globally motivate the energy transition to a sustainable energy sector requiring very high penetration level of renewable energy sources in the world energy matrix (Gallo et al. 2016). Environmental concerns are an important factor in sustainable development. For a variety of reasons, activities which continually degrade the environment are not sustainable over time, e.g., the cumulative impact on the environment of such activities often leads to a variety of health, ecological, and other problems. Improved energy efficiency leads to reduced energy losses. Most of efficiency improvement methods produce direct environmental benefits in two ways. First, operating energy input requirements are reduced per unit output, and pollutants generated are correspondingly reduced. Second, consideration of the entire life cycle for energy resources and technologies suggests that improved efficiency reduces environmental impact during most stages of the life cycle (Dincer and Rosen 1998). The chapter on waste heat recovery discusses this perspective.

In most global energy scenarios to meet stringent CO_2 constraints, bioenergy is assumed to be the dominating new energy source, displacing fossil fuels and associated CO_2 emissions. Technological details and implementation methods of bioenergy systems have been covered in two chapters. The alarming greenhouse gas perspective has led the industry to seek new more sustainable ways of meeting the need of transport. Alternative fuel vehicles and intelligent transport systems technologies may be used to promote the sustainable development in the transport sector. Even further advances in technology include the use of fuel cells, where zero emissions might be achieved in combination with substantially higher energy efficiency rates. Alternative fuels are being used today in place of gasoline and diesel fuel made from petroleum, e.g., biodiesel, electricity, ethanol, hydrogen, methanol, natural gas, propane. Penetration of any new transport technology is fundamentally dependent on broad availability of the fuel (Lindqvist 2002). Wireless power transfer (WPT) provides the prospect of new opportunities for electric vehicles (EVs) to enhance sustainable mobility (Bi et al. 2016).

The energy sector constitutes a fundamental element of industrial economies and supports all economic activities. Economic growth is strongly linked to increased energy consumption. One of the main roads to sustainable development is a reduction of demand for energy through implementation of better technologies in the residential and tertiary sectors, transport, industry, decisive shift from fossil fuel to renewable fuels, and decarburization of fossil fuel. Economic policies should concentrate on creating the right conditions for the efficient use of productive and natural resources and for their enhancement over time. In particular, they should contribute to improved market functioning by addressing market imperfections or failure due to the existence of externalities, market power, imperfect information, or the regulatory environment. We hope that this volume will stimulate the integrated thought process for ensuring fulfillment of energy requirements in sustainable manner with due consideration of various energy usage sectors and applicable technologies in those sectors.

References

Ahuja D, Tatsutani M (2009) Sustainable energy for developing countries. SAPI EN S. Surv Perspect Integrating Environ Soc (2.1)

Bi J, Zhang R, Wang H, Liu M, Wu Y (2011) The benchmarks of carbon emissions and policy implications for China's cities case of Nanjing. Energy Policy 39:4785–4794

Bi Z, Kan T, Mi CC, Zhang Y, Zhao Z, Keoleian GA (2016) A review of wireless power transfer for electric vehicles: prospects to enhance sustainable mobility. Appl Energy 179:413–425

Dhakal S (2009) Urban energy use and carbon emissions from cities in China and policy implications. Energy Policy 37:4208–4219

Dincer I, Rosen MA (1998) A worldwide perspective on energy, environment and sustainable development. Int J Energy Res 22(15):1305–1321

Gallo AB, Simões-Moreira JR, Costa HK, Santos MM, dos Santos EM (2016) Energy storage in the energy transition context: a technology review. Renew Sustain Energy Rev 65:800–822

https://www.iea.org/roadmaps/. Accessed 01 Aug 2017

Lindqvist O (2002) Technology and policy for sustainable development. Centre for Environment and Sustainability, Chalmers University of Technology and the Göteborg University

Milner J, Davies M, Wilkinson P (2012) Urban energy, carbon management (low carbon cities) and co-benefits for human health. Curr Opin Environ Ment Sustain 4:398–404

Rosen MA (1995) The role of energy efficiency in sustainable development. In Foundations and Applications of General Science Theory. Knowledge Tools for a Sustainable Civilization. Interdisciplinary Conference. Canadian Conference on 1995, IEEE, pp. 140–148

Song Q, Li J, Duan H, Yu D, Wang Z (2017) Towards to sustainable energy-efficient city: a case study of Macau. Renew Sustain Energy Rev 75:504–14

Shifting Focus from Supply to Demand—The Changing Face of Transportation Towards Sustainability

Ashish Verma

Abstract Urban form and transport system have an enormous impact on the way people travel. With rapid growing economies and population typically seen in developing countries, there is an increasing trend of expansion of urban sprawl and auto-based mobilization. This has a direct effect on the level and form of transport demand and pattern. In the absence of the implementation of proper policy measures, like parking charges, congestion charging, fare revisions, pedestrianization, it also leads to an increased additional cost for transportation infrastructure and its operation, while at the same time, creating many environmental, economic and social problems. Sustainable transport systems are those which aim to reduce emissions, fossil fuel consumption and the consumption of natural land, while providing easy access to people. This chapter throws light on various issues and challenges related to achieving sustainable urban transportation solutions in Indian cities and how the fundamental focus has shifted from traditional supply centric approaches to demand centric approaches. A case study of steel flyover project in Bangalore, India, is also presented to emphasize this point.

Keywords Sustainability · Transportation · Demand · Supply · India

1 Introduction

Ability of both persons and goods to move, safely and quickly, from one place to the other, i.e. mobility, is the basic need for the well being and prosperity of any individual, society, or a nation and nobody can argue on this basic requirement. However, the larger question to answer in present local and global context is what is good for us, is it just mobility or 'sustainable' mobility? 'Sustainable Mobility'

A. Verma (✉)
Department of Civil Engineering and Centre for Infrastructure, Sustainable Transportation and Urban Planning (CiSTUP), Indian Institute of Science, Bangalore 560012, India
e-mail: ashishv@iisc.ac.in

A. Gautam et al. (eds.), *Sustainable Energy and Transportation*,
Energy, Environment, and Sustainability,
https://doi.org/10.1007/978-981-10-7509-4_2

means improving our ability to move faster and safer in such a way that it does not compromise the ability of our children to do so in future. For any sensible person, the answer will definitely be 'sustainable' mobility. But it is important to think and answer, whether we are really choosing or tending towards sustainable mobility or not? Unfortunately, looking at the present state of transport infrastructure development and policy making in majority of Indian cities and also cities from many other developing economies like BRICS, the answer is no, to a large extent.

This is evident from the road infrastructure-based projects dominating the cities in the present times as a solution to address the transport-related externalities; the question is while such projects are one possible solution to solve mobility problems but is it a sustainable solution? In other words, as also observed with many such past road-based infrastructure projects, the effectiveness of such measures has proved to be for a very short time (sometimes just a year or couple or even less). From past experience, if has been observed that flyovers and underpasses (no matter of short or long length) does not solve the congestion problem but only shift the point of congestion from one place to the other. If we look around our cities, we can identify many examples of such unsustainable measures. Unfortunately, such measures are often perceived by the political class as 'populist' measures that can fetch votes for them during elections. To a large extent, citizens are also responsible for the same, after all 'populist' is all about what is popular and acceptable among the masses. This is also due to the aspirational aspects of large Indian middle class to own and use a car.

Ironically, many Indian cities are currently trapped in the 'Vicious Circle of Congestion' as shown in Fig. 1. It started with the economic boom in the country that began in early 1990s and which resulted in exponential growth of car ownership, particularly since car is largely seen as 'The' symbol of wealth in the Indian society (from small cars to SUV/MUVs now and to luxury cars in future). This resulted in more congestion and delay while travelling on our city roads and which then started the vicious circle of congestion with public pressure to increase road capacity, which resulted in adding of new flyovers/underpasses, road widening, etc., because of which movements by car become more easy and fast, which further favoured urban sprawl as people can travel longer distances with car in the same time, which then lead to further increase in number and average length of trips, eventually getting us back to square one, i.e. more congestion and delay on the roads and further forcing us into the same loop again. The ongoing predominance of road-based infrastructure development will do nothing but add fuel to this already existing fire and trap our cities deeper into this vicious circle of congestion.

It is interesting to note that Indian cities are not experiencing anything new that cities in other countries have never experienced. In fact, we are only reinventing the wheel, as many cities in developed economies have experienced the same vicious circle of congestion earlier, and some of which have been able to break this vicious circle through strong political/people's will and policies and planning that took them on the path of sustainable mobility. For example, Netherland was largely in

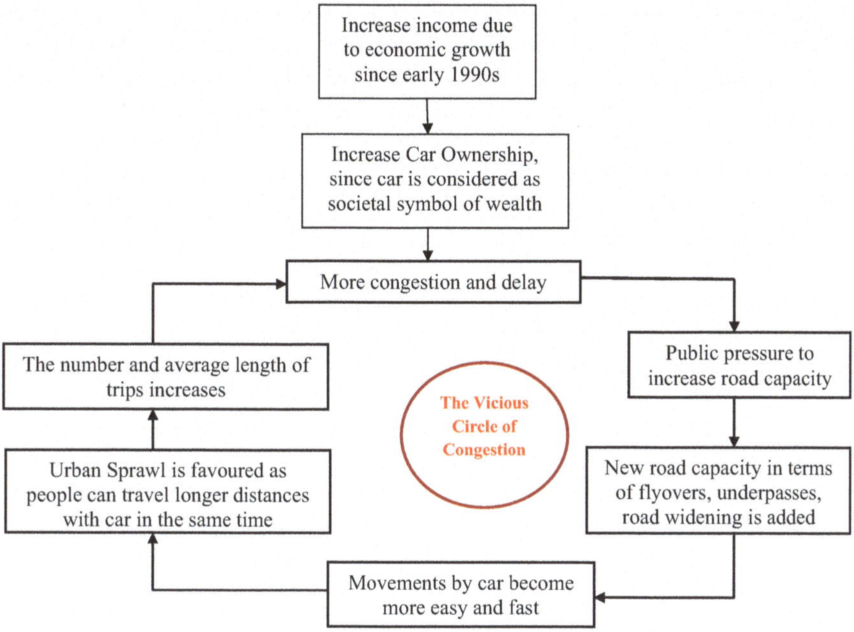

Fig. 1 Vicious circle of congestion in which many Indian cities are trapped

the same mess of congestion about 40 years before, but they chose a sustainable mobility path and today almost 60% commute trips daily are made by cycle in their cities like Amsterdam (Verma et al. 2016). Historically, if we compare the mobility development in India with that of USA and Europe, we will realize that there is a time lag of almost 50–60 years with respect to USA and 40–50 years with respect to Europe (Ecola et al. 2014). This is specifically in terms of saturation levels of motorization rates and infrastructure development which happened much earlier in USA, Europe and many other developing economies as compared to India where it is still at a very low level. However, it is interesting to understand that when infrastructure development, particularly of roads, happened in these countries, it was totally untouched from any global concern related to energy, climate change, carbon emissions, etc. The terms like climate change, carbon footprint and their possible consequences were hardly known to people at that time. Unfortunately, these global issues are much more prominent, known, and pressing in today's context, which means that India and other fast developing economies of BRICS, unfortunately, does not have the luxury of being untouched from these issues as compared to USA and Europe. This then leads us to think, whether India and other BRICS countries should follow the same path of infrastructure, particularly highway and other road-based development, that matured economies of USA, Europe, Australia, etc., followed historically or should India and other BRICS countries leap frog in terms of infrastructure development that is much more sustainable than the matured economies. More importantly, do we really have a choice today to leap frog? The answer is yes, only if Indian cities stop following the Western way of Highway or in other words only if we stop the race of constructing/expanding more and more roads, flyovers, elevated roads, etc. Unfortunately, this is happening in spite of clear message from our National Urban Transport Policy that says ‘focus on moving people rather than vehicles’ (MOUD 2006).

Today the road length per unit population or per unit area in Indian cities is much lower compared to the cities in developed world and so as the motorization levels, at the same time the length of reserved/priority public transport and cycle tracks per unit population is also very less in Indian cities as compared to many cities in European countries (Verma et al. 2016). If, as a policy, we decide today that we will not reach the levels of road indicators as high as what they are in Western world today and instead focus on increasing the density of sustainable modes then we certainly have the opportunity to leap frog. As a first step to leap frog, the government should think ways to decouple economic growth from the motorization rate and decide a desired modal share of trips for our cities and country overall that each type of mode/infrastructure should carry, by considering sustainability as the prime criteria. Based on this, the governments at national and state level, as well as civic bodies should allocate funds in such a manner that we achieve this desired modal share. On the other side, it is equally important to be aware that no commuter will adopt sustainable modes like public transport, walking, bicycling unless they are attractive travel options for them as compared to their personal vehicles. In other words, the question is; do we have high-quality public transport for us to travel safely and comfortably from our point of origin to point of final destination; do we have well connected and complete network of high-quality walking and bicycling infrastructure (also well integrated with public transport) that provides us safe end-to-end connectivity; do we have right kind of land-use transport planning in place that naturally encourages usage of sustainable modes? Unfortunately, the answer is no. So, to achieve this level of sustainable infrastructure, a complimentary set of policies, like congestion charging, parking restrictions, priority for public transport, pedestrianization of city core, various incentive schemes to promote usage of sustainable modes, etc., are required along with funding infrastructure development for sustainable modes.

In fact, this is the Western way of ‘sustainable’ mobility that many European countries like, Netherlands, Germany, Switzerland, etc., have adopted and which we should actually follow rather than the Western way of highway. This will also make the Indian population much healthier and will reduce the health expenditure of the country.

2 Case Study—Proposed Steel Flyover Project in Bangalore, India

This section presents a case study from Bangalore, India, of a recent road-based infrastructure project, namely the 'Steel Flyover Project' that was proposed by the state government of Indian state Karnataka as a measure to solve problem of congestion on a major corridor of Bangalore city (about 11 km) leading to the Bangalore airport and passing through some major road junctions that are major bottlenecks for the traffic moving on this corridor (Fig. 2). This section evaluates the sustainability of proposed solution particularly in terms of longevity of proposed solution to alleviate traffic congestion on this corridor. As a guiding principle, transportation planning for a city is done based on technical parameter called as 'Level of Service' (LOS), which is rated in six categories from A to F, and is based on the value of the ratio of traffic volume to capacity (V/C), expressed in passenger car units (PCUs) per hour (h), on individual links of the transport network. The level A (V/C tending to zero) corresponds to the best situation when vehicles move at free speed and level F (V/C tending to one or above) corresponds to the worst situation when the volume of traffic reaches or is above capacity, i.e. bumper-to-bumper traffic with only stop-and-go movement (in other words, complete traffic jam). The aim of transport planning is always to improve the transport system to bring down the LOS tending towards 'A' and as per Indian Roads Congress (IRC) guidelines, the design LOS is taken as 'C' which corresponds to V/C ratio of 0.7 (IRC 1990). In the context of the case study road corridor (airport road) leading to the Bangalore airport, it is important to note that some sections of the road (like, starting from Mekhri Circle towards Hebbal) are already operating at LOS 'F' (stop-and-go situation) during peak hour period based on the present traffic trends. While debating on merits of steel flyover or other proposed measures to reduce traffic congestion, the question really is: how; for how long; at what economic, financial and social cost; with what impact to the environment and ecology of the city we achieve the benefits of a proposed solution.

Below is a detailed technical analysis of the impact of added road capacity on airport road with steel flyover, in terms of LOS and 'Sustainability' criteria (development that not only serves today's needs but also serve the next generation in future). The aim of this technical analysis is to highlight the facts and figures which stress on the level of scientific and sustainable nature of the proposed steel flyover. Three scenarios have been considered to show the prevailing traffic situations in those conditions. The capacity of the roadway was obtained from IRC, 106: 'Guidelines for capacity of urban roads in plain areas' (IRC 1990). For the steel flyover, and for the at-grade road, the capacity is assumed as 2700 PCUs/h for three lanes (arterial category). The three scenarios are described below:

Scenario 1: Considering average vehicular growth rate of 4.75% as mentioned in the Detailed Project Report (DPR) prepared by STUP Consultants Pvt. Ltd. for steel flyover (*Source*: Bangalore Development Authority).

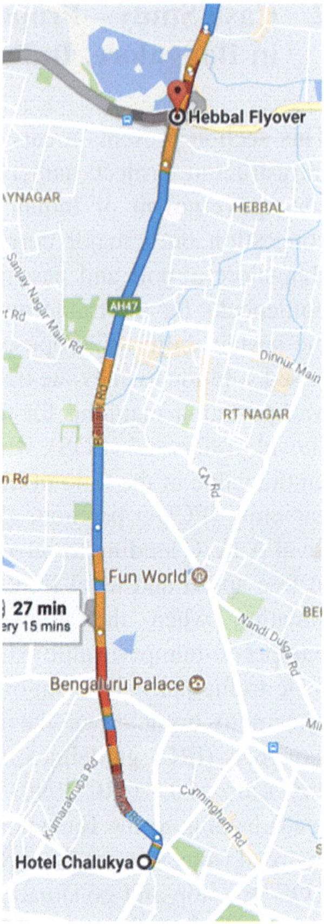

Fig. 2 Proposed steel flyover corridor in Bangalore. *Source* Google maps

Scenario 2: Considering average traffic growth rate of 10.6%, replicating the average annual growth rate in airline passenger traffic for Bangalore international airport in last 9 years (*Source*: Association of Private Airport Operators).
Scenario 3: Based on the coarse-grained assumption that car ownership (number of cars owned per 1000 population) in Bangalore has been doubling in every five years and so as the traffic (*Source*: Verma et al. 2013).

Also, since the amount of width (three-lane on each side) proposed for steel flyover and also on toll way road can easily accommodate two tracks in each direction of a hypothetical metro rail system on the same corridor, we have taken two scenarios of metro rail for analysis, which is with single metro rail track (ST) in each direction and with double metro rail tracks (DT) in each direction. This is further topped up with a scenario of only 50% people moving on the corridor shifting to metro, and a best case scenario when all are shifting to metro rail system.

These scenarios have been applied to two at-grade junctions along the proposed flyover namely; High Grounds Junction and Mekhri Circle. The traffic in consideration is the airport-bound traffic for which the capacity is 2700 PCUs/h each for the proposed flyover and at-grade roads. The traffic comparison (in PCUs) has been done for current and future stage (at the time of completion of project). Also, mode comparison is done to determine the saturation stage of existing traffic and a possible metro rail along this route. The capacity of a metro rail system is taken as 69,000 pphpd (persons per hour per direction); *Source*: (Verma and Dhingra 2001), and average car occupancy is taken as 1.5 persons per vehicle. Also, the peak hour traffic volume figures are taken from the DPR report of STUP Consultants for steel flyover (*Source*: Bangalore Development Authority).

2.1 Scenario 1: Growth Rate of 4.75%

High Grounds Junction

Time stage	2014	Current condition (2016)	Project completion (2018)
Traffic volume (PCUs)	6234	6840	7505
Capacity	2700	2700	2700 + 2700 = 5400
V/C	2.30	2.50	1.40

Mekhri Circle

Time stage	2014	Current condition (2016)	Project completion (2018)
Traffic volume (PCUs)	8884	9748	10,696
Capacity	2700	2700	2700 + 2700 = 5400
V/C	3.29	3.61	2.0

Summary: Though the growth rate is underestimated and undermines the realistic trends, the volume at Mekhri circle junction appears to be twice the capacity at the time of completion of steel flyover project, which means that it will only operate at LOS 'F' even on the first day of its operation. Even though, the total capacity of the existing at-grade road and grade-separated (flyover) comes up to 5400 (2700 + 2700), the current traffic has already overshot the combined capacities and situation only worsens by the time of completion of project (10,696 PCUs in 2018).
Metro in scenario 1: Now, evaluating the effectiveness of a possible metro from the High Grounds Junction for the same scenario (taking same growth factor of 4.75%) 2018 onwards.

Case 1: When all passengers switch to metro.

Time stage	2018	2028	2038	Saturation ST (2057)	Saturation DT (2072)
Passenger volume (based on average car occupancy of 1.5)	11,257	17,904	28,478	68,774	137,955
Metro capacity (single track (ST))	69,000	69,000	69,000	69,000	–
V/C (single track)	0.16	0.25	0.41	0.99	–
Metro capacity (double track (DT))	138,000	138,000	138,000	138,000	138,000
V/C (double track)	0.08	0.13	0.20	0.49	0.99

Case 2: When half of the passengers switch to metro.

Time stage	2018	2028	2038	Saturation ST (2072)	Saturation DT (2087)
Passenger volume	5628	8951	14,237	68,971	138,350
Metro capacity (single track)	69,000	69,000	69,000	69,000	–
V/C (single track)	0.08	0.12	0.20	0.99	–
Metro capacity (double track)	138,000	138,000	138,000	138,000	138,000
V/C (double track)	0.04	0.06	0.10	0.49	1.0

Summary: Using the traffic growth rate of STUP Consultants, the metro rail system can serve the airport road corridor for another 70 years easily with even half of the passengers shifting to metro rail, which is sustainable solution by any definition.

2.2 Scenario 2: Growth Rate of 10.6%

High Grounds Junction

Time stage	2014 traffic	Current traffic (2016)	Project completion (2018)
Traffic volume (PCUs)	6234	7625	9327
Capacity	2700	2700	5400
V/C	2.30	2.8	1.7

Mekhri Circle

Time stage	2014 traffic	Current traffic (2016)	Project completion (2018)
Traffic volume (PCUs)	8884	10,867	13,292
Capacity	2700	2700	5400
V/C	3.3	4	2.5

Summary: The assumed traffic growth rate here is in compliance with the growth in passenger traffic at BIAL. While the LOS at High Grounds Junction may still seem hovering around C&D; the volume at Mekhri Circle is more than five times the capacity by 2018.

Metro in scenario 2: Now, evaluating the effectiveness of a possible metro from the High Grounds Junction for the same scenario (taking same growth factor of 10.6%) 2018 onwards.

Case 1: When all passengers switch to metro.

Time stage	2018	2028	2030	Saturation ST (2034)	Saturation DT (2041)
Passenger volume	13,990	38,315	46,868	70,130	141,966
Metro capacity single track	69,000	69,000	69,000	69,000	–
V/C single track	0.20	0.55	0.67	1.0	–
Metro capacity double track	138,000	138,000	138,000	138,000	138,000
V/C double track	0.10	0.27	0.33	0.50	1.02

Case 2: When half of the passengers switch to metro.

Time stage	2018	2028	2037	Saturation ST (2041)	Saturation DT (2048)
Passenger volume	6996	19,158	47,445	70,900	143,715
Metro capacity (single track)	69,000	69,000	69,000	69,000	–
V/C (single track)	0.13	0.38	0.94	1.0	–
Metro capacity (double track)	138,000	138,000	138,000	138,000	138,000
V/C (double track)	0.06	0.19	0.47	0.50	1.04

Summary: Using the 10.6% traffic growth rate, the metro rail system can serve the airport road corridor up to 2048 easily with even half of the passengers shifting to metro rail, which is sustainable solution by any definition.

2.3 *Scenario 3: Double in 5 Years*

High Grounds Junction

Time stage	2014 traffic	Project completion (2019)	(2024)
Traffic volume (PCUs)	6234	4192	8384
Capacity	2700	5400	5400
V/C	2.3	0.77	1.6

Mekhri Circle

Time stage	2014 traffic	Project completion (2019)	(2024)
Traffic volume (PCUs)	8884	17,768	35,536
Capacity	2700	5400	5400
V/C	3.29	3.29	6.6

Summary: For this scenario, the LOS at High Grounds Junction is already D at the beginning of the project (2019), and the LOS at Mekhri Circle show horrific results where the volume has reached three times the capacity.

Metro in scenario 3: Now, evaluating the effectiveness of a possible metro from the High Grounds Junction for the same scenario (doubling the passenger every 5 years) 2014 onwards.

Case 1: When all passengers switch to metro.

Time stage	2019	2024	2029	Saturation ST (2034)	Saturation DT (2039)
Passenger volume	12,576	25,152	50,300	100,608	201,216
Metro capacity (single track)	69,000	69,000	69,000	69,000	–
V/C (single track)	0.18	0.36	0.72	1.45	–
Metro capacity (double track)	138,000	138,000	138,000	138,000	138,000
V/C (double track)	0.09	0.18	0.36	0.7	1.45

Case 2: When half of the passengers switch to metro.

Time stage	2019	2024	2029	2034	Saturation ST (2039)	Saturation DT (2044)
Passenger volume	6288	12,576	25,152	50300	100,600	201,216
Metro capacity (ST)	69,000	69,000	69,000	69,000	6900	–
V/C (ST)	0.09	0.18	0.36	0.72	1.45	–
Metro capacity (DT)	138,000	138,000	138,000	138,000	138,000	138,000
V/C (DT)	0.045	0.09	0.18	0.36	0.7	1.45

Summary: Even with a assumption of very high traffic growth rate, the metro rail system can serve the airport road corridor up to 2044 easily with even half of the passengers shifting to metro rail, which is again sustainable solution by any definition.

Further, with the growth in traffic on airport road in future and the high speed at which the vehicles will travel on the steel flyover and below, the air (from exhaust emission) and noise pollution can only increase adding further to the externalities due to transport.

So, if not steel flyover then what is the sustainable mobility solution for the airport road and also for the Bangalore city? From the analysis above, a metro rail system on the airport road corridor right up to the Bangalore airport and physically well integrated with airport terminal building itself may serve as an alternative solution which will also be sustainable on many grounds. This metro rail line to the airport should get connected and integrated at one or couple of transfer metro stations so that people can easily connect to the city network of metro rail system and then travel to any part of the Bangalore city. While doing so, the government should plan feeder bus systems for each metro rail station within 2–3 km influence zone on each side of metro station along with pedestrian/cycling-friendly infrastructure within 500–1000 m influence area around metro stations. On the metro stations in sub-urban locations of city, park-and-ride facility at metro station can be created and integrated so that people leave their car in park-and-ride and travel to the core area of city by metro. Further, buses and metro rail system being there for the same purpose of public transport, they can be operated under a single public transport agency rather than separate (BMTC and BMRCL), to eliminate redundancy of capacity and competition within public transport modes and to achieve the highest level of efficiency and network-wide connectivity. Worldwide, in cities which are very high on liveability index and have excellent and sustainable transport system, all public transport modes are operated by a single agency.

3 Summary and Recommendations

This chapter throws light on various issues and challenges related to achieving sustainable urban transportation solutions in Indian cities and how the fundamental focus has shifted from traditional supply centric approaches to demand centric approaches. A case study of steel flyover project from Bangalore, India, is also presented to emphasize this point. To conclude, the following questions should always be asked while deciding any transportation improvement in any Indian city:

- What is the amount of person capacity added per unit of investment made?
- What is the amount of person capacity added per unit of space width created?
- By what amount the fossil fuel consumption will change?
- By what amount the air and noise pollution will change?

If the above questions are answered in an unbiased manner than such transport improvements are likely to be sustainable. Following is the recommendation approach for road ahead in Indian cities:

- Development choices to decouple economic growth from private motorization growth.
- A multi-tier approach that includes comprehensive urban planning, improved policy making, effective economic instruments among others.

- Creating opportunities for cities to use existing transport infrastructure wisely, while generating funds to improve sustainable transport options.
- Developing cities such as from India have a challenging opportunity to build a system in which public transport and NMT become the first choice for mobility.
- Long run endeavour of 'Avoid-Shift-Improve' approach to build resilient and sustainable cities.

References

Ecola A, Rohr C, Zmud J, Kuhnimhaf T, Phleps P (2014) The future of driving in developing countries. ifmo project report, ifmo, Germany

Indian Roads Congress (IRC) (1990) Guidelines for capacity of urban roads in plain areas. IRC, 106, India

Ministry of Urban Development (MoUD) (2006) National Urban Transport Policy. Available at http://www.urbanindia.nic.in/policies/TransportPolicy.pdf. Accessed on 04 Dec 2010

Verma A, Dhingra SL (2001) Suitability of alternative systems for urban mass transport for Indian cities. TrasportiEuropei, Q J Transp Law Econ Eng Trieste 18:4–15 (Italy, Anno VII)

Verma A, Velmurugan S, Singh SK, Gurtoo A, Ramanayya TV (2013) Global mobility monitor network—India mobility final report. Project Report, IISc Bangalore, India

Verma A, Bhat PB, Rai N (2016) Urban mobility trends in india and other BRIC countries and their policy implications. Urban Transp J IUT (India) 15(1):1–26

Sustainable Transport Solutions for the Concept of Smart City

Minu Joshi, Ajay Vaidya and Manali Deshmukh

Abstract In the year 2008, the world faced one of the worst economic recessions. At this point in time, the idea of a smart city was floated by IBM which was a part of the smarter planet initiative (Paroutis et al. 2013). As per various definitions, a smart city is an urban development that takes the help of technology to manage its assets like its infrastructure, important buildings, community services, transportation. The concept of smart city was welcomed by China, UAE, Korea and others. There are six indicators of a smart city, namely "*smart economy, smart governance, smart citizen, smart living, smart environment and smart mobility*" (Kumar and Dahiya 2017)—mobility or transportation takes the most important position due to its influence on all the other indicators. Statistics show that millions of dollars are invested in research and development of new sustainable modes of transportation systems all over the globe. There are innumerable examples of sustainable smart transport all over the world like Paris, Boston, Germany, Singapore. Each one of these examples lays emphasis on the fact that a well-designed and efficient transport system contributes to economic growth and helps in refining the quality of life of the people, thereby becoming one of the most important sectors of urban development. Post-studying these examples, one can realize that sustainable transportation system contributes to

M. Joshi (✉)
Ecour Studio, Pune, India
e-mail: ecour.ar@gmail.com

M. Joshi
Theory and Practice of Sustainable Design, Cardiff University, Cardiff, UK

A. Vaidya
AVA, Pune, India
e-mail: ajayvaidya@outlook.com

A. Vaidya
Urban Design, Cardiff University, Cardiff, UK

M. Deshmukh
Disha Architects, Pune, India
e-mail: manalicdeshmukh@gmail.com

M. Deshmukh
Construction Management, YCMOU, Nashik, India

A. Gautam et al. (eds.), *Sustainable Energy and Transportation*,
Energy, Environment, and Sustainability,
https://doi.org/10.1007/978-981-10-7509-4_3

all the different indicators or aspects of a smart city. It is projected by the United Nations that in the coming 15 years the population of top 100 cities of India will grow by 60%. Also with growth in opportunities and increase in the buying capacity, personal motor-vehicle ownership doubles every decade. Land is a precious resource, and utilizing it for making roads to meet the needs of the motorized vehicles is not a solution. This chapter examines the Indian scenario and smart city dream and various successful models of sustainable transportation addressing issues similar to the existing Indian cities, and makes recommendations for the same.

Keywords Sustainable transport · Smart city transport system · Indian smart cites
Case studies in smart transport · Important lessons for sustainable transport

1 Smart City, Sustainable Smart Transport and Indian Scenario

1.1 Indian Scenario—Transportation

Transportation is one of the key contributors to the GDP as per the Government of India. Its current share is close to 6.7%, and it is projected to grow to about 26% by 2026 (Mehra 2016). The Government of India has framed and adopted various policies and schemes to transform the poor state of urban transport system into an effective system contributing to the smart city vision. The issues like excessive traffic, poor public transport, pollution, poor road conditions, lack of intelligent transport systems, poor infrastructure are some of the inherent challenges and threats in shifting towards smart transportation in India. In spite of advances in transportation technology as well as urban planning, there is a lack in understanding of how cities work and how the infrastructure affects day-to-day activities. Every city in the Indian context has a unique character, unique complexity and a unique set of challenges that need to be addressed before proposing any kind of transport system. It would be wise to implement smart sustainable transport only post a thorough understanding of the need of the city, community and the quality of life that the system will offer. With advancement in opportunities and the smart city mission, Government of India has taken more initiatives towards encouraging smart fuel vehicles and smart technology. However, it will be difficult for India to keep the pace with other countries in terms of application of technological advancements for smart transportation due to the slow pace of development, rapid growth in population, as well as time taken in decision making and implementation. Prime Minister Narendra Modi's vision "Digital India", has set an ambitious plan to build 100 smart cities (http://smartcities.gov.in/content/innerpage/what-is-smart-city.php). These cities will be designed on the following key principles or indicators: "*smart economy, smart governance, smart citizen, smart living, smart environment and smart mobility*". Apart from the 100 smart cities, 500 other cities in India will be rejuvenated and transformed by the provision of better connectivity and infrastructure.

1.2 Smart City Concept and Indian Cities

As per various definitions, a smart city is an urban development that takes help of technology to manage its assets like its infrastructure, important buildings, community services, transportation. However, a smart city would have a slightly different association and meaning in India than Europe or the USA. Even in India, there are many ways of defining a smart city. Various aspirations of citizens define the concept of smart city limited to that particular city. Connectivity and mobility along with services and infrastructure are among the primary aspirations of the citizens for a smart city development. The holistic approach of the Government of India through policies like Urban Transformation Policy 2014, Make in India Scheme, Faster Adoption and Manufacturing of Electric/Hybrid Vehicles, Digital India, AMRUT, etc., will support the smart city mission primarily by transforming the urban transport system (Mehra 2016). In 2016, 13 cities have been selected to be taken up for development as smart cities in the fast track round (http://www.smartcities.gov.in). There are eight pointers given by the Government of India as the interpretation of the smart city in Indian context. These talk about some typical features of holistic development needed in Indian smart cities.

1. Efficient use of land by promoting mixed-use development and making provisions in the building bye-laws to adapt the change statewise.
2. Providing housing opportunities for all.
3. Promoting walkability, reducing traffic, improving air quality, promoting public transport, promoting local economy, providing environment that encourages interaction and ensures safety and designing the road network that provides opportunities to cyclists and pedestrians.
4. Preserving and increasing recreational spaces and green pockets like parks, playgrounds, river banks in order to enhance the quality of life of citizens, reduce the urban heat island effects in areas and maintain a comfortable microclimate as well as ecocity balance.
5. Promoting a variety of smart public transport options—transit-oriented development (TOD), public transport, last mile para-transport connectivity, pods, etc.
6. Measures to bring accountability and transparence in the governance by the means of introducing "e-governance" which is both time saving and cost effective. Forming e-groups for discussions and monitoring of programmes as well as activities.
7. Promoting the local economy by encouraging activities like local cuisine, health, education, arts and craft, culture, sports goods, furniture, hosiery, textile, dairy, which also gives an identity to the city.

8. Promoting smart solutions for infrastructure and services in the area-based development in order to safeguard them from disasters as well as reduce the use of resources and provide affordable services.

From the above pointers, one can conclude that the aim of smart city mission is to promote development of cities that provide adequate infrastructure, clean environment, sustainable transport solutions, respectable housing facilities for all, promote economic growth, etc., and by doing all this, improve the quality of life of its citizens.

2 Learning from Good Examples Around the World

Public transportation in cities is considered as a key aspect to relieve traffic overcrowding and promote environmentally friendly transportation modes. Various cities around the world have developed excellent transportation systems that promote use of more sustainable means of transport over personal motorized vehicles. In such cities, policies and systems both are designed to encourage people to choose more sustainable options for travelling. One of the key aspects noticed in most of the examples discussed here is that planning contributes to reducing private car traffic and increasing the use of alternative forms of transport, namely car pooling, cycling, walking and public transport, by making them more desirable (Duchene 2000).

2.1 Paris

The spirit of the Paris Region mobility plan/urban travel plan (PDU—Plan de Déplacements Urbains) is to provide support for the development of the Paris Region through the enhancement of its facilities. The PDU defines the principles for the organization of transport of goods and people, control of vehicular traffic and parking. The need of a better and smarter transport in Paris was felt way back after the Second World War. At that time, the fast growth led to an unrestrained sprawl of the core area of Paris towards the fringes. Population of the core area rose by 3.3 million from 1945 to 1975, touching the 10 million mark (Trevien and Mayer 2013). The Central Government set up policies to restructure the region and control its growth, viz. redistribution of district boundaries, construction of a new airport, highway and train system, rehabilitation of population from the core area to new towns. To connect these various places, Paris required setting up an integrated public transport network. The lack of evolution of the existing commuter rail network and its poor connectivity within and across Paris created the need of improving the existing transport system. The new or improvised transport system promoted the following:

- The harmonious and controlled development of the territory.
- The emergence of a common culture on urban and intermunicipal travel.

Along with this, certain policies were also devised so as to ensure efficient operation of transport networks, and increased sharing of publicly accessible spaces. In order to reduce private car traffic and promote public transport, the PDU reduced differences in travel time in-between the various forms of transport used, taking due account of their corresponding zones of significance. This strategy also looks at a tiered system of transport networks, in particular, the rail and expressway network that provides effective conveyance on an urban scale. The primary road network, on which the assorted modes coincide, provides for transport within the high-density zone, while organizing transport in both the inner city and the fringes. Measures were taken to integrate the network so as to make it a more convenient for the passengers. Key actions (STIF 2003) listed below have transformed an inadequate network to one that may be looked upon as a model for other cities to follow.

- With a view to promote intermodality between modes of transport, parking and ride facilities like large car and bicycle parking facilities are set up near metro stations.
- To keep travellers informed with up-to-date information, the STIF (Syndicat des Transports d'Ile-de-France) created an online resource in collaboration with all public transport companies for the region.
- With the introduction of the Navigo smartcard in 2001, for the Paris train, metro, RER (Réseau Express Régional), tram and bus system, travelling within Paris has become easier because separate tickets do not have to be purchased while changing between transport modes.
- The STIF approves future projects before they are implemented to ensure that upcoming infrastructure is integrated well with existing ones.
- Recent initiatives have been launched to integrate public transport with car rental services. This is integrated with the Navigo pass allowing users to rent cars at reduced rates thus renouncing the need to own one.
- The cycle track system is planned to connect various scales across the city and suburban area including long- and short-distance travel within the inner city and the fringe areas.
- Access to public transit stations is given priority by developing pedestrian infrastructure equipping pedestrians to easily walk within their districts and access transit stations.

The collaboration between the STIF and regional council and cooperation from the Ministry of Transport have enabled an integrated transit (urban mobility plan) and land-use plan (regional plan). This ensures that good access to public transport is linked to all future developments. For example, the business district of La

Defense provides good connectivity to inner as well as the suburban Paris Region and 85% of the employees use public transport to get to work (STIF 2003).

Once major action was the decision to implement the 1965 plan and build five new towns outside Paris and decongest the inner city. This plan also ensured good connectivity to Paris via rail services. New smaller centres within these towns would be developed to make them attractive for people and to reduce the need for travel across the city for work, recreation or daily essentials. The extension of two metro lines ensured good connectivity to these counties. To provide good connections with the city, two metro lines were extended to serve these counties.

On the basis of the above-mentioned pointers, three main factors can therefore account for the increase in favour of public transport post-1975:

- Increase in the supply of public transport services
  Long-distance commute has become more attractive via public transport with the suburban rail systems built after 1975 (know as the RER).
- Improved integration of services
  The integrated ticketing system in Paris enables travel on multiple public transport modes with an unlimited number of trips across different companies.
- A reduction of alternative modes of transport
  Integrated season passes for transit has changed some people's decisions to take the bus instead of other transport modes (e.g. private car).

Multiple modes of public transit are being used by a numerous and ever-increasing quantum of persons due to consistent advances in the integration of passes. These enable people to use them across the bus, metro, RER services, etc., and public transport modes. This is in spite of having more than eighty operators.

The following projects could further improve public transport systems in Paris in the future:

- Enabling ticket/pass integration with park and ride facilities adjacent transit stations to make it easier to access to station. The success of the present system will be strengthened with a cohesive tariff system.
- A quality bus network which has communication patterns under a single new agency independent of operators will greatly improve the transport networks. This project, termed "Mobilien", has been mentioned in the urban mobility plan framework.
- Prioritizing interchanges for people, bicycles, taxis, etc., to enable easy access.
- Setting up a centralized information system along with a single point of contact for addressing complaints for public transport users.
- Improving synchronization between taxis and public transit systems. Currently, an independent body is responsible for taxis.
- Concerning space organization on streets, priority is given to pedestrians, cyclists and buses.
- Re-organizing the metro and RER systems to better cater to inner and outer Paris during the day.

- Strengthening the bus network with bus rapid transit systems extending to outer fringes of Paris to ensure connectivity for every area.
- Increasing the usability of public transit between places of interest within Paris.

The city council along with the Paris transport operator has initiated the "Bus 2001" project to be merged with the existing bus network. Amendment of the existing system will be essential for its extension beyond the "Boulevard Peripherique" (ring motorway) in unison with the metro lines (Duchene 2000). For pedestrian's safety and comfort, work on footpaths and road crossings will be enhanced for them. The introduction of new cycle tracks will be considered. Pedestrian paths and cycle tracks between Paris and the adjacent municipalities will be extended. Pedestrian paths and cycle tracks between Paris and the adjacent municipalities will be extended resulting into enhanced urban conditions in the new towns and suburban complex. Pedestrians, bicycles, motorcycles and buses will have easier access to the metro and railway stations. The main bus network will offer an efficient alternative to the private car. The cycle track system, including establishment for bicycle bays, will provide for easy access to stations and public facilities. This network will be interconnected, uninterrupted and clearly identified.

2.1.1 Paris—Lessons Learnt

- The public transit system in Paris is a high-quality case with integration across multiple modes on a wide region (approximately 12,000 km) and with a large number of transport operating companies. The impact of integrating multiple modes of transport on one network resulted in growing patronage of transport systems. Numerous factors are responsible for this high level of integration.
- Due to the presence of several operating companies, it was necessary to offer passengers with integrated tickets among them.
- Integration has been incremental (It took twenty-five years to achieve a single ticket accepted by all transport companies). The positive results of each step encouraged local authorities to take the next one.
- The public transport authority (STIF) was legally given administrative power, to impose its decisions on companies regarding fare policy, travel routes, etc. This ensured compliance from all private companies to a common goal for the betterment of the public transport system.
- Integration of services did not cause losses of incomes to operators since STIF always compensated for the price decrease in ticket fares which it made obligatory on them.
- The two main transport companies (RATP and SNCF) are state-owned enterprises played a positive role in the integration of public transport services at the beginning. Today, the policy framework established by the STIF applies to all private companies as well.

From the Paris example, it is clear that building up an integrated smart transport network is a long-drawn process. But with proper planning strategies, we find a positive measurable outcome is possible over a period of time.

2.2 Germany

The case of Germany is very interesting as it has managed to stabilize high levels of car ownership with sustainable modes of commute like walking, cycling and public transport. As per statistics, Germany has one of the world's highest car ownership rates along with a very powerful lobby of car manufacturing industry. In spite of immense popularity of cars (Wolf 1986; Schmucki 2001 as cited in Buehler and Pucher 2009), Germany has managed to design a transport system that is sustainable and smart by adopting the right policies and strategies. The policy adopted by Germany was to manage public transport, cycling and walking and to discipline the use to cars rather than completely eliminating it. Freiburg, the environmental capital of Germany, is considered as the most sustainable city.

Car ownership grew in Germany from 1950 to 2006 resulting into a 42-fold rise in motorization rate (Buehler and Pucher 2009). Germany made certain policies as well as infrastructure revolutions to promote sustainable transport. The Government policies can be classified into five categories as follows. These have been very important for transport sustainability in Germany.

- Increased taxes and restrictions on car usage
  The strategy adopted by Germany was essentially to make "cars" the least preferred alternative to travel. One of the ways adopted to do so is much higher fees and taxes on car ownership and use along with motor fuel taxes making cars an expensive mode of transport. The increased taxes have generated enough revenue to cover government expenditures on maintenance and construction of roadways. Other strategies apart from taxes, adopted by most of the German cities, are the speed limits and reduced parking spaces as well as car-free zones, restricted entries, deliberate dead ends, turn restrictions, one-way street networks, etc., making it very difficult to travel by cars. Most of the road networks in German cities have speed limits as low as 30 km/h (Beatley 2000; Newman et al. 2009). Most of the developments use strategies like road narrowing, raised intersections, extra curves or zigzag routes as well as speed humps. At many places, ultimate restrictive measures are taken by banning the car altogether.
- Ensuring the provision of high-quality, low-priced, well-synchronized public transport services as a viable alternative to the car
  In recent times, Germany has made considerable improvements in the public transport realm as a result of which more and more people prefer to use it. German metros, trams and buses are reasonably new and do not require as much maintenance or repair cost. German buses and trams are often articulated, carrying more passengers and requiring fewer drivers per passenger (Buehler and

Pucher 2009). All these reasons make public transport an affordable choice. Along with this, public transport is further enhanced by signal priority at the intersections, well-planned spacing of bus and tram stops to reduce the time taken per trip, making it a faster option to travel. This increases the number of people using the public transport resulting in higher revenues. Much of the success of the German public transport is due to its multimodal coordination of public transport services, fares and schedules within the metropolitan areas (Buehler and Pucher 2009) allowing an uninterrupted journey for the passengers. The planning is such that it ties together public transport, cycling facilities as well as walking, giving priority to the people using these means to commute. Safe pedestrian walkways and bike parking and renting facilities near bus and rail stops encourage people to choose more sustainable means.

- Improving pedestrian walkways and well-connected cycle routes that are safe and convenient
  Great improvements have been made to pedestrian walkways and bicycle routes since the 1970s. The facilities created for pedestrians include creation of car-free zones in most of the cities along with well-lit, well-maintained walkways, well-designed street furniture, pedestrian refuge islands, clearly marked zebra crossings and pedestrian-activated crossings (Pucher and Dijkstra 2003). To encourage use of bicycles, well-connected continuous cycle routes, extensive parking for cyclists, shortcut connections for cyclists, etc., are designed. The facilities enable the cyclists and pedestrians to cover any trip without interruptions or issues created by traffic. Government policies are also designed to provide utmost safety to the pedestrians and cyclists.
- Promoting high-density mixed-use development and discouraging urban sprawl, keeping distances walkable
  To decrease the average trip distances, a high-density mixed-use development is planned. This not only enables more people to take advantage of the public transport but also makes it more economically viable by generating higher passenger volumes. The urban development policies are designed to curb urban sprawl.
- Coordination of various policies to ensure their combined effect
  It is difficult to restrict car usage and make it a more expensive choice without offering alternative solutions to car use. Germany, along with its policies to make "car" the least preferred option to travel, has also made sure to provide world-class public transport which is not only cheap but also less time-consuming. The car-restrictive measures have been successful as a result of improved conditions for walking, cycling as well as buses and trams. This made the policy reforms more acceptable to the people. In the majority of the German cities, transport and land-use planning are done through the same department and hence there are more opportunities to coordinate these two (Schmidt and Buehler 2007).

2.2.1 Germany—Lessons Learnt

Germany is a good example of car-restriction policies backed by excellent alternate sustainable means of transport. However, the German scenario was not always sustainable as it is now. Decades ago, German policies as well as development supported the use of cars. Soon more and more people started using cars and issues like traffic congestion, pollution worsened resulting into poor quality of life in most of the neighbourhoods. As a result of this, new land-use policies and reformations in public transport triggered. The 1970s witnessed an era of "car-restrictive" policies that became widespread in most of the German cities. People started realizing the benefits of car-free zones, better walkways, cycling routes over the last four decades. In short, the German situation was not always like it is now. Reformations in policies at various levels and provision of better alternatives for travel have made it possible for people to choose a more sustainable mode of transport. The overall framework for sustainable transport was provided by the German Federal Government, which included not only a hike in petrol taxes but also reduction in money spent on roads and increasing investments on public transport. People played a very important role in the shift from a car-dominated system towards more sustainable options.

2.3 Boston

Boston serves as a feeder, connector and common point of contact due to its strategic location as the centre of six New England states. Boston is a unique example as it uses rail, water, road and air as the part of its transportation infrastructure to move people and goods. The integrated national and international transportation of Boston supports the city's role as a international economic player by promoting expansion in varied economic sectors (bostonindicator.com). The quality of life of people who live and work in Boston depends largely on their capability to travel from one place to another for increasingly assorted purposes. Keeping this in mind, Boston has improved the efficiency of its public transport enabling the people to travel economically. Like Germany and Paris, the policies adopted in Boston were to promote use of a range of transportation options from walking and bicycling to driving or taking public transit for activities like work, educational advancement, shopping, socializing. This not only enabled people to move from one place to another, but also supported the city's environment by reducing pollution, congestion and greenhouse gas emissions. Boston drafted certain sets of policies based on the following indicators to develop an efficient competitive edge in its transportation system after understanding the commuter's needs (bostonindicator.com).

- Transportation system that enhanced the national and global competitiveness of Boston

Boston's Logan airport is one of the busiest airports catering to both domestic as well as international travellers. The well-developed infrastructure to move people, goods and services to national and international destinations reinforced Boston as a port of entry for new immigrants and thus enhanced its reach as a metro city within the national and worldwide economies. As a result, the port of Boston ranked high in total container traffic and hence in volume growth among all North American ports.

- Traffic patterns and travel options
  Traffic congestion in Boston was considered a reason of loss in the number of economic, social and environmental aspects of the metro city. This was because of the economic activity loss which was due to increased journey time, increased pollution levels and social seclusion. Commuters spent long hours in delayed traffic congestions reducing their productive time. By developing and integrating a wide range of available travel options, Boston government helped to offset some of these negative impacts and improve the overall growth of the region.
- Household income spent on transportation
  Transportation costs constituted a major expenditure for most households in Boston, often second only to housing (Hass 2006). Cars were expensive to purchase, and a recurring cost to cover fuel, maintenance, and repair was far-fetched. Thus like housing and medical care, transportation costs affected the decisions people made about where to live, where to start a business or bring up a family. The efficient public transit systems and the infrastructure developed by Boston enabled residents to prefer integrated transport systems, thus reducing the individuals cost of travel.
- Boston's integrated transport system
  Greater Boston's public transit system—the Massachusetts Bay Transport Authority (MBTA)—provides a variety of options for getting around the region. The efficient planning of the rapid transit lines operates along major stations and the light-rail streetcar lines, link surface stops at subway or elevated stations. The rapid bus transit line operates as a feeder for commuters from Dudley Square to downtown and from South Station to the South Boston Waterfront and Logan Airport. A passenger rail network and some local and express bus routes aid the residents to reach their communities comfortably. Para-transit facilities such as "The Ride" are especially provided for elderly and people with special needs. A water transportation system is also provided for travellers to Boston's Inner Harbor and between several Inner Harbor docks, including Logan Airport, Charlestown Navy Yard, Rowes Wharf and Long Wharf.
- Allocation of daily trips and trends in mass transit use
  A strong relationship between annual vehicle miles travelled per household and the availability of public transit systems was observed in Boston. Greater Boston cities and towns with the lowest vehicle mile travelled rates had access to the commuter rail. Multiple modes of travel were made available for the cities and towns in the inner core at fair frequencies. The MBTA being one of the largest

transit systems in ridership; a large number of trips were made using the MBTA on an average working day, thus promoting the use of public transit systems.

- Equitable and high-quality transportation access for all
  People using public transport in Boston were dependent on the availability of amenities like heating/air conditioning, shelters at stops, and comfortable seats as well as provision for persons with special needs in the public vehicle. Measures were taken to improve the quality of these amenities that greatly affected the rider's satisfaction. Sidewalks were also well maintained, thus improving the experience of the commuters walking to catch the transit system, and this in turn improved the equality of urban life. This high level of comfort and safety provided in the transit systems in Boston brought comfort to elderly, families with children and people with special needs. Transit quality and safety also ensured secure travelling of school-going children, who depended on public transportation for mobility and access to school. As streets were now safe, there were more walkers and cyclists, resulting in less traffic, better quality living and more "eyes on the street". As the public transportation was of high quality and served both residential and employment centres, more people opted for it. This created a virtuous cycle attracting more passengers and increase in service frequency, thereby further improving service.
- Access to Healthy and Environmentally Sustainable Transport Systems
  Boston believed that an environmentally sound transportation system can help to maintain health, reduce traffic congestion and improve the quality of life. A variety of sustainable transportation options that can fit different needs, including walking, bicycling, public transit and greener cars were thus provided for. Along with promoting physical activity for children, by walking to school, some of the benefits also included reduced transportation costs for the district and a sense of connection and community between the school and the neighbourhood. Boston known as a bicycling and walking city, released the New Balance Hub—way bicycle sharing system, which logged a number of rides to include annual members and casual riders. Boston governance implemented a policy in which public schools reserves 50% of each school's seats for children within the walk zone which extends 1 mile for elementary schools, 1.5 miles for middle schools and 2 miles for high schools.

2.3.1 Boston—Lessons Learnt

There are four key lessons to be learnt from the Boston example as discussed below.

- Public Transit Systems
  Understanding the numbers involved in transit use, car ownership and transportation systems helped Boston to effectively plan for population growth and a smooth movement pattern. However, Boston's work places still remained challenging to access for individuals without cars, and time taken by public

transit systems remained long and frustrating. Addressing these issues, the MBTA recently completed purchasing sufficient line trains to be built and working by the beginning of 2018. In addition, track advancement will assist in keeping the line free of ice and snow in winter.

- Alternative transportation options to help decrease the city's greenhouse gas emissions
  Due to an increase in automobile ownerships, the scattered development pattern is unsustainable and contributes to air, noise and water pollution. As a solution, in Boston most buses, trucks and construction vehicles use diesel fuel, which produces particulates and nitrogen oxides. City of Boston has an alternative fuel vehicle procurement policy for city vehicles. The city's diesel vehicles run on a cleaner blend of biodiesel and ultra-low-sulphur fuel, reducing emissions by 12–17% (bostinno.streetwise.co). Boston is the largest municipal purchaser of biodiesel in New England. The city has undertaken a large-scale retrofit of its school bus fleet (MIT report 2005), using ultra-low-sulphur diesel, and is being equipped with pollution control technologies, reducing tailpipe emissions by more than 90%. The Boston Bikes initiative seeks to make Boston a world-class bicycle city by creating safe and inviting conditions for all residents and visitors by expanding bicycle lanes, offering Ride-Along Fridays and sponsoring the annual Hub on Wheels event. The new complete streets approach puts pedestrians, bicyclists and transit users on equal footing with motor-vehicle drivers. The initiative aims to improve the quality of life in Boston by creating streets that are both great public spaces and sustainable transportation networks. It embraces innovation to address climate change and promote healthy living. The objective is to ensure Boston's streets are multimodal, green and smart. The City of Boston also launched EV Boston, part of the complete street initiative designed to promote the use of electric vehicles in Boston.
- Adequate public funding
  Transportation infrastructure to include roads, flyovers, public transit, airports and seaports is the supporting systems of economic connectivity and dynamism for Greater Boston and Massachusetts. The existence of a economically sustainable multimodal system is necessary to guarantee a state of maintained system and also to provide future expansions that allow residents, goods and services to remain exceedingly mobile. Hence, a lot of funding is utilized in research and implementation of better and more sustainable transportation technology.

2.4 Singapore

Singapore is an island city-state of 640 km^2 with a population of 3 million (1996). It once used to be a small fishing village of 150 occupants and later expanded to be British regional trading post in 1819, and since independence in 1965 into a major Asian metropolitan centre, Singapore has witnessed significant changes in the evolution of its economy (Yuan 1997). With limited land, a large populace and an animated economy, the stress on Singapore's roads has been extreme (especially during peak hours). This has resulted in a deteriorating traffic situation, particularly in the city centre. Many of the roads built are too narrow to accommodate the increasing volume of traffic and growth in car population. The ratio of cars to households has risen from 1:4 (i.e. 26 per 100 households) in 1980 to 1:3 (Yuan 1997) (i.e. 31 per 100 households) in 1990. Easing urban traffic congestion can be done in numerous ways. At a conceptual level, one can identify and categorize these measures according to three broad groups. The first set of measures are related to the development of making more urban structures conducive to the distribution of economic activities and improved physical integration between work, recreational facilities and housing with transportation. The second category comprises measures that address the service of transport facilities; i.e. they are basically directed at increasing urban transport capacity by providing better services. Measures in the third category are aimed at creating a more inclusive environment, specifically, in terms of encouraging effective use of existing transport facilities. Singapore has integrated all three types of measures within a single urban land transportation policy. Additionally, Singapore has taken steps towards restricting car possession through a quota scheme executed in 1990, thereby limiting the amount of new cars. Therefore, for cities to function efficiently and provide a liveable environment, urban planning and transportation planning/management must be coordinated and integrated within one strategic package (Yuan 1997). This has been adapted in Singapore. In 1972, a long-term extensive plan was formulated for the complete urban landscape, with the intention of integrating land use with transportation. The plan fulfilled this by focusing on the enhancement of new towns around the core; the decentralization of manufacturing, retail and institutions to the new towns; and the creation of a strategic road and rail network. The objective is to ensure free-flowing traffic such that mobility in all forms within the city is efficient. The Masterplan outlines the pursuit of a "people-centred" transport system (Sun 2013).

- Integration of public transport with land use
  The first strategy aims at integrating transport planning and urban development by strongly relating the development of urban transport facilities with other measures for supporting desired patterns of land use (Yuan 1997). This includes strategies for an appropriate development and focussed high building densities around mass rapid transit stations including more connections where required so as to ensure good accessibility for travellers to their destinations. Additional efforts include plans to decentralize commercial activities from the central area to four new regional centres in suburban areas to minimize the threat of congestion in the central area. It will also incorporate urban development into the current transit bringing work spaces closer to homes. This will result in a less transport-intensive, reasonable and efficient and harmonious urban environment. The Singaporean Masterplan does not absolutely address the issue of land use/transport integration in any detail. Though it is an outstanding document with strong policy and project directions across a range of transport issues, it does not

directly discuss transit-oriented development concepts such as better neighbourhood design, the structure of public spaces with respect to public transport, or typologies for the mixed-use developments that might surround mass transit stations (Hale and Charles 2008). Singaporean planning generally sees a close relationship between mass transit (in particular) and major urban development initiatives.

- Better services
  Singapore's Masterplan gives priority to public transport over private transport. It is formulated over the strategy that expansion of road networks cannot meet the needs of the community at large. The target of 70% of peak hour journeys by public transport is a case in point (Hale and Charles 2008). Another important target is the goal of doubling daily public transport trips to 10 million by 2020 (LTA 2008). The transport system is largely based on the mass rapid transit (MRT) rail network, which appears set for further emphasis under the expansion plans outlined in the LTA (Land Transport Authority) 2008 Masterplan. The bus service receives special attention in the plan for targeted improvement. However, it is functioning efficiently on the ground at the moment. The level of integration between bus and rail provides an important case study.

(a) Mass Rapid Transit: Singapore's Metro
Singapore has invested in a mass transit system in order to provide citizens and permanent residents with a fast and effective means to commute. Introduced in the late 1980s and first operating on a single line only, the Singaporean metro system or mass rapid transit (MRT) has now become an effective system integral to the city. Within the past three decades, the MRT has evolved into a vast and reliable system of public transport with an objective of 8 out of 10 families within a 10-min walk of a rail station. As is the case with other methods of public transport in Singapore, the MRT network is not owned and operated by a single company. It is rather a joint effort of SBS transit, a subsidiary of the leading transportation company ComfortDelGro, and SMRT trains (Yuan 1997). This does not affect one's commute since tickets and fares are calculated by the distance travelled and are independent of the means of transportation or the operator.

(b) The Bus system
In 1925, the Singapore Traction Company (STC) was set up to run a bus system based on trolleys. Within 10 years, motor vehicles became sought after and the STC started its motorized omnibuses. The lack of rules for implementation, however, led to privately owned bus companies that provided bus services outside the STC routes. The privately owned buses became a preferred transport mode as it offered routes that the STC were not covering. In 1935, the many small operators were consolidated into 10 companies with 144 buses. These buses ran regular services with bus schedules and routes in the different parts of Singapore. However, the bus companies suffered operational difficulties and poor management. Each of these companies operated from a different territory without incorporation of timetables, routes or fares making it

complicated for passengers. To solve this problem, the government decided to re-examine the public transport system of Singapore. The government reorganized the bus system in the 1970s, renovating it from one which was uncontrolled with a free hand, to one that had a centrally planned approach. In 1973, the remaining bus companies fused to form the private entity, Singapore Bus Service (SBS). The formation of SBS resulted in the integration of all bus routes and the standardization of fares (Cheong and Loh 2015). In 2009, the Land Transport Authority (LTA) took over the bus planning function from operators in order to plan Singapore's land transport network more holistically. In 2012, the government increased bus services quickly while awaiting for the new rail projects to be completed by launching the Bus Service Enhancement Programme (BSEP). In this S$1.1 billion (US$815 million) scheme, the government partners would introduce 1,000 new buses into the current bus fleet (Cheong and Loh 2015). When the BSEP is fully implemented by 2017, 100% of feeder services will be required to operate at eight-minute intervals during peak periods. The MRT, with its route of 83 km and forty-eight stations, in conjunction with public buses, serves three million users every day and constitutes some 51% of all daily motorized trips (Yuan 1997)

- Providing people centric environment and policy

(a) Facilitate Walking and Cycling
As per the Masterplan, walkers and bicyclists can anticipate increased and improved services in the future. For pedestrians, more than 200 km of new sheltered walkways around transport nodes will be available by 2018 under the Walk2Ride programme, four times more than today's 46 km (Sun and Yew 2015). Currently, all MRT stations and bus interchanges have minimum one barrier-free access route, and most bus stops are barrier-free. The remaining bus stops are to be upgraded where feasible. Additionally, key pedestrian foot overbridges will be gradually installed with lifts. For walkers, a longer green man time at 500 pedestrian crossings is complete around about 2015 under the Green Man + Programme. For cyclists, a cycling network of over 700 km island wide will be completed by 2030 (Sun and Yew 2015). To encourage cycling to the stations, more bicycle racks are installed. A total of 5,800 tracks were installed from 2013 to 2014, at 34 MRT stations. A bicycle sharing pilot is also being planned.

(b) Promote Flexi-Travel
LTA is also encouraging flexi-travel. The Free Pre-Peak Travel on MRT Scheme, launched in June 2013, encourages commuters to travel early and avoid the peak periods. This scheme was extended to June 2015. They are also launching a Travel Smart Network providing companies grants to encourage employers to support flexi-travel arrangements for their employees.

(c) Maintaining Affordable Public Transport
Sustaining an affordable public transport (PT) is an important socio-economic component of sustainable development. Ensuring inexpensive PT for the

general public is also a key factor in supporting the push for greater PT mode share. The Public Transport Concession Scheme is available for select groups or individuals to enjoy reduced travel rates. These groups include children under the age of seven, students from primary to tertiary (polytechnics and universities), full-time national servicemen, senior citizens, persons with disabilities (PWD) and the adult monthly travel card (LTA 2015). Approximately, half a million commuters enjoy the concession schemes.

Future plans

The LTA plans to extend the rail network to about 360 km by 2030 (up from 182 km today). Dozens of new mass rapid transit (MRT) stations shall be built. Additionally, with the distribution of new trains and signalling system upgrading, MRT commuters can expect faster train arrivals and better connectivity (Land Transport Authority 2015). The BSEP will add 80 new bus services and 1,000 new buses, for public bus by 2017 to ease traveller crowds and reduce waiting times. Three more integrated transport hubs will be developed by 2019 on top of the existing seven. Travellers will be able to make more comfortable transfers and enjoy the greater convenience offered by such hubs.

2.4.1 Singapore—Lessons Learnt

In considering best-practice principles for transport planning, it is important to be mindful of the different contexts and operating environments, and other characteristics. Singapore's people-centred land transport system is delivered through a clear set of important "universal values" presents itself, and these should be pursued in any location that is serious about a planned transport future. Singapore has achieved an effective transport system through integrated planning and development, commuter-centric land transport strategies and demand management. In doing so, it has proven sustainable transport to be a key element in the development of sustainable cities.

3 Conclusions and Recommendations for Indian Cities

Transport is a basic need. Urban transit is diversifying. Economic prosperity and growing car use are not necessarily linked together like they were till a decade ago (Newman and Kenworthy 2011). It is expected that public transport trips will increase. Cycling and walking are visibly expanding in many developed countries (Pucher et al. 2010). These developments are supported by substantial supply-side interventions and policies (Santos et al. 2010). Sustainable transport is one that does not rely on natural resources, rather it uses even renewable energy in the best possible way. Based on the above-discussed cases, the hierarchy for sustainable transport would be as follows:

1. Walk.
2. Bicycle.
3. Public Transport (Bus, train, etc.).
4. Electric/solar/hybrid vehicles (pooling is even better than using individual vehicles).

Sustainable transport focuses mainly on conservation of resources and provision of pollution free safe environment. Indian cities are densely populated and can be transformed into powerhouses of economy by planning them to support sustainable mobility. Technology and problem solving can be used to aid to various sustainable means of transport making it a smart city in a very holistic way. According to Shanghai Declaration on Better Cities, Better Life, it is mentioned that cities should integrate environmental issues into urban planning and administration and accelerate the transition to sustainable development. One of the greatest challenges is to plan and invest smartly in the infrastructure for sustainable smart transport. As mentioned earlier, a blanket smart city strategy may not work in Indian scenario. We come across many cases (including the case studies in this chapter) where smart transport strategies are planned based on the needs of community, post analysing and understanding the anthropology of that community. These are the cities where such strategies have worked best and are endorsed by its people. Transport solution for a smart city has to be a right mix of "smart" and "sustainable". Below are a few suggestions for transforming Indian cities towards a more sustainable smart transport system.

1. Walk
   There are various strategies proposed around the world to promote walking. Safe and well-designed footpaths are a prime requirement. The sidewalks and footpaths should be clean, welcoming and comfortable for all age groups of people. Surveys and studies show that people prefer walking if the basic amenities like a grocery store, school, post office, ATM, etc., are located at a walking distance. Barton et al. (2010), sets criteria for accessibility to promote active travel according to which amenities like school, health centres and local shops must be located within a radius of 600 m. While amenities like library, post office, community centre, leisure centre can be located in the radius of 800–1900 m. Similar strategy is put forth in ecocity proposal where a bus stop, emergency centres, service points, shops, larger public parks, youth centre, schools should be located within 300 and 500 m (Skala et al. un-known; Barton et al. 2010). In short, it should not take more than 15 min of walk for people to reach their desired destination. This encourages them to walk or cycle. Combining smart technologies with this further makes it a more preferable choice. These could be sensor-based solar lighting, seating areas with Wi-fi and charging points, generating electricity by walking on footpaths, information boards, variable messaging signs and interactive information panels.

2. Bicycle
   Most of the population in India, walk or use a bicycle due to low affordability. There are also few who use these means for health benefits or leisure purposes. Studying the condition of roads and basic infrastructure in India, there are issues like absence of proper footpaths and bicycle tracks, absence of bicycle parking spaces, low safety on roads, pollution, etc., that discourages people from using bicycles as a daily means of commute. These are over and above the issues posed by urban density and longer travel distances. In the year 2015, Columbia brought together more than 4000 dignitaries from across the world to discuss the challenges and opportunities of urban cycling. Using a bicycle not only offers a health benefit but also reduces the need of car and indirectly the issue of pollution. Like walking, designing a safe and dedicated infrastructure for cycling would encourage people to choose to cycle. A well-knit network of bicycle lanes with proper parking facilities that connect basic amenities and recreational areas is necessary for ensuring safe and uninterrupted cycling conditions as seen in the case of Paris and Germany. Other very good examples for such a network are Copenhagen, Amsterdam, Utrecht, Strasbourg, etc., that fall under top 10 cities designed for cycling. Ideas like bicycle rental and bicycle sharing, digital bicycle parking, glow in the dark bicycle lanes, signages, real-time traffic information for cyclists etc., are some of the smart solutions to promote cycling.
3. Public Transport (Bus, train, etc.)
   As said by Gustavo Petro—"A developed Country is not where the poor have cars…. it is where the rich use public transport". Various public transport modes are the backbone of a Smart city. The smart city mission -India has proposed smart transport proposals like metro, smart bus stops, real-time traffic updates, last mile para-transport connectivity, etc. Various cities like Pune, Ahmedabad, Jaipur, etc., have been selected for the provision of smart transport solutions. Making public transport frequent, comfortable and easy to use will promote people to use it. To add to this electric buses and CNG hybrid buses will help reducing air pollution and result in 30% less consumption of fuel as seen in the Boston Case. Solutions like large-scale retrofit of school/office buses using ultra-low-sulphur diesel, and equipped with pollution control technologies, reducing tailpipe emissions by more than 90% as done in the case of Boston, can also be applicable in India. Real-time update of location of the bus and the time it will take to reach the stop, further promotes its usage. Most of the Western cities follow this system. Another convenient aspect is the smart card for payment of the bus/metro fare as seen in the case of Paris and Singapore. The smart bus stops and metro stations can have solar powered lighting fixtures, fans, etc., as well as charging points for mobile and laptop. Wireless Internet can also be provided for the commuters. Safe bicycle parking near the station encourages people to use bicycle to reach the bus stop/metro station if they are not at a walkable distance [Bicycle Metro]. This kind of facilities encourages people to use public transport systems over motorized vehicles. If more and more people start using the public transportation facilities, need for building wider roads and

car parking facilities will reduce. Air and noise pollution on the road will also reduce.

4. Electric/solar/hybrid vehicles (pooling is even better than using individual vehicles)
   Apart from using bicycle, walking or using public transport, one can also choose to use personal motorized vehicles. Plug-in electric vehicles that run on battery, solar vehicles or hybrid vehicles are some of the sustainable solutions. With the advancement in the technology and higher efficiency in the electric motors, one can consider switching to the EVs or HEVs (electric vehicles or hybrid electric vehicles) to achieve better fuel economy. Hybrid and plug-in vehicles have considerably lower emissions as compared to conventional vehicles. Of course for an EV, one must understand that the life cycle emissions depend on the sources of electricity used to charge it. The countries having abundant solar power or wind/tidal power should rely on the same for powering the EVs to further minimize the emissions. The countries which depend on fossil fuels for electricity generation may not really benefit from switching to EVs. Several car manufacturing companies such as Ford, Mahindra, Toyota; provide options like eco-boost, hybrid, plug-in hybrid, etc., which are not only eco-friendly but also budget friendly. Apart from personal vehicles being EVs, even public vehicles like taxi, auto-rickshaw can be electric or hybrid. These can provide last mile para-transport and play a major role in enhancing the public transit access (Kaing 2012). Another sustainable option is to carpool. There are various mobile applications as well as NGOs that facilitate car pool. This method is economic, sustainable as well as time saving. If people carpool, the need of parking will automatically reduce. For everyday commuters, availability of shared cars around major transit stations can allow them to reach their final destination quickly and conveniently. As seen in the case of Germany, the policies should discourage people from using cars and the public transit infrastructure should be world class.

To conclude, Indian cities will need a solid sustainable transport system that can cater to the needs of growing population and promote a better lifestyle by providing a better environment. This will happen only if city-specific urban planning and policies related to transport go hand in hand in favour of more sustainable measures for the future.

References

Barton H et al (2010) Shaping neighbourhoods for local health and global sustainability, 2nd edn. Routledge, London

Beatley T (2000) Green urbanism: learning from European cities. Island Press, Washington, D.C.

"Bike Metro" Los Angeles County Metropolitan Transportation Authority

Buehler R, Pucher J (2009) Sustainable transport that works: lessons from Germany. World Transp Policy Pract 15(1):13–46. Retrieved 10 June 2017, from http://www.eco-logica.co.uk/pdf/wtpp15.1.pdf

Cheong C, Loh N (2015) Singapore's bus reform experience. Journeys, pp 17–28. Available at: https://www.lta.gov.sg/ltaacademy/pdf/J15May_p18Choi_SingaporeBusReform.pdf. Accessed 18 June 2017

Duchene C (2000) Paris region mobility plan (PDU): a strategy for the durable development of the region and the capital. Urban Transportation and Environment

Hale C, Charles P (2008) Visions for a sustainable transport future—a comparative analysis of transport planning approaches in Singapore, Vienna and Brisbane. In: Australian Transport Research Forum. Transport Tasmania, pp 17–37. Available at: http://atrf.info/papers/2008/2008_Hale_Charles.pdf. Accessed 15 June 2017

Hass PM (2006) Housing and transportation cost trade-offs and burdens of working households in 28 metros. Center for Neighborhood Technology. Retrieved from http://www2.nhc.org/media/documents/chp-pub-hl06-cnt-report.pdf

http://www.smartcitiesdive.com/ex/sustainablecitiescollective/measuring-life-quality-boston-indicators-project/150386/

http://www.bostonindicators.org/indicators

http://bostinno.streetwise.co/channels/how-boston-is-reducing-carbon-emissions-greenhouse-gas-emissions/

http://smartcities.gov.in/content/makeinindia.com

https://www.internations.org/singapore-expats/guide/16089-transport-driving/public-transport-in-singapore-16095 (2017) Public Transport in Singapore. Available at: https://www.internations.org/singapore-expats/guide/16089-transport-driving/public-transport-in-singapore-16095. Accessed 18 June 2017

Kaing EK (2012) The taxi: friend or foe? Understanding planners' perceptions of the taxi industry. University of California, Los Angeles

Kumar V, Dahiya B (2017) Smart economy in smart cities. https://www.researchgate.net/publication/306924920_Smart_Economy_in_Smart_Cities. Accessed 17 June 2017

LTA (2008) Land transport masterplan: a people-centred land transport system. Land Transport Authority, Singapore

LTA (2015) Sustainability update 2014. Singapore Land Transport Authority, Singapore, pp 3–6. Available at: https://www.lta.gov.sg/content/dam/ltaweb/corp/PublicationsResearch/files/ReportNewsletter/SustainabilityUpdate2014.pdf. Accessed 15 June 2017

Mehra S (2016) Grant Thornton: an instinct for growth. Smart transportation-transforming Indian cities. Available at: http://www.grantthornton.in/globalassets/1.-member-firms/india/assets/pdfs/smart-transportation-report.pdf. Accessed 7 June 2017

MIT report, Vehicles and Air Quality (2005). In summer air quality symposium series at Endicott house. Retrieved 15 June 2017, from https://energy.mit.edu/wp-content/uploads/2005/08/MIT-LFEE-06-003-RP.pdf

Newman P, Kenworthy J (2011) 'Peak Car Use': understanding the demise of automobile dependence. World Transp Policy Pract 17(2):31–42

Newman P, Beatley T, Boyer H (2009) Resilient cities. Responding to peak oil and climate change. Island Press, Washington, D.C.

Paroutis S et al (2013) A strategic view on smart city technology: the case of IBM smarter cities during a recession. Technol Forecast Soc Change. http://dx.doi.org/10.1016/j.techfore.2013.08.041

Pucher J, Dijkstra L (2003) Promoting safe walking and cycling to improve public health: lessons from the Netherlands and Germany. Am J Public Health 93(9):1509–1517

Pucher J, Dill J, Handy S (2010) Infrastructure, programs, and policies to increase bicycling: an international review. Prev Med 50:106–125

Santos G, Teytelboym A, Behrendt H (2010) Part II: policy instruments for sustainable road transport. Res Transp Econ 28:46–91

Schmidt S, Buehler R (2007) The planning process in the US and Germany: a comparative analysis. Int Plan Stud 12:55–75

Skala F et al (un-known) Encouraging walking, the role of urban design experiences of the EU-ECOCITY Project. Walk 21. Available at: http://www.oekostadt.at/root/img/pool/files/walk21_toronto_text.pdf. Accessed 4 Sept 2017

STIF (2003) Integration and regulatory structures in public transport case study—Paris-ile de France. Syndicat des Transports d'Ile-de-France, Paris, pp 1–20. Available at: https://www.emta.com/IMG/pdf/Surveys_Paris_case_study.pdf?17/ac1ebd88f513ebac00762db1ab967e19f820ca3d. Accessed 9 June 2017

Sun G (2013) Sustainable land transport system. Singapore's experience. Available at: https://sustainabledevelopment.un.org/content/documents/3756sun.pdf. Accessed 25 June 2017

Sun G, Yew S (2015) Sustainable transport in Singapore. Journeys, (May), pp 29–37. Available at: https://www.lta.gov.sg/ltaacademy/pdf/J15May_p30Sun_sustainabletransportinsingapore.pdf. Accessed 15 June 2017

Trevien C, Mayer T (2013) Urban public transportation and firm location choice evidence from the regional express rail of Paris Metropolitan area. In: 53rd congress of the European regional science association: regional integration: Europe, the mediterranean and the world economy, 27–31 Aug 2013, Palermo, Italy. European Regional Science Association (ERSA) Louvain-la-Neuve. http://hdl.handle.net/10419/123954

Yuan LL (1997) A case study on urban transportation development and management in Singapore. In: Second international expert panel meeting on urban infrastructure development, Bangkok, Thailand, pp 8–9

Role of Electric Vehicles in Future Road Transport

Punit Kumar, K. N. Srivastava and Atul Dhar

Abstract With increasing pollution in the environment, there is a need of sustainable generation and utilization of energy. Transport is the technologically most challenging major contributor to environmental pollution. With increasing global adaption of renewable energy generation, if somehow, the transportation energy supply can be shifted to renewable sources; then, it will be major leap in mitigating the impact of pollution on human beings. Battery electric vehicle provides the most feasible solution right now, which has a developed and demonstrated infrastructure in terms of technology development and fuel, in this case electricity, distribution network. With decreasing tariffs of renewable energy and new and cheap battery technologies, it is self-evident that the future has electric vehicles and renewable power generation. This chapter provides the insight into the battery electric vehicle technology, needed resources and the economics which is developing around this technology.

Keywords Electric vehicle · Lithium battery · Sustainable transport technologies

1 Introduction

With food, cloth and house, there happens to be urge for mobility on which every human being has relied. It is not just the case of present but also of the past. Humans have been roaming around the planet, and doing so, transport has become a necessity. While there is a possibility for reducing the need to travel, it remains an integral part of our life until we invent robots so sophisticated that can become our surrogate. For now and for foreseeable future, the human commute has increased

K. N. Srivastava
ABB Corporate Research, Västerås, Sweden

P. Kumar · A. Dhar (✉)
School of Engineering, IIT Mandi, Mandi 175005, India
e-mail: add@iitmandi.ac.in

A. Gautam et al. (eds.), *Sustainable Energy and Transportation*,
Energy, Environment, and Sustainability,
https://doi.org/10.1007/978-981-10-7509-4_4

compared to past and even going to increase in future reason being the globalization.

Humans first started travelling on their feet. Later, they started using animals for this purpose. In the next stage, the wheel was invented. The wheel was a great invention and changed the life of humans forever. The wheel was then used with animals, and thus, cart was invented. These carts were used to travel from Europe to Asia via silk route for transport of goods and people. Horse carts were in use till the nineteenth century in USA. After the invention and popularization of electric motor, the electric cars were favourite among people owing to the easiness in driving and silence it had. However, with the advancement in the gasoline engine technology, gasoline cars started picking pace in the race. Starter motor made hand cranking obsolete, ignition system, rubber engine mounts and carburettor made easy to drive gasoline cars. In 1910, Ford was selling their model T car at half the price of the any electric car available at that time. This became the end of electric vehicles for that time. Being economical as well as reliable, the gasoline cars became favourite to the people and usage of gasoline cars started to increase. With all the advancement and excitement, however, the nature began to respond to the changes. With all the carbon that was till then buried under the ground, started moving to the atmosphere, our planet began to warm and climate began to change. It did not stop there but with time has increased to a level that it needs a solution; otherwise, we will get regular natural crisis and devastations.

In 2014, humans have released 32,381 tons of CO_2 in the atmosphere at global level; out of which, 42% is from electricity and heat generation and 23% is from transport sector (IEA 2017). China and USA are leading with 27.4 and 15.7% of this emission, which in total accounts for more than 40% of total global emissions. This gives an indication that the developed countries like the USA are the main source because of which all humans have to suffer. In the European Union (EU), transport sector accounts for 33.2% (Reece 2016) of overall energy consumption which ultimately emits the carbon fuel burned in the engines to atmosphere causing global warming. This global pollution is causing the ischemic heart diseases, strokes and lung cancer and acute lower respiratory infections in children. An estimated 12.6 million deaths and 7 million premature deaths are accounted to air pollution in the year 2012 (WHO 2017a, b). So while the causes have been identified and transport is one important source of these emissions, options to curb the emissions from transport need to be explored. These options are summarized briefly in subsequent sub-sections.

1.1 Internal Combustion (IC) Engines

IC engines are the most popular and widespread propulsion system in the world used in transport vehicles. Due to their good reliability, range and power, these have been used from scooters to buses and trucks. In IC engine, the air fuel comes together in the engine cylinder and ignites. The way the fuel and air comes into the cylinder and mechanism of ignition classify the engine in two types: spark ignition engine (SI) and compressed ignition (CI) engine.

In SI engine, air and petrol are mixed together and sent into the engine cylinder. In cylinder, the piston compresses the mixture to a compression ratio of about 9–14 and combusted by using the spark plug. The SI engines are the highest power density energy converters. Though they are reliable, their overall efficiency is not more than 40% for gasoline at optimum driving conditions and produces about 411 g of CO_2 per mile (Greenhouse Gas Emissions from a Typical Passenger Vehicle 2014). New technologies like direct fuel injection, air management system, thermal energy recuperation, precise fuelling, lean combustion and higher compression ratios are being used to alleviate the efficiency problem. SI engine also produces air pollutant like CO, NO_x and UHC and greenhouse gas CO_2.

In CI engine, the air is brought into the cylinder and compressed by the piston. At a certain point, the fuel is brought into the chamber which is ignited due to the combustible conditions in the chamber. CI engine has higher efficiency due to the higher compression ratio, un-throttled air intake and lean operation but produces about 411 g of CO_2 from each mile driven due to higher carbon content of diesel than gasoline (Greenhouse Gas Emissions from a Typical Passenger Vehicle 2014).

Some novel technologies are also under development like homogenous charged compression ignition (HCCI), reactivity controlled compressed ignition (RCCI), premixed charge compression ignition (PCCI) and cold air intake (CAI). These all technologies rely upon the low temperature lean and efficient combustion to limit the emissions produced.

Researchers around the globe are trying to limit our dependency on gasoline and diesel and are trying to use other fuels in the same engine as an attempt to contain the global warming and ensure the fuel supply after the exhaustion of these fuels (diesel and petrol). The other fuels have some degree of satisfaction in terms of use and resource development but mostly are in the development stage.

1.2 Battery Electric Vehicles (BEV)

BEVs are fully electric vehicles designed and developed to run entirely on electricity. Energy is stored in battery packs usually underneath the seats and is supplied to high torque electric motors, which drive the car. BEVs provide zero emissions if measured from tank to wheel emissions. But the electricity production remains dependent highly on fossil fuel plants which forbids battery electric vehicles to be called zero emission vehicles from the perspective of well to wheel. However, the global commitment and policies to reduce emissions have increased the adaption rate to solar power generation which, if adopted successfully, will make battery electric vehicles to be truly zero emission vehicles.

Electric Vehicles Program (Sperling 1995) was started before the IC engine counterpart and was even popular in USA in the year 1904. IC engine vehicles of the early twentieth century were tough to start, it needed cranking, noisy, smelly and horses were frightened by it. Whereas the electric vehicles were silent and easy to drive. But then, the gasoline car technology development made electric car obsolete in few years. Gasoline powered cars could be started by motor, rubber engine mounts reduced vibrations, and advances in carburation and ignition made gasoline cars easier to drive. Electric vehicles were restricted because of the limited development in battery technology. Gasoline cars became cheap, were available for half the price of electric vehicles, and became reliable. After Ford started selling model T, the middle-class families could afford the cars (History 2017). That was the end of electric vehicles for that time. After this, no noticeable development was seen in the electric vehicle sector. In 1980, the General Motors announced to produce electric vehicles commercially. It was the time of oil crisis, and the idea was later abandoned as the prices of oil came down again. The present interest in electric vehicles is because of the environmental concerns. After this, renewed interest General Motors again announced the electric vehicle plans and even unveiled a sporty electric vehicle in 1990. But later, the plans never materialized properly, and electric vehicles were never progressed. The reason for these dramatic events was probably the battery technology that never reached a point of satisfaction. When the lithium-ion batteries came into existence, the interest in the electric vehicle was again raised. After the General Motors planned to drop the electric vehicle program, Tesla Motors came into existence. Elon Musk, CEO of Tesla Motors, invested heavily in the lithium-ion battery run EVs and himself designed the Tesla Roadster which demonstrated the lithium battery technology to be feasible (Vance 2015). At present, they are continuously decreasing the price of the cars and also launching cars for every sector. Currently, the cheapest Tesla car Model 3 has the range of 346 km and costs around 35,000 USD (TESLA 2017), whereas the cheap gasoline and diesel car costs around 10,000 USD.

1.3 Fuel Cell Electric Vehicles

Parallel to battery electric vehicles, the fuel cell technology has been invented to replace the IC engine vehicles. A fuel cell (Sperling 1995), in the present case hydrogen fuel cell, is a device which converts the hydrogen and oxygen into useful electricity. Fuel cell is similar to battery in case that it has electrodes, electrolyte and positive–negative terminals. But unlike battery, it does not store electricity. It produces electricity continuously from the hydrogen gas just like the IC engines. The electricity produced by the fuel cell is used to drive the electric motor powertrain, which propels the vehicle. The fuel cells are twice as efficient as compared to the IC engine and do not have a problem of range like the BEVs. These vehicles can simply be fuelled at fuel station like gasoline vehicles, and vehicle can have increased range capacity in few minutes. The reaction of hydrogen and oxygen produces water as the product and does not emit any pollutants.

Initial fuel cells were limited to space application and were first used in mid-1960s for Gemini spacecraft and later in Apollo spacecraft and space shuttles.

This NASA research initiated the interest of this new technology for vehicles. However, the technology was not yet ready for the adaption. The fuel cells were heavy, bulky and costly. In the nineties, after an extensive research in the area, several buses were built in Europe and United States. At present, the fuel cell technology has reached a position that it can satisfy the customer's expectations (Greene 2013). However, the cost of the technology is not yet compelling for the customers, and thus, this technology is not in the mainstream till now. The hydrogen production and distribution infrastructure are the roadblocks to the successful implementation of the hydrogen fuel cells. US, Germany, Japan, South Korea and Scandinavian countries all have plans to bring fuel cell electric vehicles to the market and deploy the needed hydrogen production and distribution infrastructure (Greene 2013; Fuel Cell Electric Vehicles 2013).

1.4 Hybrid Vehicles

In the simplest terms, hybrid vehicle (Sperling 1995) is the combination of IC engine and battery electric vehicles. It generates power from the IC engine running at constant, more economical engine operating mode. This engine then produces electricity which charges the batteries. The power from this battery ultimately drives the electric motors. The IC engines in the hybrid vehicles are small and are used as an electricity generator to extend the range of a battery vehicle. Engines are run at low speeds which are more efficient and use all electronically controlled system to maintain the efficiency as in the advanced IC engine vehicles. In terms of complexity, these systems are more complex than IC engine or battery electric vehicles and also require flywheel or ultra-capacitor for peak power demand in some configurations.

Hybrid technology has been in the minds of inventors from the early twentieth century, but the development of IC engine vehicles never gave chance to hybrids. Hybrid technology interest was invoked again in 1970s over concern of oil dependency and in 1980s over concern of climate change. After that, there has been significant development in the hybrid vehicles and different configurations have been designed to satisfy the customers. Toyota has been a clear lead player in this technology covering 66% market of hybrids in 2014 (German 2015). Toyota with its technology is able to decrease the fuel consumption by 30% followed by the Ford and Hyundai/Kia with 14 and 8%, respectively. Total hybrid share of the vehicles was maximum for Japan with over 20% followed by Netherlands and USA with around 6 and 3%, respectively, in 2013 (German 2015). Increased complexity, weight and need to fuel two different power systems rather limits the interest in the hybrid technology and thus the future of hybrids looks not so promising.

With recent high growth in popularity of battery electric vehicles, next section discusses the technological and other logistics details of BEVs in detail.

2 Battery Electric Vehicle Technology

In its simplest form, the battery electric vehicle powertrain has two components: (1) batteries and (2) electric motors with controllers. Batteries store the energy. Controller controls the energy supplied to motor from battery. With this supplied energy, the motor rotates and drives the vehicle.

2.1 Batteries

Battery is a device in which several cells are connected in series to provide required energy output in terms of volts and ampere (Larminie and Lowry 2013). The cells convert chemical energy into the electrical energy as DC power. While recharging these batteries, the reactions are reversed by reversing the current and thus bringing the battery back to charged state. Depending upon the types of chemical used in these cells, there are different batteries available. Lead acid batteries are most readily available and used batteries in the market. In these lead-acid batteries, the negative plates have spongy lead and positive plate have lead oxide. The electrolyte used in these batteries is sulphuric acid. The most important advantage of these batteries is the extremely low internal resistance. Internal resistance reduces the voltage at the battery terminals and increases while charging. This is also a waste of energy; thus, it is desired to have a low resistance for batteries to be used in electric vehicles. Hence, lead-acid batteries are good candidate when it comes to have low resistance. Specific energy of these batteries, however, is very low 20–35 Wh/kg. For comparison, the energy density of petrol is around 10,000 Wh/kg. To have a similar energy content as 1 kg of petrol, 330 kg of lead-acid batteries is needed. Although, the EVs have better efficiency, which reduces the energy requirement for the same range and thus reduces the weight of the batteries. To have a range of 100 km, the petrol needed is around 5 kg where for the same range, batteries needed are of around 300 kg. However, if the weight of the engine is considered, then the engine alone weighs around 150 kg in the car. In addition, the petroleum fuel car has flywheel, gear box and transmission system which is omitted in the EVs and motors are connected directly to the wheels. This compensates some weight problem for the EVs.

In the past, many of the cars were made using the lead-acid batteries, but recent advancement in different batteries has reached to a level that the new batteries are showing promises regarding the specific energy for the use in the EVs. Table 1 shows the properties of different batteries. After using Nickle metal hydride batteries for some years, lithium-ion batteries are currently showing promises and presently are in use for EVs at industrial level. As mentioned in Table 1, lithium-ion provides 3 times the specific power than the lead-acid batteries and thus reduces the weight of the electric vehicle considerably. In addition, it has low charging time of 2 h and in which also, 60% can be charged in 30 min. While this

Table 1 Comparison of commercially available batteries

Battery	Specific energy (Wh/kg)	Energy density (Wh/L)	Specific power (W/kg)
Lead acid	30	75	250
NiCd	50	80	150
NiMH	65	150	200
Li-ion	90	150	300
Zinc–air	230	270	–

does not remove the charging time problem completely, it reduces this significantly. These batteries also do not have memory effect and thus need not be discharged fully. The main problem with these batteries is the safety concerns. These batteries can burst into flames in case of overcharging or short circuit. To avoid these circumstances, various mechanical and electronic measures are taken for controlling the temperature and pressure of each cell. The charging and discharging are also controlled through charge controller. Tesla Motors are providing the warranty of 8 years for battery packs which is rather sufficient for a car. These have proven the reliability of lithium-ion batteries for the electric vehicles.

These lithium-ion batteries came into market in nineties and have not changed since then. These batteries have comparatively high specific energy, but it still lag behind petrol and diesel. There is a need for a battery with high specific energy and reliability. Metal air batteries are offering some of these qualities right now. Metal air batteries have shown to have specific energy of 225 Wh/kg in early 2000s (Larminie and Lowry 2013) and now have reached around 4000 Wh/kg for lithium–air batteries (Lee et al. 2011). This is a completely new type of battery in which the metal electrode is consumed in the discharging process and needs to be changed every time. These electrodes can be thought of as the fuel like gasoline for the vehicle. Due to this reason, charging time is also less than 10 min, time to change the consumed metal electrodes. But this battery is under development as of now and cannot be commercialized before further development (Larminie and Lowry 2013; Dromantien and Tripolskaja 2009).

At present, the leading electric car manufacturers are focusing on lithium-ion batteries, which have shown the reliability and have been used at industrial level to make electric cars.

2.2 *Electric Motors*

Electric motors are the machines which convert the electrical energy into speed and torque. The current supplied by the batteries in the coil when comes in the magnetic field of the magnet, coil feels a force which drives the axle connected to the coil (Larminie and Lowry 2013). The torque and the speed of the motor are controlled by the supplied voltage and the strength of the magnetic field. In DC brushed motors, the direction of the current is changed periodically in a way that the force is exerted on the coil in a single direction. This produced power is increased with increasing the supplied voltage. But it cannot be done indefinitely. This is restricted by the heat generated through the losses in the motor. These brushes of this motor are a big problem because of the friction and wear. These brushes are removed and replaced by the electronic switches in a brushless motor. Unlike brushed motor, heat generated in these motors is not concentrated at the centre but is generated in the stator. So it is comparatively easy to dissipate. More advanced motor is the induction motor. It is powered by alternating current. In electric motor, as we only have DC available, this DC needs to be converted into AC by use of the inverter. In a three-phase induction motor, the winding is such that the coil constantly gets force exerted in one direction. In this way, the motor continuously rotates. As mentioned above, brushless DC motor dissipates heat in the outer stator shell from which it is much easier to remove heat. It also reduces the size and mass of the motor which is preferred in electric vehicles. Brushless DC motors also have high power density compared to other motors which is an added advantage.

3 Resources Needed for Sustainable Transport

Petroleum-based vehicles, with all the benevolent characteristics, have only one drawback. They are driven on a fuel, which is increasing the emissions and causing the climate change. To avoid similar situation in future, future transport needs to be sustainable in terms of both, operation as well as production. In case of EVs, the electricity needs to be generated in a sustainable manner and sufficient lithium should be available to have an electric vehicle future.

3.1 Lithium Consumption and Reserves

Increased popularity of electric vehicles demands lithium to be mined at a great scale. With transport as a service, decreasing prices of driving an EV per mile will attract more customers in future (Arbib and Seba 2017). Each lithium battery needs about 2% by volume lithium. With that, it is estimated that 1 million tons of lithium will be needed per year till 2030. With transport as a service, there is an expectation that the model of individual car ownership will be disrupted and replaced with fleets of cars providing transport service to the customers. In that case, few cars will be used more efficiently to deliver transport service to the customers, which will also be cheap. With this model, the number of cars on the road will also decrease. In the USA, the cars on road are expected to decrease to 50 million in 2030 from the current 250 million (Arbib and Seba 2017). This decrement in cars will limit the requirement of lithium at some point of time.

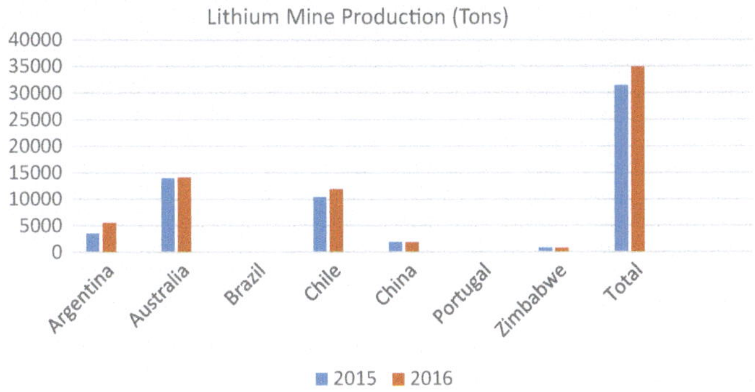

Fig. 1 Global lithium production for 2015 and 2016 (Jaskula 2017)

Fig. 2 Global sector wise lithium consumption (Jaskula 2017)

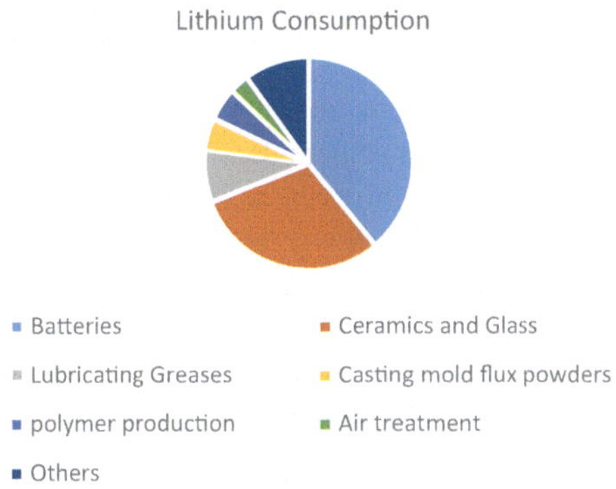

The current production of lithium (Fig. 1) (Jaskula 2017), without the data of US lithium production (as withheld to disclose company proprietary data), shows around 30,000 tons/year. Australia leads the way with over 14,000 tons of production each year followed by Chile with over 10,000 tons. If we look at the consumption of the lithium (Fig. 2), a major share is taken by battery production. This fraction of battery is going to increase drastically in near future.

Figure 3 shows the lithium resources whose economical extraction is currently and potentially feasible (Jaskula 2017). In total, around 45 million tons of lithium resources are currently available. This value might increase in future as new methods are derived to extract this commodity economically. According to the predictions, the demand for lithium is going to increase to just over 100,000 tons of lithium (534 k ton of lithium carbonate) in 2025 (Hocking et al. 2016). Considering these numbers, it seems there is enough lithium available for electric vehicles and these reserves will increase further when new technologies are invented to extract lithium from more resources in the future.

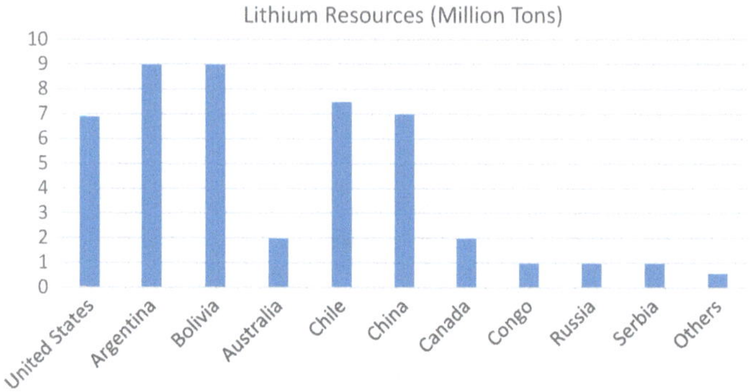

Fig. 3 Global lithium resources available (Jaskula 2017)

3.2 Renewable Electricity Generation

The tank to wheel emissions from an electric vehicle are zero; thus, it can be called a zero emission vehicle. But if the electricity itself is produced from fossil fuel, then it might result in even a worse system. It is worse because in conventional IC engine vehicles, power is being produced in vehicle itself and was consumed in vehicle. But if the power is produced at a particular place and then transferred to a vehicle in different place, then there is a transmission loss occurring in the system which reduces the efficiency of the system. Hence, the production of the electricity needs to be in a sustainable manner for EVs to be sustainable.

Electricity is produced from several ways: thermal steam plants, gas turbine plants, hydro plants, nuclear plants, solar, wind etc. Thermal and gas plants are energized by coal, natural gas or other fuel gases. These conventional methods of electricity generation are non-renewable with very little efficiency. Lower efficiency means that a lot of fuel is directly going waste and in the process contributing to climate degradation, loss from both sides.

World total renewable power capacity is on the rise in recent decade (Fig. 4). In that also, solar, wind and biomass have seen drastic growth and support. Solar has reached to 300 GW from being insignificant a decade before. Wind is also following the same trend. While developed countries are focusing and increasing their renewable share, India is not lagging behind. India is adding renewable resources at even higher rate (Fig. 5). India has a total capacity of 90 GW from renewable resources currently. Solar and wind are the most focused at present while reaching to 10 and 30 GW, respectively. This growth in solar has been seen in just last five years. The efforts made by India are appreciable, and India is fulfilling the commitment made in the Paris climate deal.

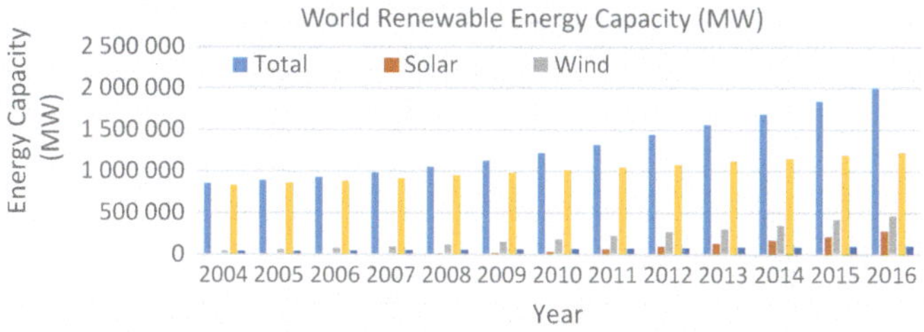

Fig. 4 World renewable energy capacity (Renewable Capacity Statistics 2016)

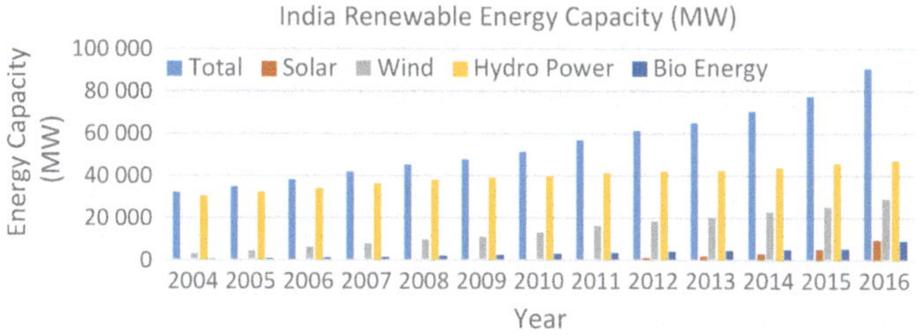

Fig. 5 India renewable energy capacity (Renewable Capacity Statistics 2016)

4 Economics of EVs

4.1 Evolution of EV Market

Historically, the electric vehicles were invented as early as nineteenth century. However, low range and high battery cost remained as the hurdle in the adaption of EVs. With the invention and industrialization of IC engine cars in early twentieth century, the cars were much cheaper and were in reach of global population. Thus, battery electric vehicle remained untouched, and it took some time and technological development to realize its potential.

After 2010, when Tesla Motors went public, there was a rapid rise in EVs adaption globally. Below are the countries where a mass adoption of EVs was seen (Outlook 2016). The USA has seen continuous rise in the registration of electric cars from 2010 to 2015 and was the largest market in the world until China surpassed it in 2015. In European countries, there has been a constant interest in the electric vehicles. Especially, Norwegian people love Tesla. Norway is the largest market in the world for electric vehicles in terms of market share. They have 23% of electric cars. Following are the Netherlands and Sweden with 10 and 3%. While being the largest market in terms of sales, USA and China have around 1% in market share for electric cars (Fig. 6).

Popularity of electric vehicles is increasing worldwide. There is number of reasons for this trend, including the plummeting battery prices and continuously

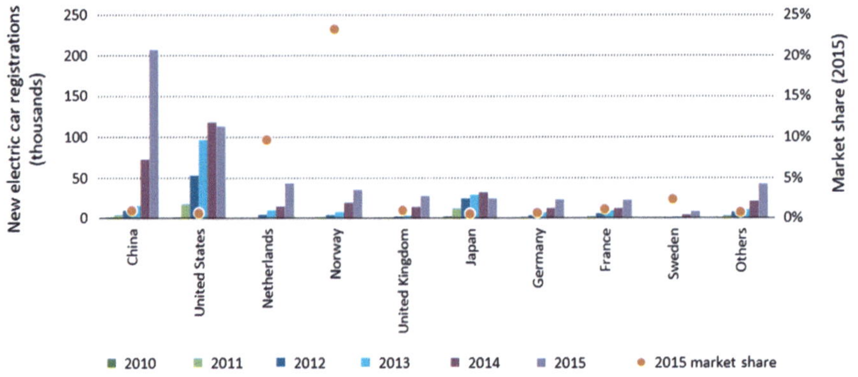

Fig. 6 Electric vehicle registration in countries (Global EV Outlook 2016)

spreading recharging network. The main problem with the electric vehicle due to which electric vehicle had not seen the roads in past is the range anxiety concerns of the drivers. With increased energy density of the lithium-ion batteries, this anxiety is somewhat controlled. To address this problem further, Tesla Motors started making supercharging networks all over the USA and later on other countries to let drivers charge their cars for free, even if the car is not Tesla. This made driving electric vehicle completely free. With more charging station addition to the network, the range anxiety will end.

In every investment, the payback time is of high importance. For comparative price tag of an EV and a diesel car, driving an EV costs 1000 USD cheaper than diesel car for normal miles driven in a year (Idaho National Laboratory 2017). With faster recharging time, longer lifetime and low cost, electric vehicle cost is going down and with life of 20 years, the EVs are going to be favourite to people. New technology like metal air batteries (Dromantien and Tripolskaja 2009) when developed completely, will topple this anxiety problem.

The trend of adapting the electric cars is on the rise and expected to be so for the years to come with the support of the technological development, renewable energy development, competitive economic development, government policies and intensive support.

4.2 *Global Policies*

Figure 7 shows the government policies and incentives on electric vehicles around the globe (Outlook 2016). Countries are giving tax incentives and tax credits to relaxed roadways for driving the electric vehicles. China, the largest market, has nationwide policies of rebates at registration and sales tax exemption for electric car owners. While USA has targeted policies, it gives tax credits. European countries are leading the way in policy making and implementation with Norway providing

VAT and sales tax exemption, and waivers on parking and toll fees, access to bus lanes to drive the vehicle. With all the incentives and policies, government is also forcing manufacturers to shift to electric vehicle programs by tightening the emission norms. India does not have much electric vehicle market share currently, but still imposing strict emission norms. Recently, in 2017, India has imposed Bharat stage IV norms nationwide on automakers. Now, India is deciding to skip the Bharat V and impose Bharat VI directly in 2020. In application side, the Indian government is focusing on pool taxi services as well as electric buses in some states (The New Indian Express 2017; The Times of India 2017; Electric Buses in India 2015). Indian government is suggesting to lower taxes and loan interest rates on electric vehicles while capping sales of petrol and diesel cars, and it has been considered as a radical shift in policy (The Indian Express 2017). From this chart, it is clear that these policies have a much larger effect on the adaption of the electric vehicle.

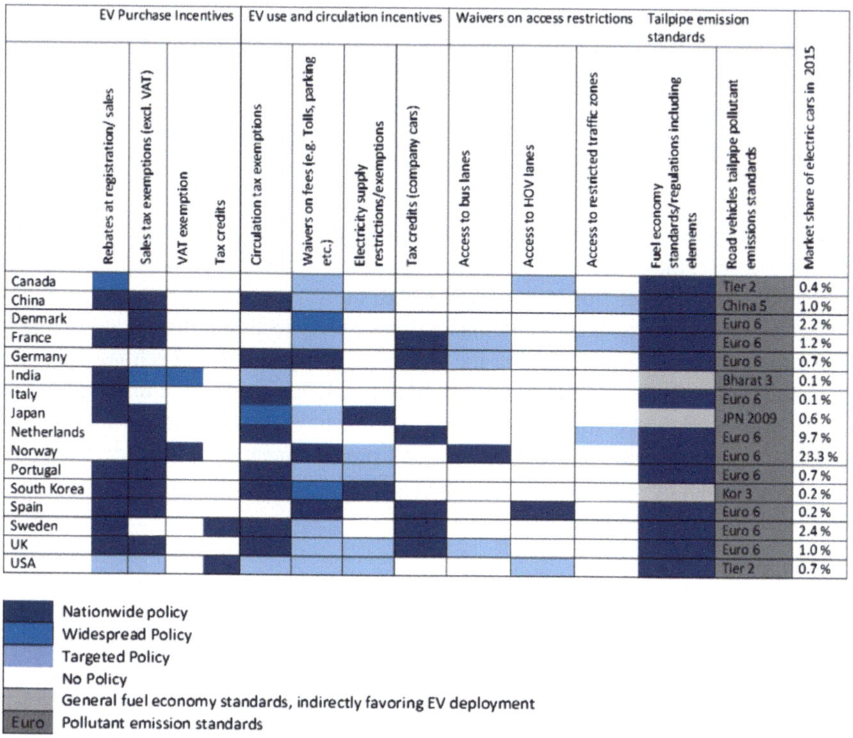

Fig. 7 Global policies on emissions and zero emissions vehicles (Outlook 2016)

4.3 Business Models for Sustained Electric Vehicle Growth

4.3.1 Charging Stations

While Tesla Motors is installing supercharging networks to various places covering the long distances, there are many start-ups forming businesses around the charging problem of electric vehicles such as in Canada SWTCH (The Globe and Mail 2017), in USA PlugShare (Scientific American 2017) and EVmatch (Clean Technica 2017). These have similar business model of connecting the private home owners to drivers for renting the electrical outlet. They have made an online platform facilitating the aggregation, bookings and payments all at one place. These online platforms even take into account the charging rate and the time of charge in the day to smartly calculate the bill. Airbnb has now partnered with Tesla to provide such facilities and attract EV drivers as customers (Scientific American 2017).

A new start-up has gone a step further and is providing the electricity for free. Volta is a star-tup which is combining the recharging of the electric vehicles and advertising. After charging the cars at the Volta charging station, the driver does not need to pay for the electricity. Instead he needs to sit in his car and watch the advertisements, which company shows on the display (The Globe and Mail 2017).

4.3.2 Vehicle to Grid (V2G)

Despite being the second biggest investment after the house, on an average of about 95% of the time, the cars are parked. There are times in a day at which the grid is on high loads. To facilitate this high demand grid, these vehicles can be used to pump electricity into the grid at peak demand times. V2G technique provides demand response to the grid via discharging the electric vehicle batteries (Technavio 2017). To enable this, a bi-directional electricity flow facilitator needs to be installed in the cars. There is also a communication between vehicle owner and utility grid to control the power drain from the vehicle and precise metering of the electricity to calculate the money exchange at later stage.

4.3.3 Sharing of Vehicle

In present time, people are using their assets more effectively and efficiently. Houses are shared; computational power is shared; data are shared; and even jewellery is shared peer-to-peer, then why not cars! (Hoegh Autoliners 2017). As mentioned earlier also, cars are the most unused asset we have. 95% of the time, cars are parked. Unused asset is waste. To use this till now wasted asset, cars could be shared between persons when not in use or when there are empty seats in the car.

Despite the fact that the global car sales are decreasing, car ownership is an old concept. People want to use their investment. This opportunity is being seized by many start-ups and OEM car companies to establish shared economy for cars. Finland has been the front runner in this, and capital Helsinki is making it unnecessary to own a car for its citizens. In Finland, this concept is called as Mobility as a Service (MAAS) (Goodall et al. 2017). Advancement in mobile phone communications, cashless transactions and integration of many such systems is fuelling the growth of shared vehicle economy.

5 Near Future Prospects

What is the future of transport? What will be a future car would be like? How shall we travel from long and short distances? These are some of the questions that come to mind immediately when we think of pollution due to the transport sector and cannot find a solution in conventional vehicles. In this chapter, some technologies which are worked upon in recent times to address these questions have been discussed.

Despite technological development in lithium-ion battery, unless metal–air batteries are developed, range anxiety is going to affect the adaption of electric vehicles. To address this problem, Sweden recently experimented with a new road that charges the vehicle driving upon itself. If implemented successfully, this will completely remove the need to stop for charging. It will even be better than the conventional fuel cars. In Sweden, ELWAYS (2017) has used a technology called conductive feeding in which car uses a movable arm to connect to the conductor fixed in the road to charge itself. In Korea, Korea Advanced Institute of Science and Technology (KAIST) has made a demonstration with their inductive feeding technology online electric vehicle (OLEV) (Popular Science 2017). OLEV gets its power feed magnetically from the electric strips buried in the roads. OLEV, now in service in Seoul, pulls three buses behind it.

After successfully applying the electric vehicle technology to the private passenger vehicles, now companies are starting to try this on high-density passenger and goods vehicles as well. Goods carrying vehicles need more power and torque at times, which is seen as a problem in this sector. But Tesla recently has announced to make an all-electric truck Tesla Semi (Electrek 2017). They have completed a prototype and will be launching it in September, 2017.

Bikes are not different than cars and mass transport vehicle. Start-ups and big companies globally are covering from scooters and bicycle to sports bike to be transformed into an electric vehicle. Emflux is one of such start-up that is focusing on electric sports bikes in India. Ather energy is focusing largely on scooters.

For addressing the intercity and long travel distances, Hyperloop (Musk 2013) is the concept which is floating around for some time now. The capsules are moved in low-pressure tubes at high speed of 700–800 km/h. Due to very less air in the tube, it is possible to get very high speed overcoming minimal drag thus saving energy. Elon Musk first proposed the idea of Hyperloop and published a paper on SpaceX and Tesla website. From then on, several companies are actively pursuing the idea.

Besides that, several teams around the world are also working on designing the hyper pods to participate in the competition organized by SpaceX. In the proposed design of Hyperloop, it works completely on solar PV and in some cases, it is also claimed that it will generate excess electricity which can be supplied to grid.

Imagine a future where all the energy we need is in some way or another comes largely from the sun. We have solar PV panels, solar thermal collectors, wind mills which are continuously generating energy in your neighbourhood and may be at your roof. This completely clean energy is then transferred to all of your appliances which are talking to each other including your panels and cars. You take your car to office. There is no noise, no vibration, no pollution, high amount of power coming from zero emission sources. Your car is continuously getting charged directly from the roads with wireless charging. Your car drops you and then drives itself to the man standing on the next corner. Picks him up and works as a taxi earning bucks for you. You track your car and in the evening see the total revenue earned that day. In more long routes, the underground tunnel infrastructure will also be used. Your car will enter into an underground tunnel system in which the car is placed on a car skate and then transferred to destination at high speeds of 200 km/h skipping the traffic. There if you need to go to another city, the Hyperloop could be used.

6 Conclusion

The sustainable future of transport depends upon the advancements of technologies. As per the current status of the technologies, battery electric vehicle clearly wins and dominates among the considered transportation alternatives. The fuel cell technology is still too costly, and complexities in the hybrid vehicle technologies described earlier in the chapter limits its capabilities. To produce the lithium-ion batteries of the electric cars, the steps have been already taken by the companies. The lithium reserves are identified to be sufficient to meet the targets. If the global target of shifting to renewable energy generation is met, the battery electric vehicle will certainly be the future for sustainable transport.

References

Arbib J, Seba T (2017) RethinkX: rethinking transportation 2020–2030

Clean Technica (2017) EV charging startup EVmatch connects private charging station owners with EV drivers [online]. Available at: https://cleantechnica.com/2017/04/03/evmatch-launches-pilot/. Accessed 14 June 2017

Dromantien RĖ, Tripolskaja L (2009) Metal air batteries. European Association for Storage of Energy

Electrek (2017) Tesla Semi [online]. https://electrek.co/guides/tesla-semi/. Accessed 16 June 2017

Electric Buses in India (2015) Technology, policy and benefits 2015. Global Green Growth Institute, Center for Study for Science, Technology and Policy

ELWAYS (2017) Charging technology [online]. Available at: http://elways.se/charging-technology/?lang=en. Accessed 16 June 2017
Fuel Cell Electric Vehicles (2013) The Road Ahead 2013. FuelCellToday
German J (2015) Hybrid vehicles: trends in technology development and cost reduction. International Council on Clean Transportation
Goodall BW, Dovey T, Bornstein J, Bonthron B (2017) The rise of mobility as a service. Deloitte Rev 20:112–129
Greene DL (2013) Worldwide status of hydrogen fuel cell vehicle technology and prospects for commercialization 2013. Oak Ridge National Laboratories
Greenhouse Gas Emissions from a Typical Passenger Vehicle 2014. United States Environmental Protection Agency
History (2017) MODEL T [online]. Available at: http://www.history.com/topics/model-t. Accessed 10 June 2017
Hocking M, Kan J, Young P, Terry C, Begleiter D (2016) Lithium 101. Deutsch Bank Markets Research
Hoegh Autoliners (2017) Sharing economy disrupts car ownership model [online]. Available at: http://www.hoeghautoliners.com/news-and-media/blogs/sharing-economy-disrupts-car-ownership-model. Accessed 14 June 2017
Idaho National Laboratory (2017) Comparing energy cost for per mile for electric and gasoline fueled vehicles [online]. Available at: https://avt.inl.gov/sites/default/files/pdf/fsev/costs.pdf. Accessed 25 June 2017
IEA (2017) CO_2 emissions from fuel combustion [online]. Available at: http://www.iea.org/bookshop/729-CO2_Emissions_from_Fuel_Combustion. Accessed 15 June 2017
Jaskula BW (2017) Lithium. United States Geological Survey
Larminie J, Lowry J (2013) Electric vehicle technology explained
Lee J, Kim ST, Cao R, Choi N, Liu M, Lee KT (2011) Metal—air batteries with high energy density: Li–Air versus Zn–Air. Adv Energy Mater 1:34–50
Musk E (2013) Hyperloop alpha. PDF). SpaceX. Retrieved 13 Aug
Global EV Outlook (2016) Beyond one million electric cars. International Energy Agency
Popular Science (2017) Korea's online electric vehicle gathers power from the road through wireless induction [online]. Available at: http://www.popsci.com/cars/article/2010-03/koreas-online-electric-vehicle-gathers-power-road-wirelessly. Accessed 16 June 2017
Reece G (2016) EU energy in figures—statistical pocketbook 2016. European commission
Renewable Capacity Statistics (2016) International Renewable Energy Agency
Scientific American (2017) Electric car charging could follow Airbnb model [online]. Available at: https://www.scientificamerican.com/article/electric-car-charging-could-follow-airbnb-model/. Accessed 14 June 2017
Sperling D (1995) Future drive—electric vehicles and sustainable transportation
Technavio (2017) What is vehicle-to-grid (V2G) and how is it improving the EV market? [online]. Available at: https://www.technavio.com/blog/what-is-vehicle-to-grid-v2g-and-how-is-it-improving-the-ev-market. Accessed 14 June 2017
TESLA (2017) Model 3 [online]. Available at: https://www.tesla.com/model3. Accessed 25 June 2017
The Globe and Mail (2017) Toronto startup seeks to boost electric-vehicle charging options [online]. Available at: https://www.theglobeandmail.com/globe-drive/culture/commuting/toronto-startup-seeks-to-boost-electric-vehicle-charging-options/article35084045/. Accessed 13 June 2017
The Indian Express (2017) India's electric vehicles push likely to benefit Chinese car makers [online]. Available at: http://indianexpress.com/article/auto-travel/cars/indias-electric-vehicles-push-likely-to-benefit-chinese-car-makers-4670851/. Accessed 13 June 2017
The New Indian Express (2017) E-vehicles push: govt to focus on public transport system to tackle pollution [online]. Available at: http://www.newindianexpress.com/states/telangana/2017/may/18/tsrtcs-electric-bus-plan-hits-roadblock-as-costs-too-high-1606115.html. Accessed 12 June 2017

The Times of India (2017) Soon, metropolitan transport corporation to operate electric buses [online]. Available at: http://timesofindia.indiatimes.com/city/chennai/soon-mtc-to-operate-electric-buses/articleshow/59507846.cms. Accessed 12 June 2017

Vance A (2015) Elon Musk: how the billionaire CEO of SpaceX and Tesla is shaping our future

WHO (2017a) An estimated 12.6 million deaths each year are attributable to unhealthy environments [online]. Available at: http://www.who.int/mediacentre/news/releases/2016/deaths-attributable-to-unhealthy-environments/en/. Accessed 15 June 2017

WHO (2017b) 7 million premature deaths annually linked to air pollution [online]. Available at: http://www.who.int/mediacentre/news/releases/2014/air-pollution/en/. Accessed 15 June 2017

Supplementing the Energy Need of Diesel Engines in Indian Transport and Power Sectors

Achinta Sarkar, Maryom Dabi and Ujjwal K. Saha

Abstract The progress and prosperity of a nation like India depends largely upon the fossil fuel-based power production sectors. However, the use of these fuels leads to detrimental effects on the ecosystem because of various pollutants emitted from combustion of fossil fuel. Being the second largest populist country in the globe, India is heavily dependent on imported fossil fuels, thereby making it the major source of global warming and pollutant emission. The fossil energy in India is primarily used in transport and stationary power production sectors. Another problem with these fossil fuels is that it is located in the certain part of the globe which makes oil-deficient countries to depend on them. Therefore, an alternative arrangement is necessitated to reduce these dependencies and oil imports. Alternative renewable energy sources in the form of vegetable oil, biodiesel, biogas, producer gas, and alcohols have good prospects to replace or supplement fossil fuel. Oils derived from vegetables show the most promising fuel for diesel engines. However, vegetable oils have the lower calorific values along with the high viscosity and density as compared to diesel, which are not suitable properties to run the engine. Hence, these properties are improvised through blending, preheating, transesterifications, and emulsifications. Transesterification of vegetable oil yields biodiesel which is the most prominent and popular among the processes. Gaseous fuels such as biogas and producer gas have also been successfully implemented in diesel engine through the dual fuel mode. Stressing the importance of alternative fuel sources for diesel engines, the present study focuses on the effects of these arrangements on engine performance and emission characteristics.

Keywords Fossil fuels · Pollutant emissions · Ecosystem · Alternative fuels Diesel engine

A. Sarkar · M. Dabi · U. K. Saha (✉)
Department of Mechanical Engineering, IIT Guwahati, Guwahati 781039, India
e-mail: saha@iitg.ernet.in

A. Sarkar
e-mail: s.achinta@iitg.ernet.in

M. Dabi
e-mail: maryom.dabi@iitg.ernet.in

A. Gautam et al. (eds.), *Sustainable Energy and Transportation*,
Energy, Environment, and Sustainability,
https://doi.org/10.1007/978-981-10-7509-4_5

1 Introduction to Energy Scenario in India

Production, conversion, reservation, and efficient utilization of fossil and biorenewable energy sources are the key challenges to the researchers throughout the globe to keep a balance between the supply and demand in transport and power production sectors. The growth of the population and the advancement of the lifestyle of the societies in the both developed and developing nations in the world enhance the exponential utilization of energy. The energy being supplied to meet the demand at different transport and power production sectors is mainly from the limited fossil reservoirs. The British Petroleum (BP) Energy Outlook 2035 (BP Energy Outlook 2017) reported that the fossil fuels (oil, coal and gas) will remain the dominant source of energy and will supply almost 75% of total energy till 2035. Consequently, the depletions of inadequate fossil fuel reserves are fast growing which is challenging the country's energy security. Besides, the burning of these fossil fuels brings the globe at the carbon age with the consequence of global warming and alarming threat to the green ecosystem. India among the realms of the world is the fastest growing economy and has the second largest population in the world. Therefore, it has become the largest energy consumer in the world. India, China, and the USA have the world population of 18, 19, and 4%, respectively, and according to International Energy Outlook (IEA) report on India Energy Outlook (India Energy Outlook: IEA Special Report 2015), the nations consumed 6, 22, and 16% of world energy, respectively. However, according to this report, in 2040, India will be the largest populous nation in the world with respect to the energy consumption and will be the major driving force in global trends in all aspects. However, to sustain the progress of the nation, the crucial challenge is to ensure the balance of demand and supply of energy. The country has been meeting its commercial energy demand largely by imported fossil fuels such as coal, petroleum oil and natural gas. The scenario of these fuels has been briefly described below.

1.1 Petroleum Oil

Fossil oil is known as the highly traded commodity in the world. In India, oil is considered as the second largest primary source of energy for the commercial power production sectors. However, the country relatively has the poor resources of fossil oil reserves. Currently, India has the total reserve of petroleum oil of 5.7 billion barrels in offshore and onshore which is only 0.5% of the world reserves (India Energy Outlook: IEA Special Report 2015). However, the resources are insufficient to meet the rigorous budding demand. Figures 1 and 2 show the production, import, and import dependency of fossil oil in India (National Energy Map for India: Technology Vision 2030; World Energy Outlook 2009). They clearly indicate that

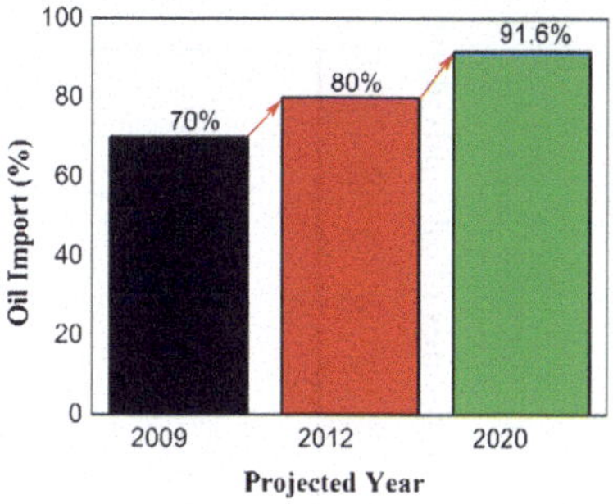

Fig. 1 India's oil dependency on imported crude petroleum oil. *Source* IEA, World Energy Outlook (2009)

the import and import dependency of petroleum oil increase with times (year) and the increment is projected as very high beyond 2030. It is evident from the data published in different reports that India need to proceed with concrete meaningful policy to ensure the energy security of the country and to balance the supply and demand to sustain the increasing trends of the development of the nation. Presently, the country is importing the crude petroleum oil from the most politically unstable Persian Gulf countries accounting a 65% of total import. In the new policy of energy, India is looking for the new regions beyond the Gulf for seeking the new sources and investments in oil fields and makes the deal with Sudan, Nigeria, Syria, Vietnam, Russia, North America, etc. According to the energy information agency (EIA), India will be the fourth largest oil importer in the world by 2025 after USA, China, and Japan (http://www.eia.doe.gov/emeu/cabs/India/Oil.html).

1.2 Natural Gas

As the reserves of the natural gas are abundant and combustion of this low carbon fuel produces less pollutant in comparison to higher carbon gaseous fuels, hence it is considered as a "bridge" fuel to "low-carbon future" (Ahuja and Tatsutani 2009; Kirkland 2010). In recent times, the Government of India has targeted natural gas as the fuel for future in Indian domestic and commercial purposes. However, the country is not endowed with adequate sources of natural gas and has only of 0.7% of the world reserves (British Petroleum 2013). However, the use of the gas was increased by twofolds during 2002–2012 (British Petroleum 2013). It was estimated by the World Energy Outlook 2013 that in India, natural gas had the fastest growing demands as compared to coal and oil. During 1990–2011, the compound annual growth rate (CAGR) of coal, oil, and natural gas was 4.9, 5.6, and 7.5%, respectively (IEA 2013), and the CAGR of coal, oil, and natural gas during the fiscal of 2011–2035 will be 3.1, 3.5, and 4.4 (IEA 2013), whereas the average CAGR of the world is estimated as only 1.6% (IEA 2013). Hence, due to the little reserves of natural gas and increasing demand growth, the dependency of imported natural gas is increasing. According to business-as-usual (BAU) scenario, the import

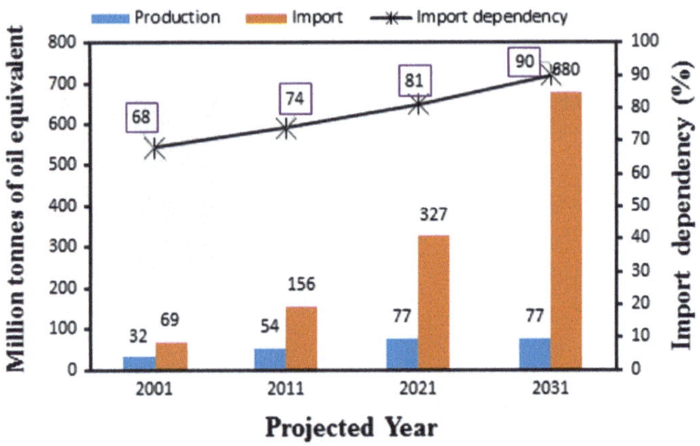

Fig. 2 Production, import, and import dependency of fossil oil in the business-as-usual (BAU) scenario. *Source* National Energy Map for India: Technology Vision 2035

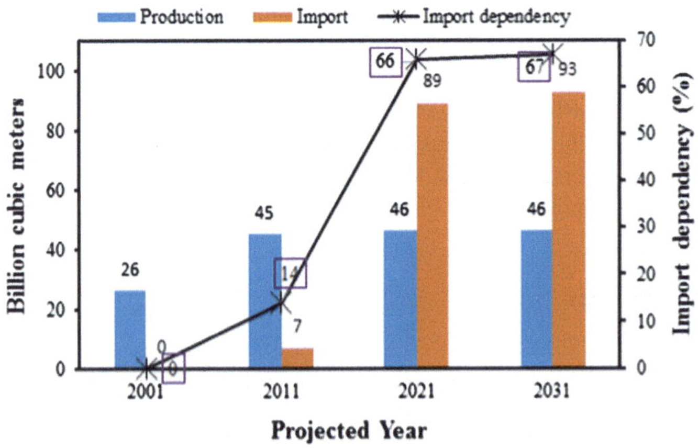

Fig. 3 Production, import, and import dependency of natural gas in the business-as-usual scenario. *Source* National Energy Map for India: Technology Vision 2035

dependency of natural gas from 2001 to 2021 will increase from almost 0 to 66% and, in 2031, the dependency will hover between 66 and 67% due to the development of infrastructure like making pipeline network to ensure the supply of gas at proper station and the bilateral agreement between the countries (National Energy Map for India: Technology Vision 2030). Figures 3 and 4 show the production, import, import dependency, and total gas demand of natural gas. It clearly shows the huge increment of import dependency and total gas demand from 2011 to 2040.

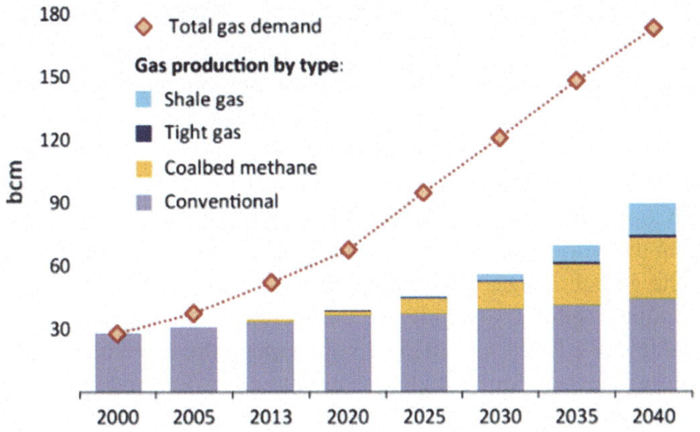

Fig. 4 Natural gas production in India in the new policy scenario. *Source* IEA, World Energy Outlook Special Report 2015 (bcm = billion cubic meters)

1.3 Inter-relations of GDP and Climate in India

The relation of the growth of the gross domestic product (GDP) to energy usage in India is shown in Fig. 5. It is shown that with the increase of GDP in India, energy demand increases with faster rate. On the other hand, meeting this energy demand has been fulfilled and is being targeted by importing the fossil fuels such as coal, petroleum oil, and natural gas. The above discussion of the production, import, and import dependency of fossil fuels (coal, oil, and natural gas) implies that the India has no other options to meet the energy demand of the country and to sustain the growth of the GDP. However, as all the fuels are carbonaceous, the country has become one of the largest greenhouse gas (GHG) emitters in the world with the global share of 6.5% (Olivier et al. 2015). As stated in World Resources Institute (WRI) data, the top 10 GHG gas emitters in the world are China, USA, European Union, India, Russian Federation, Indonesia, Brazil, Japan, Canada, and Mexico and they are responsible for the emission of 70% of total global GHG (http://www.wri.org/blog/). Among them, India is the fourth largest emitter of the GHG in the

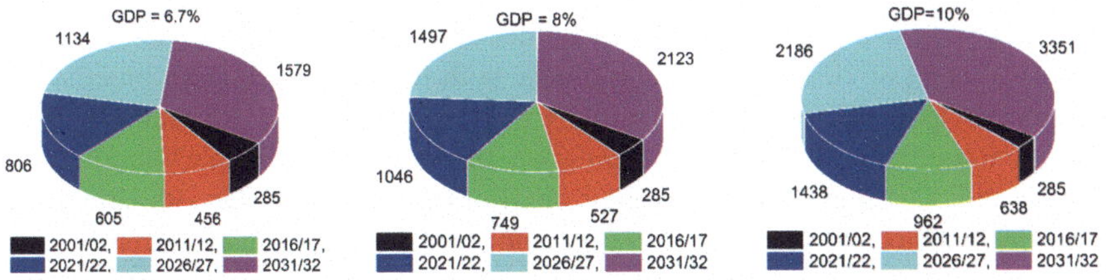

Fig. 5 Relationship between the growth of GDP and the energy dependency in India. *Source* National Energy Map of India, Technology Vision 2030

world. According to the Biennial Update Report (BUR) 2010, India released approximately 2136.84 million tonnes of CO_2 equivalent to GHG and, in 2030, the emission will climb up to 90% in contrast to present emission status (PARIS SUMMIT 2015). However, enhancing GHG emission keeps the whole world in completely devastating states which would destroy the green ecosystem, and for that reason, the historical Paris Climate Agreement (PCA) under the United Nations Framework Convention on Climate Change (UNFCCC) has been signed among 191 nations in the world. The nations except the USA have committed to hover "the increase in the global average temperature to well below 2 °C above pre-industrial levels and pursue efforts to limit the temperature increase to 1.5 °C above pre-industrial levels" (UNFCCC 2015). In this aspect, the agreement makes its own constitution for the developed and developing countries and binds them in a common platform to achieve the goals of the agreement. India is also very keen and conscious of her responsibility about the PCA. In this noble aspect, the Government of India has implemented eight national policies on long-term basis in the context of climate change in the report of National Action Plan on Climate Change (NAPCC), and they are: (i) national solar mission, (ii) national mission for enhanced energy efficiency, (iii) national mission on sustainable habitat, (iv) national water mission, (v) national mission for sustaining the Himalayan ecosystem, (vi) national mission for green India, (vii) national mission for sustainable agriculture, and (viii) national mission on strategic knowledge for climate change (http://www.moef.nic.in/ccd-napcc).

1.4 Status of Biofuels in India

The development of biofuel in India shows a slow progress as compared to other countries of the world. Concerned with the vulnerability of energy security and need of alternative fuels to substitute or supplement fossil fuel, the Government of India under the Ministry of New and Renewable Energy announced "National Policy on Biofuels" in 2009. It was observed that biofuels have "a ray of hope" in providing energy security of the country. The policy was aimed at mainstreaming biofuels and bringing about accelerated development and promotion of the cultivation, production, and use of biofuels to substitute petrol and diesel in transport and other sectors. The policy sets a goal to ensure that a minimum level of biofuels become readily available in the market to meet the demand at any given time. An indicative target of 20% blending of biofuels, both for biodiesel and for bioethanol, is proposed by 2017. The policy targeting of exploiting over 400 species of trees bearing non-edible oil seeds depending on their techno-economic viability for production of biofuel (National Policy on Biofuels 2009). In January 2003, the government launched the sale of 5% ethanol-blended petrol, and currently, this program is carried out in 21 states and 4 union territories with the target to achieve 10% ethanol blend in petrol. More than 111 crore liters of ethanol blended in the ratio of 10% were carried out in the ethanol supply year 2015–16. In the view of high demand of ethanol for blending, the production of second-generation ethanol through lignocelluloses route has been stressed now (Annual Report 2016–2017). On August 10, 2015, the government announced marketing of high-speed diesel blended with biodiesel (B-5) in selected retail outlets of oil marketing companies (OMCs) in New Delhi, Vishakhapatnam, Haldia, and Vijayawada. The government also allowed sale of biodiesel (B100) by private manufacturers to bulk consumer. Biodiesel-blended diesel (B5) is now being sold by OMCs in 6 states in more than 3621 retail outlets (Annual Report 2016–2017; Annual Report 2015–2016).

2 Energy Use in Transport Sector

According to BAU scenario, the energy use in transport sector in 2031 will be increased by 186.34% with respect to the consumption in 2016 as shown in Fig. 6 (National Energy Map for India: Technology Vision 2030). Figure 7 shows the net import, net production, net supply, and consumption by the different industrial and transport sectors and the total consumption by all the sectors which are consuming petroleum oil (National Energy Map for India: Technology Vision 2030). As observed in Fig. 7, the oil used in transport sector is increasing with time, and in the year of 2031/32, the consumption of oil according to BAU scenario will be enhanced approximately by 188.24% with respect to the present time (2016/17), and this will be the maximum consumption as compared to the consumption by the others sectors (National Energy Map for India: Technology Vision 2030).

Figure 8 shows the energy consumption in transport sectors by various modes in India (International Energy Agency, World Energy Outlook 2007). The consumption of energy (oil) in trucks and buses is found to be very high in comparison with consumption by the others modes in transport sector at all the times, i.e., in past, present, and future. On the other hand, the engines in trucks and buses are generally run with diesel. Therefore, diesel consumption will be expected to be high. Figure 9 shows the diesel consumption from 2014 to 2040. The consumption

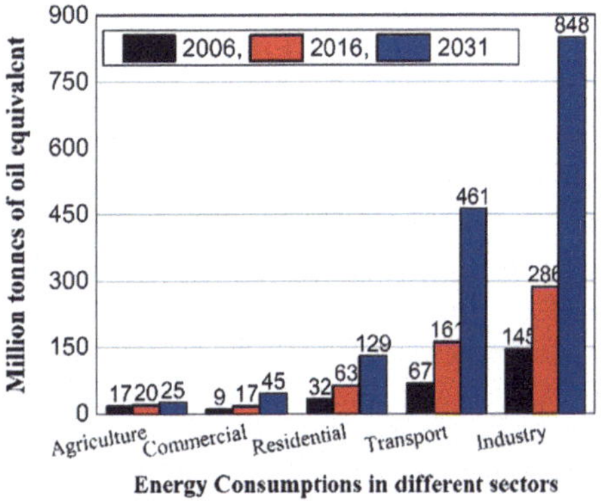

Fig. 6 Sector-wise energy consumption in the BAU Scenario. *Source* National Energy Map of India, Technology Vision 2030

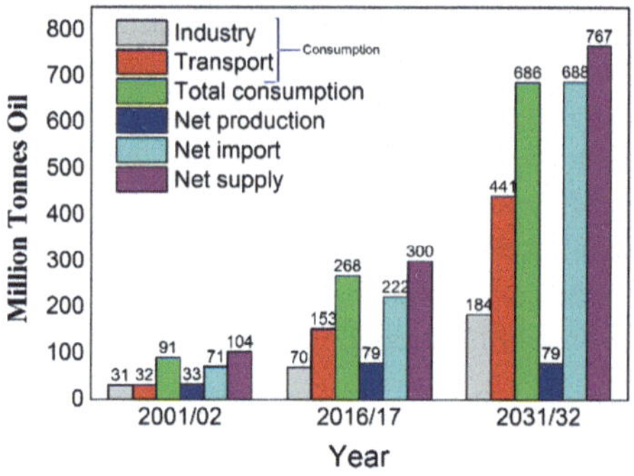

Fig. 7 Petroleum oil import, production, supply, and use in transport sector. *Source* National Energy Map of India, Technology Vision 2030

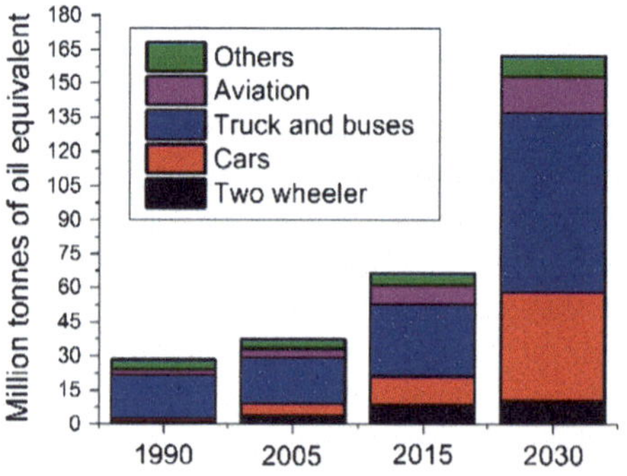

Fig. 8 Energy consumption in transport sector in India. *Source* IEA (2007) (Mtoe = Million tonnes of oil equivalent)

of diesel in transport sector is found to be dominant. In 2013, the diesel consumption in transportation sector was 1 mb/d (million barrel/day) which was 70% of the total oil consumption. This is because of a large number of freight vehicles (approximately, 60% run with diesel) and buses (around 35%). As a result, the use of total oil and at the same time the use of diesel will enhance extensively by 2040 (World Energy Outlook 2009).

The discussions in the previous sections revealed that with the enhancement of GDP, energy use in India enhances. On the other hand, India is heavily dependent on imported fossil fuels. Consequently, GHG emission in India is increasing excessively. Figure 10 shows the GHG emission in different sectors (Busby and Shidore 2017). Figure 11 shows the GHG emissions in transport sectors by different modes (India: Greenhouse Gas Emissions 2007). In the transport sector, the maximum GHG emission was found in road transport sector as shown in Fig. 11, whereas in Fig. 10 it is projected that the GHG emissions in 2030 will enhance substantially. Therefore, it is expected that in the road transport, the GHG emissions will also enhance significantly.

In the discussion of GHG emissions in the road transport, it is evident that the GHG increases due to predominant use of fossil diesel. However, it can be reduced by the use of the bioliquid and gaseous fuels. Figure 12 shows the scenario of the increment of the use of biofuels in the world (OECD/IEA 2011). The use, from the

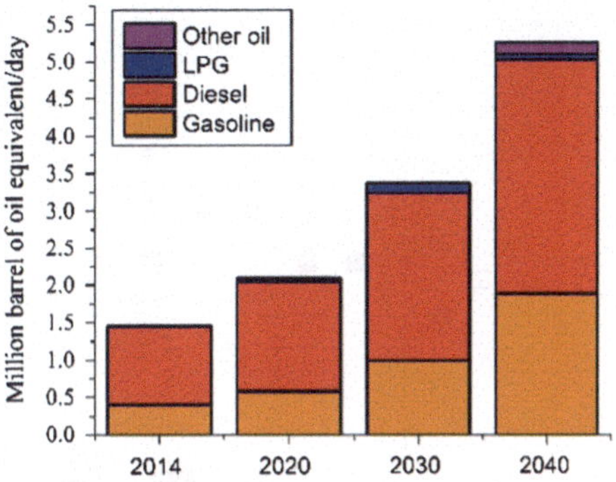

Fig. 9 Category-wise fuel demand in transport sector in India. *Source* World Energy Outlook, IEA (2015)

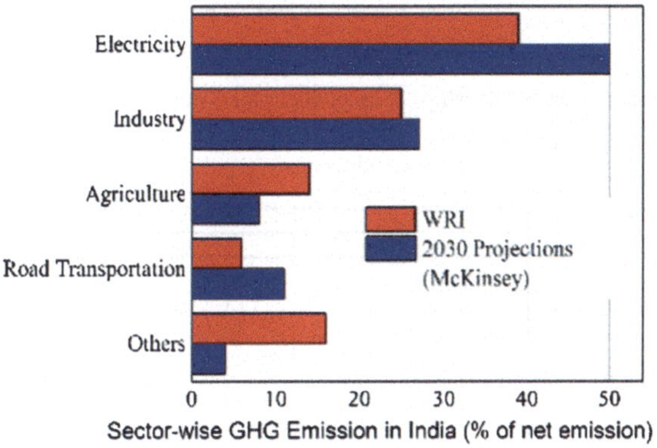

Fig. 10 Sector-wise greenhouse gas emission in India. *Source* Busby and Shidore (2017)

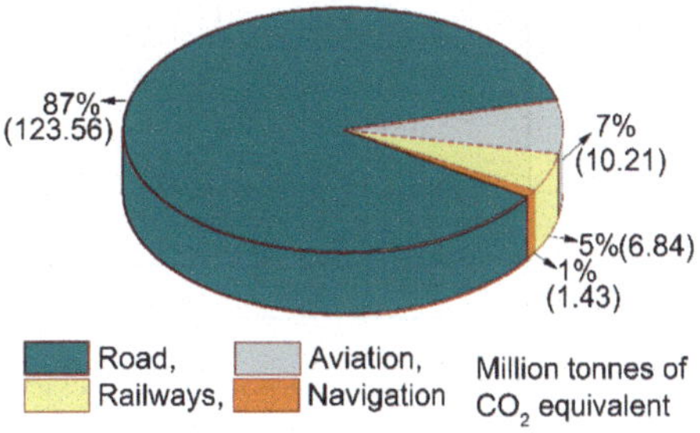

Fig. 11 Greenhouse gas emission in transport sectors in India in 2007. *Source* INCCA-2007

present to the projected future, is increasing, and the increment is found to be higher in the OECD countries than the developing countries. It may be due to the development of the new technology and the transfer of technology between the OECD countries. However, in India, the use of biofuels is also increasing (Fig. 12). Figure 13 shows the promising source of biofuel in different strategic states (National Policy on Biofuels 2009). Some of the biofuels are in commercial states, and some of them are in the research and development states. In Table 1, the different biofuels are presented and the safe use of biofuels blended with the

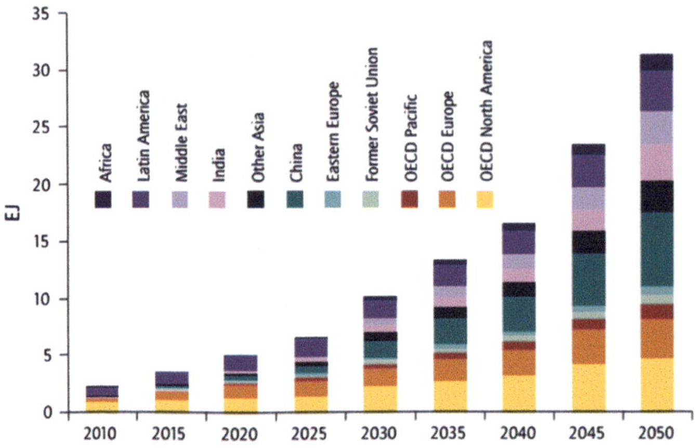

Fig. 12 Biofuel energy demand in transport sector in different regions of the world. *Source* IEA, Technology Road Map, Biofuel fuels for Transport (EJ = exajoules)

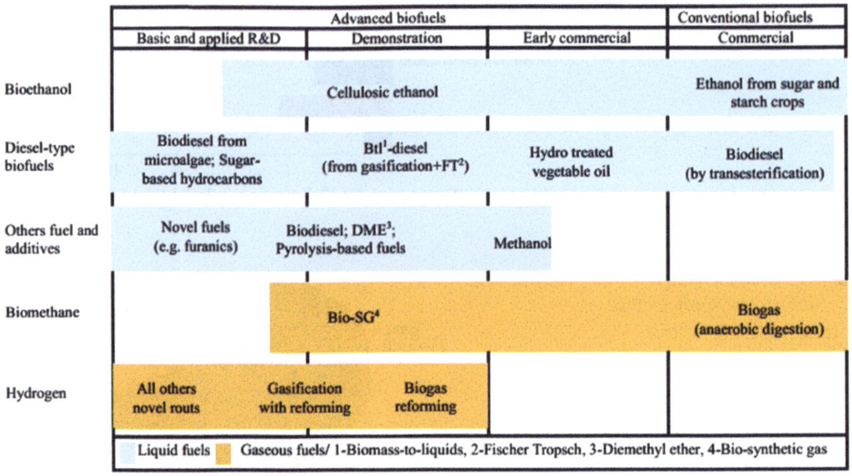

Fig. 13 Commercial status of primary biofuel technology in the world. *Source* IEA, Technology Road Map, Biofuel fuels for Transport

conventional petroleum fuels is described (OECD/IEA 2011). The substantial improvement of the diesel engine performance and emission with biofuels (liquid and gas) was noticed in published literature as well as in the investigation carried out by the authors.

Table 1 Overview of biofuel blending characteristics (OECD/IEA2011)

Biofuels	Blending characteristics
Sugar-based ethanol	E10 (Ethanol)–E15 (E25 in Brazil) in conventional vehicles; E85–E100 in ethanol vehicles
Starch-based ethanol	Same as above
Cellulosic ethanol	Same as above
Conventional biodiesel (fatty acid methyl ester)	Up to B-20 (biodiesel 20%) in conventional diesel engines
Hydrotreated vegetable oil (HVO)	Fully compatible with existing vehicles and distribution infrastructures
BtL (biomass-to-liquid) diesel	Same as above
Algae oil-based biodiesel/biojet fuel	After hydrotreating, fully compatible with existing vehicles and distribution infrastructures
Biogas	After upgrading, fully compatible with natural gas vehicles and fueling infrastructures
Bio-SG	Same as above
Biobutanol	Used in gasoline vehicles in blends up to 85%
Dimethyl ether	Compatible with LPG infrastructure
Methanol	10–20% in gasoline engine, blends up to 85% in FFVs (flexible-fuel vehicles)
Sugar-based diesel/jet fuel	Fully compatible with existing vehicles and distribution infrastructures

Source IEA, Technology Road Map, Biofuels for Transport

3 Implications of Alternative Sources

It is evident from above review that there is an urgent need of an alternative energy source to supplement or substitute the fossil energy for the future. In this perspective, biofuel can supplement/substitute the diesel fuel in the form of either liquid or gaseous mode. In liquid route, biofuels can be directly used in an engine, while the use of gaseous mode needs a minor modification of the engine. The types of biofuel are shown in Fig. 14. In the following sections, the implications of these biofuels in diesel engine have been discussed.

3.1 Biodiesel

According to ASTM D6751, biodiesel is defined as the "fuel comprised of mono-alkyl esters of long-chain fatty acids derived from vegetable oils or animal fats, designated as B100." Biodiesel can be derived from various edible and non-edible feedstocks such as soybean, coconut, sunflower, jatropha, animal fats, waste cooking oil. Subramanian et al. (2005) delineate that there are over 300 species of different non-edible trees in Indian Territory which produce oil-bearing

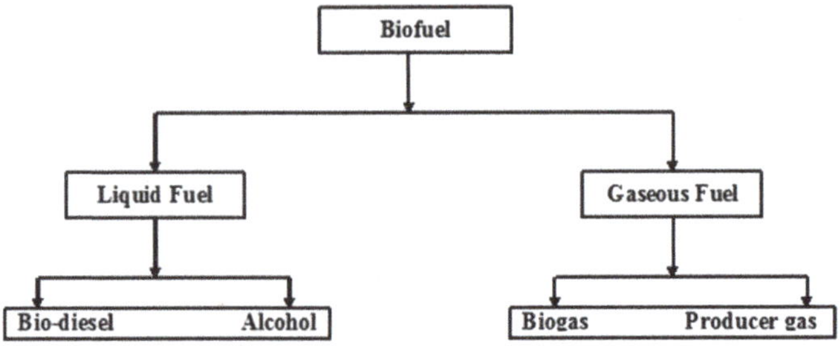

Fig. 14 Forms of biofuel

Fig. 15 Few selected non-edible seeds as the raw material to produce biodiesel (Kumar and Sharma 2011) (Reproduced with permission)

seeds. Hence, there is a significant potential to produce biodiesel from these feedstocks as alternative fuel to the petrodiesel. Some sources of non-edible feedstocks that are abundant in India are shown in Fig. 15. The production of biodiesel from oil seed consists of various processes (Fig. 16). It includes seed decortications, oil extraction, oil filtration, degumming, demoisturization, transestrification, and washing.

Transesterification is the main process, while others serve as pre- and post-processes for the production of biodiesel. In the transesterification process, the oil extracted from the seed is reacted with alcohol in the presence of catalyst to produce biodiesel. The alcohol used is either methanol (mostly used) or ethanol,

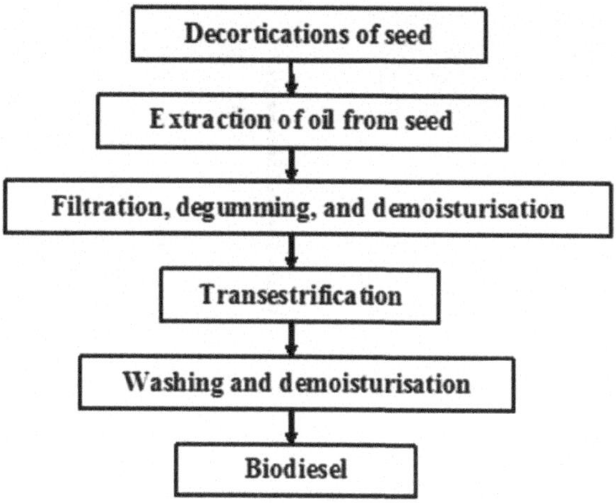

Fig. 16 Process involved in the production of biodiesel

while potassium or sodium hydroxide is used as catalyst. The yield of biodiesel depends on the free fatty acid composition of oil, proportions of alcohol and catalyst in reaction, and temperature and time of reaction. The quality of biodiesel produce is ascertained through comparing its properties with the biodiesel standards. In the USA, the biodiesel needs to confirm with ASTM D6751 standard, while EN 14214 standard is used in Europe. In India, the quality of biodiesel needs to be as per IS 15607 standard. The biodiesel is composed of saturated and unsaturated fatty acids. It is characterized by higher density and viscosity and lower calorific value as compared to diesel.

The use of biodiesel in engines has several advantages as compared to diesel such as biodegradability, higher flash point, and excellent lubricity (Knothe et al. 2005). The use of biodiesel shows lower thermal efficiency in comparison with diesel fuel. The blending of biodiesel with diesel to a certain ratio gives better thermal efficiency (BTE) than the diesel. Figure 17 shows that the variations of BTE with loads for different blends of biodiesel (derived from rubber seeds) have been tested in the engine (Ramadhas et al. 2005). The maximum BTE of 28% is obtained with 10% biodiesel (B10) which is quite higher than the BTE obtained with pure diesel. On the other hand, the BTEs obtained with B50, B75, and B100 are found to be 25, 25, and 24%, respectively. However, in most of the cases, the blending up to 20% shows a better performance (Sinha and Agarwal 2005; Ramadhas et al. 2005; Muralidharan et al. 2011; Chauhan et al. 2011; Raheman and Ghadge 2007). As shown in Figs. 18 and 19, the use of biodiesel blend decreases the emission of CO, smoke level, and hydrocarbon (HC); however, it increases the NO_x and CO_2 emission (Ramadhas et al. 2005; Chauhan et al. 2011; Raheman and Ghadge 2007; Buyukkaya 2010; How et al. 2014). Biodiesel is emerging as the most preferred alternative fuel because of its similar characteristics to that of diesel fuel.

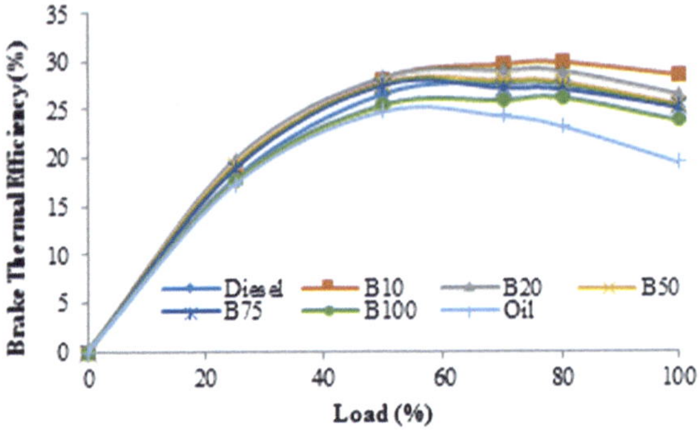

Fig. 17 Brake thermal efficiency of rubber seed biodiesel and diesel with respect to load (Ramadhas et al. 2005)

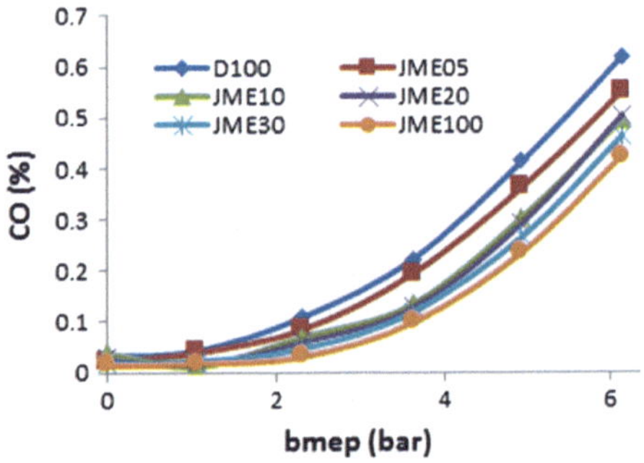

Fig. 18 CO emission at the various loads for different blends of jatropha methyl ester (Chauhan et al. 2011)

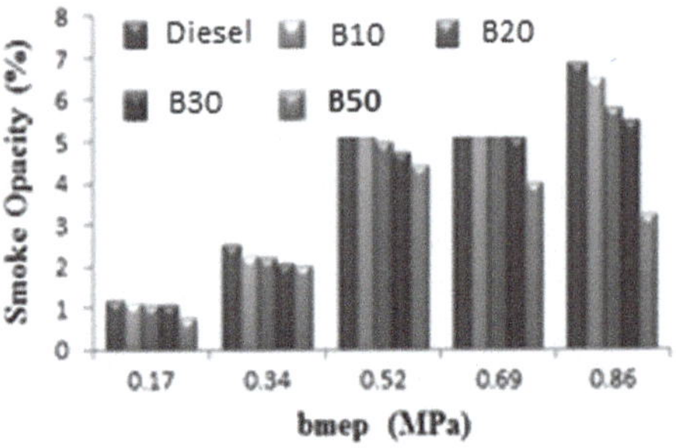

Fig. 19 Smoke emission at the various loads for different blends of coconut biodiesel (How et al. 2014)

3.2 *Alcohol*

Alcohols are used in the diesel engines as an additive to improvise overall engine performance. It is blended with diesel up to the 25% by volume. Alcohols have the characteristics of lower heating value, lower boiling points, and lower cetane number as compared to diesel (Table 2). This lower heating value of alcohol increases the brake-specific fuel consumption. However, it does not cause much impact on the engine performance. The use of alcohol reduces the emission of CO, smoke opacity, and total hydrocarbon (Sayin et al. 2010; Li et al. 2005; Muthaiyan

Table 2 Properties of the alcohols (Zaharin et al. 2017; Sayin et al. 2010; Li et al. 2005)

Alcohol	Diesel	Methanol	Ethanol	Propanol	Butanol
Chemical formula	$C_{12}H_{26}$	CH_3OH	C_2H_5OH	C_3H_7OH	C_4H_9OH
Molecular weight	170	32	46.07	60.1	74.12
Density at 20 °C (kg/m^3)	830	790	789	803.7	809.7
Calorific value (MJ/kg)	42	20.27	26.8	29.82	32.01
Heat of vaporization (kJ/kg)	260	1100	840	789.6	706.3
Boiling point (°C)	180–360	64.7	78.4	97	117
Cetane number	52	4	5–8	12	17

and Gomathinayagam 2016). The presence of oxygen molecules promotes complete combustion and produces lower exhaust emission (Zaharin et al. 2017).

Among the alcohols, ethanol and butanol have become popular as these are non-toxic and can be easily produced from the various biological feedstocks by the process of fermentation. However, alcohols have the higher heat of vaporization and lower calorific value as compared to diesel fuel. As a result, it is used as a blended fuel with diesel fuel to achieve an efficient engine performance and to lower the pollutant emission. Sathiyamoorthi and Sankaranarayanan (2017) have tested the performance and emission characteristics of a direct injection diesel engine with the blends of neat lemongrass oil–diesel and supplemented ethanol as the additive. In the investigation, ethanol is used as 2.5 and 5% by volume. It is observed that 5% ethanol in the blend contributes maximum BTE as compared to other blends. Both HC and NO_x are noticed to be higher, while CO is found to be lower with 5% ethanol. Venu and Madhavan (2017) have investigated the influence of ternary blends (alcohol–biodiesel–diesel) and use diethyl ether (DEE) as an ignition enhancer. Two types of alcohols resembling ethanol and methanol have been used in the study. Experiments were conducted with (a) pure diesel, (b) two ternary blends (TBs) with methanol (MBD) and ethanol (EBD), and (c) the ternary blends with DEE. In the TBs, methanol, ethanol, biodiesel, and diesel volume percentages are chosen as 20, 40, and 40, respectively. After the preparation of blends, 5 and 10% of DEE by volume are added and these are designated as EBD-5DEE, EBD-10DEE, MBD-5DEE, and MBD-10DEE, respectively. The results obtained in the study reveal that the DEE in EBD blends increases the combustion duration, cylinder pressure, and brake-specific fuel consumption (BSFC) with reduced NO_x, particulate matter (PM), and smoke emissions due to reduced ignition delay and higher latent heat of evaporation. Overall, EBD-5DEE and MBD-5DEE reflect an improved engine performance, combustion, and emission characteristics than EBD blends.

Paul et al. (2013) have conducted several experiments in dual fuel as well as single diesel and diesel–ethanol fuel modes. In dual-fuel mode, compressed natural gas (CNG) is injected at the intake port during the engine suction period. However, the discussion of this study is confined on the engine performance and pollutant

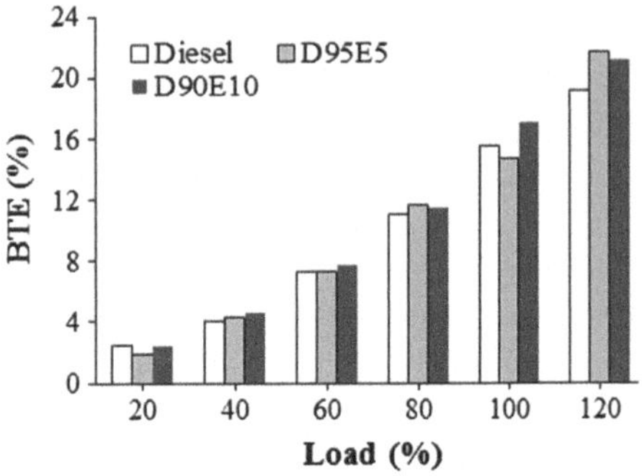

Fig. 20 Variation of BTE with load for diesel–ethanol blends (Paul et al. 2013)

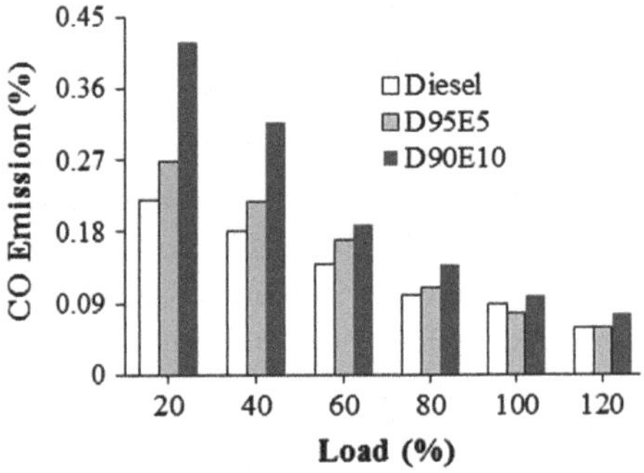

Fig. 21 Variation of CO emission with load for diesel–ethanol blends (Paul et al. 2013)

emissions in single liquid fuel mode of operation. In the single mode of operation, experiments are conducted with pure diesel, 5 and 10% ethanol by volume blended with diesel, respectively. These blends are designated as D95E5 and D90E10, respectively. The variation of BTE with applied load on engine is depicted in Fig. 20. It is observed that with the increase of ethanol volume percentage in the blend, BTE increases, and the BTE is consistently found to be higher from lower to higher loads with D90B10 blend as compared to D95B5 blend. The maximum increment of BTE by 11.04% is observed with D90B10 blend as compared to pure diesel. However, CO emission is found to increase with the increased ethanol percentage in the blend (Fig. 21). On the other hand, the CO emission decreases with the increase of applied load on the engine. The same trends are also found with unburnt hydrocarbon (UBHC) emission (Fig. 22). At 100% load, UBHC is found to be lower with the blended fuel. Again, NO_x emission is noticed to be higher with increasing loads (Fig. 23) and blends. Figure 24 shows the variation of smoke opacity with loads where the smoke opacity reduces with the blended fuel in comparison with pure diesel mode. At all loads, the investigators observed the lower smoke opacity with D95E5 blend.

Balamurugan and Nalini (2014) carried out the engine performance and emission investigation with the binary blends of butanol–diesel and propanol–diesel, respectively. Butanol and propanol are blended separately with the diesel fuel at the

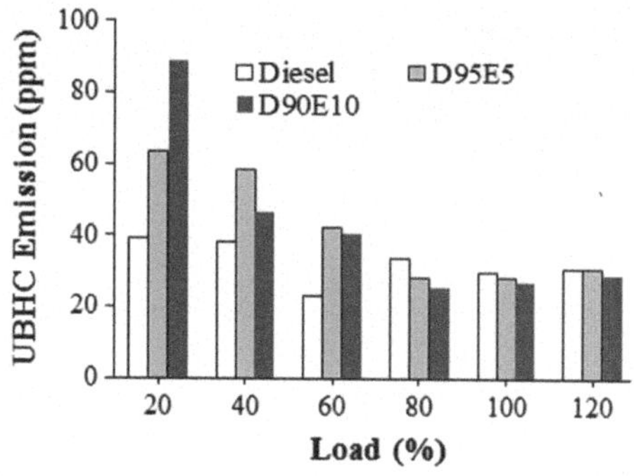

Fig. 22 Variation of UBHC emission with load for diesel–ethanol blends (Paul et al. 2013)

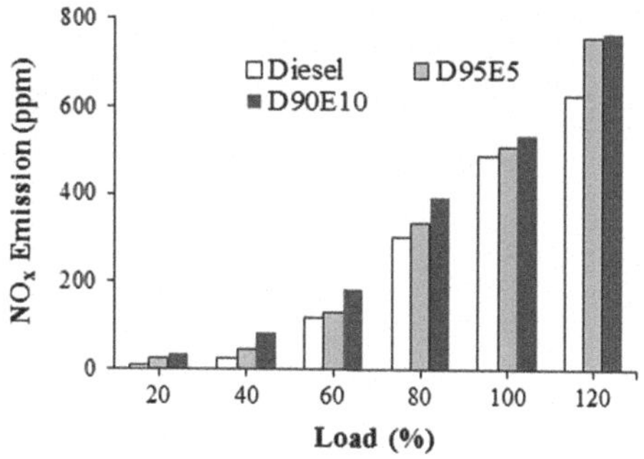

Fig. 23 Variation of NO_x emission with load for diesel–ethanol blends (Paul et al. 2013)

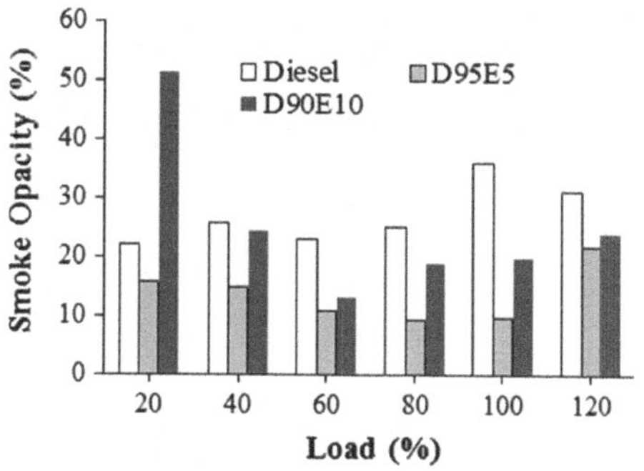

Fig. 24 Variation of smoke opacity with load for diesel–ethanol blends (Paul et al. 2013)

rate of 4 and 8% by volume. The BTE is found to be higher for the butanol blends in comparison with both propanol and diesel (Fig. 25). The maximum increment of BTE by 8.92 and 10.52% is noticed for 4 and 8% butanol in contrast to pure diesel. There has been a higher emission of CO with the alcohols blends (Fig. 26) in contrast to pure diesel. Similar trends of HC emission are also observed in Fig. 27. However, NO_x emission is noticed to be lower for all blends of alcohols (Fig. 28). Figure 29 shows the smoke density variations in the engine exhaust where an

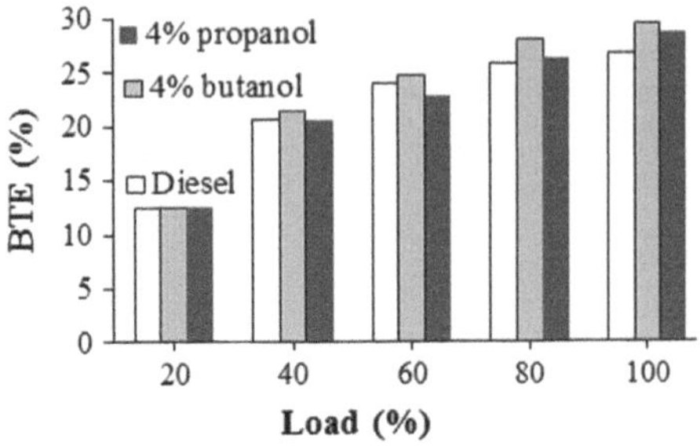

Fig. 25 Effect of adding 4% alcohols with diesel on brake thermal efficiency (Balamurugan and Nalini 2014)

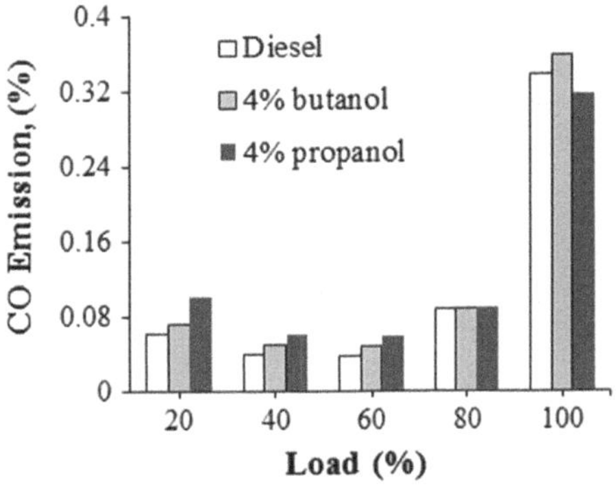

Fig. 26 Effect of adding 4% alcohols with diesel on CO emission (Balamurugan and Nalini 2014)

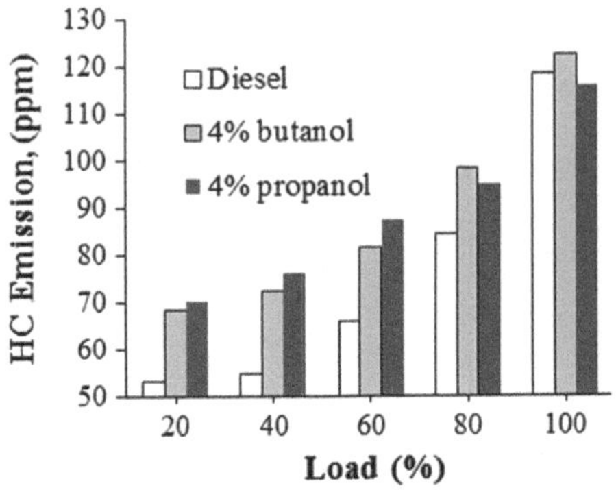

Fig. 27 Effect of adding 4% alcohols with diesel on HC emission (Balamurugan and Nalini 2014)

increasing amount of alcohols in the blends increases the smoke density. The smoke density also increases with an increase of load. It is further noticed that at higher loads, the richness of fuel increases which enhances the smoke density. The higher heat vaporization of alcohols seems to be responsible for the other adverse performance and emissions characteristics of the engine.

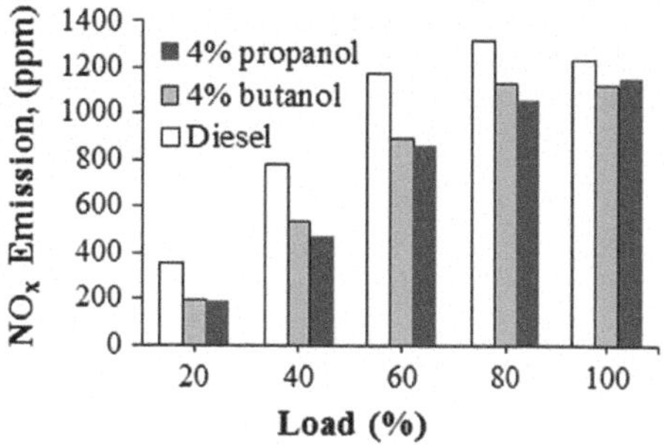

Fig. 28 Effect of adding 4% alcohols with diesel on NO_x emission (Balamurugan and Nalini 2014)

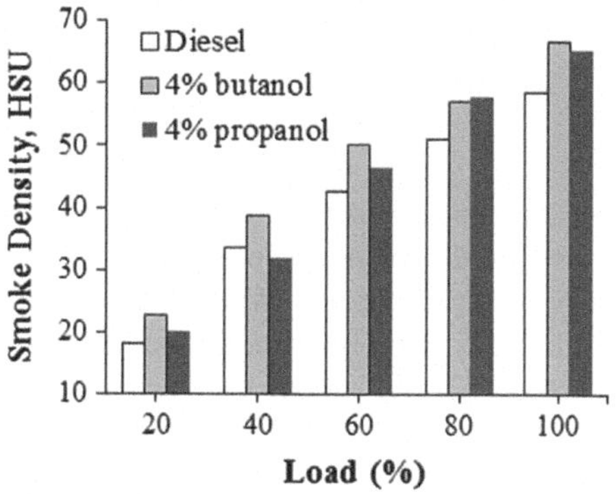

Fig. 29 Effect of adding 4% alcohols with diesel on smoke density (Balamurugan and Nalini 2014)

3.3 Gaseous Fuel

Gaseous fuel in the form of biogas and producer gas, derived from the biomass, can be used in the diesel engines through the dual-fuel mode. To facilitate the use of dual fuel, the suction line of the engine is modified to accommodate gaseous fuel as shown in Fig. 30. The gaseous fuel is inducted along with air during suction stroke which is then compressed and combusted through the injection of diesel. The gaseous fuel serves as the primary fuel and replaces the diesel, while diesel is used as the pilot fuel. The pilot fuel is needed to ignite the combustible mixture as the gaseous fuel has high autoignition temperature. Liquid fuel other than diesel such as biodiesel can also be used in this arrangement. Dual-fuel mode not only reduces the consumption of diesel, but there is an advantage of operating the engine in pure diesel mode in case of shortage of gaseous primary fuel.

3.3.1 Biogas in Dual Fuel Mode

Biogas is produced by the anaerobic fermentation of the organic materials. The organic material includes cattle waste, municipal wastes, poultry manures, and agricultural biomass. It is composed of methane (CH_4), carbon dioxide (CO_2), hydrogen (H_2), nitrogen (N_2), oxygen, and hydrogen sulfide. The composition of

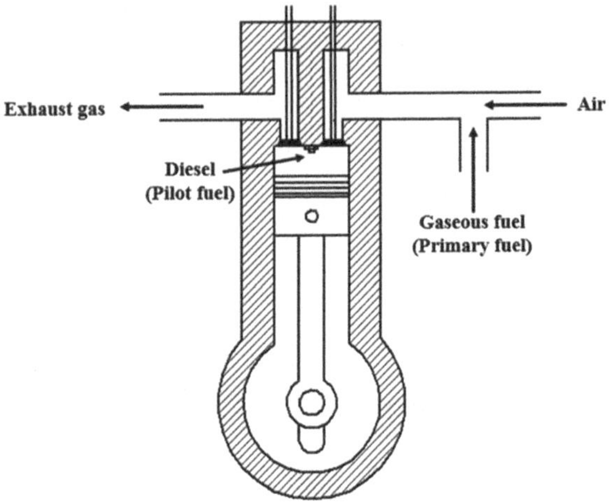

Fig. 30 Schematic diagram of dual fuel diesel engine (Dabi and Saha 2015)

the component varies with feedstock used and production conditions. CH_4 and CO_2 constitute the major components of the biogas in the range of 60–73 and 19–40%, respectively (Duc and Wattanavichien 2007; Tippayawong et al. 2007; Bedoya et al. 2012). This variation in the composition gives the heating value in the range of 26–17 MJ/kg. Biogas has been successfully used in diesel engines through the dual-fuel mode. Diesel replacement up to 90% has been achieved through this arrangement (Tippayawong et al. 2007; Sarkar and Saha 2017; Bora et al. 2016). The duel fueling reduces the brake thermal efficiency of engine due to lower calorific value of biogas as compared to diesel. Since CO_2 constitutes one of major components of biogas, CO_2 and CO emissions increase with the use of biogas. However, NO_x emission decreases because of duel fueling.

The investigation carried out by Sarkar and Saha (2017) with biogas as the primary fuel and diesel as the pilot fuel with preheating of the intake charge is shown in Fig. 31. The BTE is found to be substantially lower than the BTE obtained with pure diesel. It is believed to be due to the lower heating value of biogas. However, there is a 10% increment of BTE with the preheating of intake charge. Figure 32 shows the variations of BTE with load for 5% ethanol blended with diesel (primary fuel) and biogas (secondary fuel). The investigation with the use of oxygenated fuel such as ethanol revealed very high engine efficiency as compared to the results shown in Fig. 31. Figure 33 shows the reduction in NO_x emission with biogas as the secondary fuel. Further, there is a drastic reduction in unburned hydrocarbon (approximately 60%) and carbon monoxide (CO). The trends shown in Figs. 32 and 33 are demonstrated by the authors at Indian Institute of Technology Guwahati, India.

3.3.2 Producer Gas in Dual Fuel Mode

When the biomass is subjected to partial thermal oxidation, it produces a gaseous product, ash, oil, and tar. This thermochemical process, known as gasification, takes

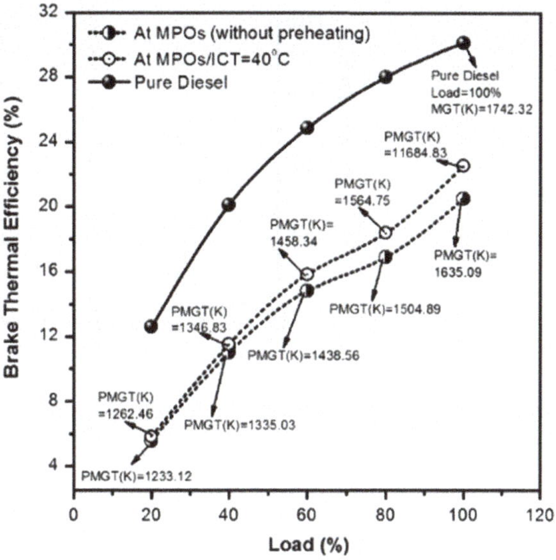

Fig. 31 Variations of brake thermal efficiency with applied loads with biogas as the primary fuel at different states (Sarkar and Saha 2017)

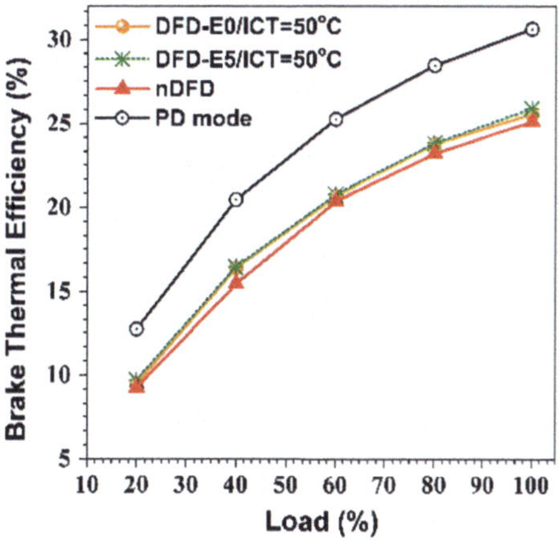

Fig. 32 Variations of brake thermal efficiency with loads and with biogas

place in the device which is called as a gasifier. The schematic diagrams of gasifier are shown in Fig. 34. In the updraft gasifier, air and biomass move in the opposite direction, while it moves in parallel and perpendicular directions in downdraft and cross-draft gasifier, respectively. Steam and oxygen are also used as the oxidizing agent. The biomass, in the form of either wood chips, briquette, or pellets, is fed to the gasifier. The gasification process consists of complex thermochemical, which has been generally classified as drying, devolatilization, oxidation, and reduction processes (Puig-Arnavat et al. 2010). The gaseous products that are produced through this process are called producer gas. Producer gas is composed of CO_2, H_2,

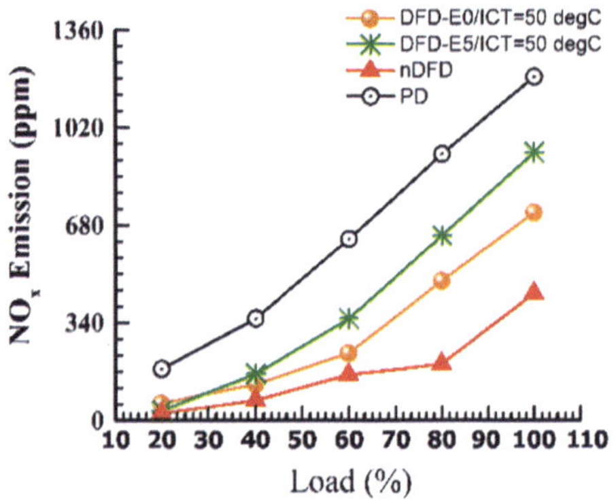

Fig. 33 Emission of NO_x with applied on engine at different conditions

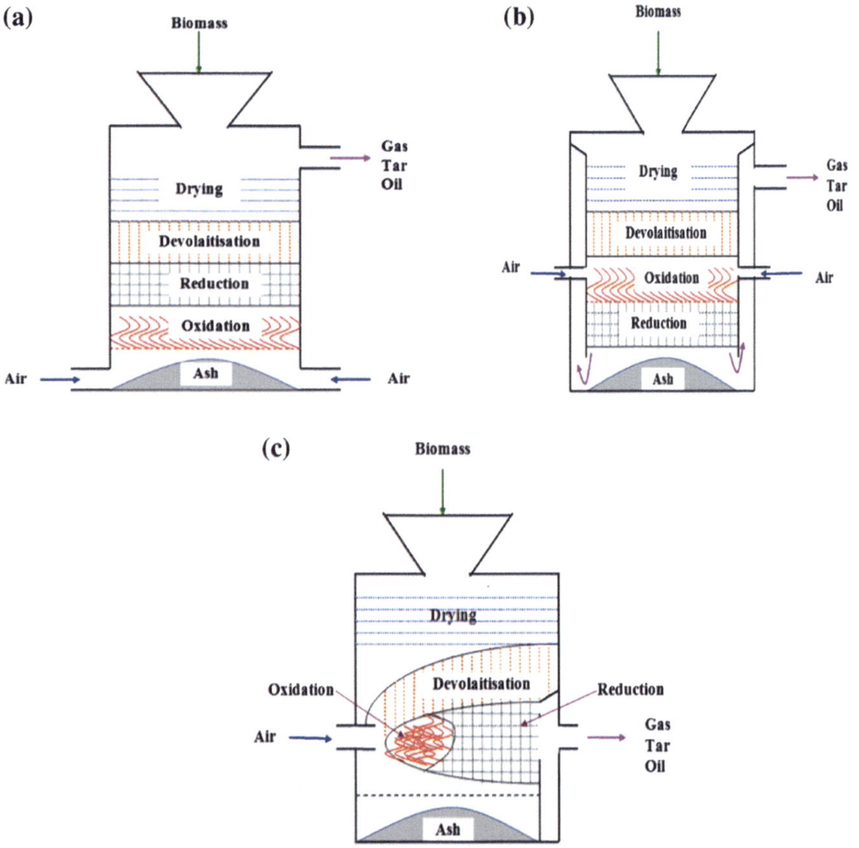

Fig. 34 Schematic diagram of gasifier: **a** updraft gasifier, **b** downdraft gasifier, and **c** cross-draft gasifier

Table 3 Composition of producer gas derived form different feedstocks (Sridhar et al. 2005; Ramadhas et al. 2006; Deshmukh et al. 2008; Raman and Ram 2013; Lal and Mohapatra 2017)

Feed stock	Volumetric composition (%)						Calorific Value
	CO	H_2	CH_4	N_2	H_2O	CO_2	
Causurina species wood and coconut shells	18.0	18.0	2–3	Rest	2.5	12.00	5.2 (MJ/Nm^3)
Wood chips and coir pith	18–22	15–19	1–5	45–55	4.0	–	4.8 and 3.5 (MJ/kg)
Hingan fruit residue	16.2	12.7	1.41	55.43	–	15.15	4.38 (MJ/m^3)
Fuel wood	21.0	23.0	0.90	46.1	–	9.00	5.6 (MJ/Nm^3)
Sawdust and cotton stalks	18.1	8.2	3.10	56.6	–	14.00	4.5 (MJ/Nm^3)

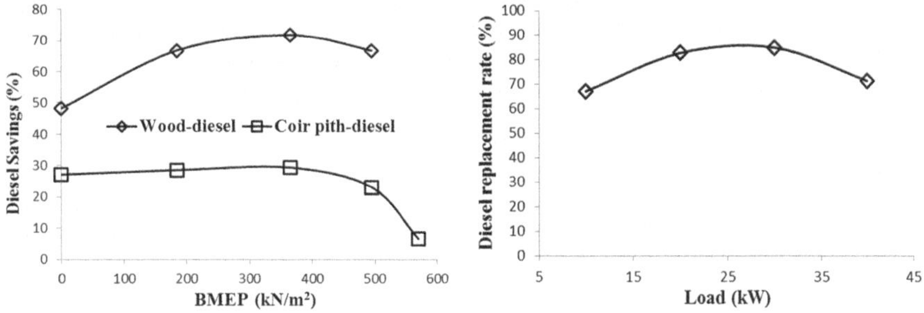

Fig. 35 Diesel savings using the producer gas derived from coir pith, wood and keekar wood (*acacia nilotica*) (Ramadhas et al. 2006; Uma et al. 2004)

N_2, CH_4, water vapor (H_2O), and carbon monoxide (CO). The composition of producer gas varies with the feedstock used. Typical composition with calorific value derived from different feedstocks is shown in Table 3. The use of producer gas in the diesel engine saves the diesel up to 80% (Fig. 35). There is a slight reduction in the engine performance due to dual fueling. There has been a remarkable reduction in the NO_x emission throughout the operating loads as compared to the diesel mode (Sridhar et al. 2005; Lal and Mohapatra 2017; Hassan et al. 2011). CO and HC emissions increase with the use of producer gas.

4 Concluding Remark

Whenever there is a negative side effect, born out of the use of the system, there is always a scope of redressing such effects through the remedial measures. In the transport and energy sectors, fossil fuel constitutes the primary energy source. It has a problem of limited reserve in earth crust and non-renewability which calls for an alternative arrangement for the future need. Such arrangement is devised and realized by exploring the available options and carrying out extensive research. The exploration and investigation of available options land up onto the use of liquid and gaseous fuel derived from biomass as a promising alternative for the fossil fuel. The best part of this alternative is that it is sustainable and renewable. It consists of fuel derived from biomass in the form of biodiesel, alcohol, biogas, and producer gas. These biomass products can directly be used in diesel engines without or with minor modification of the engine. The source of biological feedstocks to produce the biodiesel and the alcohols, namley ethanol and butanol, is abundant. At the same time, it is recommended that the blends of these biofuels with petrodiesel can improve the engine performance substantially. Therefore, practice of using of biofuel in diesel engines can substantially curtail the GHG emissions as well as dependency on imported fossil petrodiesel. The use of gaseous fuels, such as biogas and producer gas, through dual-fuel mode can substitute a substantial amount of diesel fuel. It is evident from the present review that biofuel can practically substitute or supplement diesel fuel to meet the demands of transport and power sectors of the country. A substitute or supplement of even 5% can save billions of dollars on oil import.

References

Ahuja D, Tatsutani M (2009) Sustainable energy for developing countries. SAPIENS 2:1–16

Annual Report (2015–2016). Ministry of Petroleum and Natural Gas, Government of India

Annual Report (2016–2017). Ministry of Petroleum and Natural Gas, Government of India

Balamurugan T, Nalini R (2014) Experimental investigation on performance, combustion and emission characteristics of four stroke diesel engine using diesel blended with alcohol as fuel. Energy 78:356–363

Bedoya ID, Saxena S, Cadavid FJ, Dibble RW, Wissink M (2012) Experimental study of biogas combustion in an HCCI engine for power generation with high indicated efficiency and ultra-low NO_x emissions. Energy Convers Manag 53:154–162

Bora BJ, Saha UK, Chatterjee S, Veer V (2016) Effect of load level on performance and emission characteristics of a biogas run dual fuel diesel. engine fluid mechanics and fluid power—contemporary research, lecture notes. Mech Eng. https://doi.org/10.1007/978-81-322-2743-4_143

BP Energy Outlook (2017). https://www.bp.com/content/dam/bp/pdf/energy-economics/energy-outlook-2017/bp-energy-outlook-2017.pdf

BP (British Petroleum) (2013) Statistical review of world energy, June 2013

Busby JW, Shidore S (2017) When decarbonization meets development: the sectoral feasibility of greenhouse gas mitigation in India. Energy Res Soc Sci 23:60–73

Buyukkaya E (2010) Effects of biodiesel on a DI diesel engine performance, emission and combustion characteristics. Fuel 89:3099–3105

Chauhan BS, Kumar N, Cho HM (2011) A study on the performance and emission of a diesel engine fueled with Jatropha biodiesel oil and its blends. Energy 37:616–622

Dabi M, Saha UK (2015) Experimental analysis of a dual-fuel engine fueled by producer gas derived from pine leaves and cattle dung briquettes. doi:https://doi.org/10.1115/GTINDIA2015-1263

Deshmukh SJ, Bhuyar LB, Thakre SB (2008) Investigation on performance and emission characteristics of CI engine fuelled with producer gas and esters of hingan (balanites) oil in dual fuel mode. Int J Aerosp Mech Eng 2(3):148–153
Duc PM, Wattanavichien K (2007) Study on biogas premixed charge diesel dual fuelled engine. Energy Convers Manag 48:2286–2308
Energy Information Administration. DOE, US Government. http://www.eia.doe.gov/emeu/cabs/India/Oil.html
Hassan S, Nor FM, Zainal ZA, Miskam MA (2011) Performance and emission characteristics of supercharged biomass producer gas-diesel dual fuel engine. J Appl Sci 11:1606–1611
How HG, Masjuki HH, Kalam MA, Teoh YH (2014) An investigation of the engine performance, emissions and combustion characteristics of coconut biodiesel in a high-pressure common-rail diesel engine. Energy 69:749–759
IEA, World Energy Outlook 2013. OECD-IEA, Paris
India Energy Outlook: IEA Special Report 2015. https://www.iea.org/publications/freepublications/publication/IndiaEnergyOutlook_WEO2015.pdf
India: Greenhouse Gas Emissions (2007) Ministry of Environment and Forests, Government of India 2010. http://www.moef.nic.in/downloads/public-information/Report_INCCA.pdf
International Energy Agency, World Energy Outlook (2007) China and India insights. https://www.iea.org/publications/freepublications/publication/WEO_2007.pdf
Kirkland J (2010) Natural gas could serve as 'bridge' fuel to low-carbon future. Sci Am
Knothe G, Gerpen JV, Krahl J (2005) The biodiesel handbook. AOCS Press
Kumar A, Sharma S (2011) Potential non-edible oil resources as biodiesel feedstock: an Indian perspective. Renew Sustain Energy Rev 15:1791–1800
Lal S, Mohapatra SK (2017) The effect of compression ratio on the performance and emission characteristics of a dual fuel diesel engine using biomass derived producer gas. Appl Therm Eng 119:63–72
Li D, Zhen H, Xingcai L, Wu-gao Z, Jian-guang Y (2005) Physico-chemical properties of ethanol–diesel blend fuel and its effect on performance and emissions of diesel engines. Renew Energy 30:967–976
Muralidharan K, Vasudevan D, Sheeba KN (2011) Performance, emission and combustion characteristics of biodiesel fuelled variable compression ratio engine. Energy 36:5385–5393
Muthaiyan P, Gomathinayagam S (2016) Combustion characteristics of a diesel engine using propanol diesel fuel blends. J Inst Eng (India): Series C 97:323–329
National Energy Map for India: Technology Vision 2030. http://www.teriin.org/div/psa-fullreport.pdf
NAPCC. Source: http://www.moef.nic.in/ccd-napcc
National Policy on Biofuels (2009). Ministry of New and Renewable Energy, Government of India
Olivier JGJ, Muntean M, Peters JAHW (2015) Trends in global CO_2 emissions: 2015 report. PBL Netherlands Environ Assess Agency Eur Comm Jt Res Cent 2015:1–78
PARIS SUMMIT (2015) Analysis: India's climate pledge suggests significant emissions growth up to 2013. http://www.carbonbrief.org/indias-indc
Paul A, Bose PK, Panua RS, Banerjee R (2013) An experimental investigation of performance-emission trade off of a CI engine fueled by diesel-compressed natural gas (CNG) combination and diesel-ethanol blends with CNG enrichment. Energy 55:787–802
Puig-Arnavat M, Bruno JC, Coronas A (2010) Review and analysis of biomass gasification models. Renew Sustain Energy Rev 14:2841–2851
Raheman H, Ghadge SV (2007) Performance of compression ignition engine with mahua (Madhuca indica) biodiesel. Fuel 86:2568–2573
Ramadhas AS, Muraleedharan C, Jayaraj S (2005) Performance and emission evaluation of a diesel engine fueled with methyl esters of rubber seed oil. Renew Energy 30:1789–1800
Ramadhas AS, Jayaraj S, Muraleedharan C (2006) Power generation using coir-pith and wood derived producer gas in diesel engines. Fuel Process Technol 87:849–853

Raman P, Ram NK (2013) Performance analysis of an internal combustion engine operated on producer gas, in comparison with the performance of the natural gas and diesel engines. Energy 63:317–333

Sarkar A, Saha UK (2017) Effect of intake preheating on biogas run dual fuel diesel engines. In: Proceedings of the national conference on sustainable mechanical engineering: today and beyond, 24–25 Mar 2017 at Tezpur University, India

Sathiyamoorthi R, Sankaranarayanan G (2017) The effects of using ethanol as additive on the combustion and emissions of a direct injection diesel engine fuelled with neat lemongrass oil-diesel fuel blend. Renew Energy 101:747–756

Sayin C, Ozsezen AN, Canakci M (2010) The influence of operating parameters on the performance and emissions of a DI diesel engine using methanol-blended-diesel fuel. Fuel 89:1407–1414

Source: http://www.wri.org/blog/2014/11/6-graphs-explain-world%E2%80%99s-top-10-emitters

Sinha S, Agarwal AK (2005) Performance evaluation of a biodiesel (rice bran oil methyl ester) fuelled transport diesel engine. SAE Paper 2005, 2005-01-1730

Sridhar G, Dasappa S, Sridhar HV, Paul PJ, Rajan NKS (2005) Gaseous emissions using producer gas as fuel in reciprocating engines. SAE Technical Paper 2005-01-1732. doi:https://doi.org/10.4271/2005-01-1732

Subramanian AK, Singal SK, Saxena M, Singhal S (2005) Utilization of liquid biofuels in automotive diesel engines: an Indian perspective. Biomass Bioenergy 9:65–72

Technology roadmap: biofuels for transport sectors, OECD/IEA-2011. https://www.iea.org/publications/freepublications/publication/Biofuels_Roadmap_WEB.pdf

Tippayawong N, Promwungkwa A, Rerkkriangkrai P (2007) Long-term operation of a small biogas/diesel dual-fuel engine for on-farm electricity generation. Biosys Eng 98:26–32

Uma R, Kandpal TC, Kishore VVN (2004) Emission characteristics of an electricity generation system in diesel alone and dual fuel modes. Biomass Bioenerg 27:195–203

UNFCCC (2015) Adoption of the Paris Agreement (1/CP.21). United Nations Framework Convention on Climate Change, Bonn

Venu H, Madhavan V (2017) Influence of diethyl ether (DEE) addition in ethanol-biodiesel-diesel (EBD) and methanol-biodiesel-diesel (MBD) blends in a diesel engine. Fuel 189:377–390

World Energy Outlook (2009) Executive summary. International Energy Agency (IEA), France. http://www.worldenergyoutlook.org/docs/weo2009/WEO2009_es_english.pdf

Zaharin MSM, Abdullah NR, Najafi G, Sharudin H, Yusaf T (2017) Effects of physicochemical properties of biodiesel fuel blends with alcohol on diesel engine performance and exhaust emissions: a review. Renew Sustain Energy Rev 79:475–493

Unconventional Technologies for Sustainable Coal Energy in India

Nirmal Mallick and Prabu Vairakannu

Abstract Coal is a fossil fuel and majorly satisfies our energy need. Although coal resources are depleting, it is still to be a sustainable energy source for future in terms of unutilized sources. As per the survey of "Statistical review of world energy," India stands third position in coal production after China and the USA in the year 2015 with an increase in the growth rate of 4.7% over the last year. According to the current Geological Survey of India 2016, approximately 308.8 billion tons of coal reserves are available in India up to a depth of 1200 m. In India, 70% of the electricity production and 60% of the commercial energy production depend on coal resources. Out of the total coal reserves in India, approximately 39.4% ($\sim$121.65 billion tons) of coal lies under a depth more than 300–1200 m, which is out of human reach for exploitation in an economical way by conventional method. To match the current energy demand, there are few other alternative technologies such as underground coal gasification (UCG) and coal bed methane (CBM) to utilize deep coal seams economically. Underground coal gasification (UCG) is a method of converting coal in situ into syngas, which can be used for electricity generation and synthesis of chemicals and liquid fuels. Another energy source from coal is high calorific value CBM gas, which contains 95–98% pure methane. CBM gas formed during the coalification process gets evolved during coal mining process. Capturing this methane gas prior to mining or from unutilized deep coal seams would be a sustainable energy for future demand. In this chapter, we will discuss the estimate of available energy resources and its exploitation through these unconventional technologies for sustainable energy production in India.

Keywords Unconventional technologies · Underground coal gasification (UCG) Coal bed methane (CBM) · Worldwide UCG trials · CBM sites

N. Mallick · P. Vairakannu (✉)
Department of Chemical Engineering, Indian Institute of Technology Guwahati, Guwahati 781039, Assam, India
e-mail: v.prabu@iitg.ernet.in

A. Gautam et al. (eds.), *Sustainable Energy and Transportation*,
Energy, Environment, and Sustainability,
https://doi.org/10.1007/978-981-10-7509-4_6

Nomenclature

UCG	Underground coal gasification
CBM	Coal bed methane
RCL	Reverse combustion linkage
FCL	Forward combustion linkage
CRIP	Continuous retracting injection point
TDS	Total dissolved solids
CMM	Coal mine methane
FC	Fixed carbon (%)
VM	Volatile matter (%)
M	Moisture content (%)
A	Ash content (%)
P_l	Langmuir pressure (bar)
V_l	Langmuir volume (m^3)
V_{max}	Maximum volume (m^3) of CBM gas at pressure 'P'
P	Pressure (bar)
CMPDI	Central Mine Planning and Design Institute

1 Introduction

The economic development of a country majorly depends on its available fuel resources and further use of these natural resources in a beneficial way. Coal is an abundant fossil fuel resource and possesses high heating value. It is a cheaper raw material for electricity production. India is the third largest coal producing country in the world after China and the USA. According to the current geological survey of India, it was found that 308.8 billion tons of coal reserves are available in India up to a depth of 1200 m (Geological Survey of India 2016). The production of coal in India is estimated as 639.2 Mt (Annual Report 2017) during the year 2016–17. Coal India Limited (CIL) is the largest coal producing company in India. It accounts for more than 80% of total coal production in India. Solid fossil fuel "coal" accounts for 55% of the country's energy need. The current per capita commercial primary energy consumption in India is about 350 kg of oil E/equivalent/year, which is very low as compared to other developed countries. Currently, there is energy scarcity problem due to increase in the population of India. Limited reserves of petroleum and natural gas, eco-conservation restriction on hydel power project, and geopolitical perception of nuclear power lead to increase our dependence on coal energy, which will continue to play a major role in India's energy scenario in future.

As per the recent survey, approximately 40% of total coal reserves (123.52 billion tons) of India are found below a depth of 300 m. Extraction of these coal reserves is uneconomical through conventional mining technology (Geological Survey of India 2016). Alternatively, underground coal gasification (UCG) is a suitable technology for the utilization of deep underground coal resources. UCG is an in situ clean coal technology for the conversion of deep coals into useful product gas, which can be utilized for the manufacturing of chemical feedstock. Syngas can also be utilized in power generating systems such as combined cycle (IGCC) and steam turbine cycle for clean power production (Khadse et al. 2007). UCG ensures the safety of workers as there is no coal mining process in this technology. Also, UCG avoids several problems such as ash disposal, noise and air pollution, and methane emission to atmosphere, and thus, it is an environmental friendly technology. UCG is suitable for the exploitation of deep coal seams, which has better geological conditions and less gas leakage into surrounding burden strata. However, UCG may cause environmental impacts such as land sliding and groundwater contamination. Hence, a detailed geological survey must be conducted before choosing the site for a UCG plant.

Another alternative sustainable energy source using fossil resources is coal bed methane (CBM) technology. During the coalification process, methane (CH_4) gas is produced as a respiratory by-product gas of enzymes and gets entrapped into coal seams. It has been reported that approximately 5000 ft^3 of methane gas is found per ton of coal (Longwell et al. 1995). This gas is known as coal bed methane (CBM) or sweet gas. It contains almost 95–97% of pure methane and a minor quantity of other gases such as CO_2, N_2, and higher hydrocarbons. During coal mining process, CBM gas gets liberated into underground coal mines and can cause a fire hazard. Hence, underground coal mines are required to maintain a methane concentration below 1% in air (Irving et al. 2001). As methane gas is highly explosive in nature, it needs to be captured prior to mining of a coal seam. According to a survey of Indian Network for Climate Change Assessment (INCCA) 2010, a total emission of 20.56 million tons of CH_4 into the atmosphere is calculated (Remme et al. 2011) and found that coal seams serve as a major source of CH_4 emission. A total of 1116.8 Bm^3 of CBM reserves is estimated from the available coal resources of India (Prabu and Mallick 2015). The effective utilization of this vast source of energy can serve as an efficient alternative and sustainable clean energy to fulfill the energy demand of India.

In this chapter, we will discuss unconventional technologies such as UCG and CBM for the production of sustainable energy in India. The process description, methodology, effect of process parameters, worldwide resource availability and potential reserves, and the possibility of implementation of these technologies in India are discussed.

2 Underground Coal Gasification

Underground coal gasification (UCG) is a process of in situ coal gasification, where an underground gasification chamber is constructed within a coal seam through the combustion linking techniques. This technology was originated by a German engineer, William Siemens, in 1860. He conducted a successful experiment at

Durham Coalfield in north-east England. In 1939, the USSR successfully constructed a UCG plant at Ukraine, and later, another UCG plant was established in the 1960s at Angren, Uzbekistan (Sajjad and Rasul 2014). These trials show the feasibility of in situ gasification of deep coals. Thus, several researchers were involved in UCG research to optimize the process parametric conditions.

2.1 Potential Coal Reserves for UCG

All geological fossil fuel resources are not mineable reserves. The extraction process of coal depends on several parameters such as grade and pricing, available technology for extraction, and availability of infrastructure. According to Central Mine Planning and Design Institute (CMPDI), the extractable reserves can be assessed using Eq. (2.1) (Sahu 2013).

$$\text{Extractable Reserves} = (0.9 * \text{ proved res.} + 0.7 * \text{ indicated res.} + 0.4 * \text{ inferred res.})/4.7 \tag{2.1}$$

Considering the following assumptions, the quantity of extractable reserves is estimated (Thomas 2015).

(i) A detailed exploration indicates a confidence level of 90% of the reserves established.
(ii) Regional exploration establishes the resources in indicated and inferred categories. As per the Association of German Metallurgists and Mining Engineers, 70 and 40% confidence levels to indicated and inferred resources are assumed, respectively.

The studies of CMPDI (July, 2001) show an estimate of average coal reserves to production (R:P) ratio as 4.7:1. Table 1 shows that a total of 179.79 billion tons of coal reserves, which are found at a depth greater than 300 m, is predicted. It is essential to use this huge unmineable coal deposits through UCG technology to satisfy our energy demand. An estimate of a ton of coal would produce approximately 2500 m^3 syngas. A 10% utilization of this coal provides 64.74 trillion m^3 of UCG syngas with a calorific value of 3–5 MJ/m^3. Considering India's scenario with almost 70% of the total energy receiving from coal, UCG can be reliable for fulfilling future energy demand of the country.

Table 1 Estimate of unextractable coal (below 300 m) reserves in India (Sahu 2013)

	Prove	Indicated	Inferred	Total	Extractable	Unextractable
Coal (billion tons)	103.623	66.761	9.405	179.79	30.58	149.211

2.2 UCG Methodology

UCG involves in-place gasification of coal in underground. Deep coal seam is ignited with the injection of air or pure O_2. As coal seam temperature rises, gasification is carried out with further injection of gasifying medium such as steam or CO_2 for the production of high calorific value syngas. The major components of the produced syngas are H_2, CO, CO_2, CH_4, SO_x, NO_x, and H_2S (Thomas 2015). The pressure of UCG product gas is in the range of 30–50 bar depending on the operating pressure of a coal seam (Bhutto et al. 2013). The produced syngas needs to be cleaned to a desired level, which depends on further usage.

Conventional method of UCG involves drilling of two wells, namely injection and production well. Both these wells are needed to be connected using various combustion linking techniques such as reverse combustion linkage (RCL), forward combustion linkage (FCL), hydrofracturing, and electro-linking or by explosives. After establishing a connectivity between both the wells, feed gases containing oxidizing and gasifying medium are supplied through the injection well, and the generated syngas is collected at the production well. In the RCL method, oxidizing agent is supplied through the injection well and the coal seam is ignited at the production well. Thus, the flame front would propagate toward the source of oxidizing agent and forms a narrow channel connectivity. However, in the FCL method, coal seam is ignited at the injection well with the supply of oxidizing agent. As a consequence, the connectivity is formed in the shape of a broad cavity with a lesser rate of flame propagation toward the production well (Blinderman et al. 2008).

Other techniques of UCG include continuous retracting injection point (CRIP), blinding-hole UCG, long and large tunnel gasification, and two-stage gasification well (Blinderman et al. 2008; Andrianopoulos et al. 2015; Molenaar and Bruining 1995). CRIP method is more suitable for thin and deep coal seams. In this method, the injection point of the oxidants can be shifted to appropriate locations as soon as the coal gets consumed in the existing location. This technique is highly suitable for the extraction of high-ash Indian coal seams. In this technique, a horizontal injection pipe is inserted into the borehole connectivity, the injection point is relocated with the movement of flame front, and the burning zone grows toward upstream direction. This method produces better quality gas as compared to conventional UCG method (Thorsness and Britten 1989). Two-stage gasification is a method of supplying oxidizing and gasifying agents in a cyclic manner. In the first step, the oxidizing agent is introduced into the coal seam for efficient combustion. In the second step, the gasifying agent is injected by cutting the supply of oxidizing agent. During the second stage, the exposure of high surface area of coal in the cavity to the gasifying medium produces a better quality gas than conventional method (Hongtao et al. 2011).

In the blinding-hole technology, coal seam is drilled with the insertion of concentric pipes. The inner tube is used to supply air/gasifying agent for gasification, whereas the outer tube acts as a production well. This technology is beneficial for those areas where the coal resources are under construction sites or water bodies. It has advantages over other UCG methods. As the outlet gases are flowing out at a high temperature in the concentric pipe, it heats the inlet oxidant gas passing through the inner tube (Yang 2003). The major disadvantage of this technology is that the syngas produced during gasification reduces the contact area between the injected gas and coal (Yang et al. 2007). Long and large tunnel gasification method is used in the pre-mined coal seams, where injected well is connected to production well via mined tunnels (Roddy and Younger 2010). Typically, there are two auxiliary holes drilled in such a way that it connects all the mined channels (Green 2008).

2.3 *Chemical Reactions Involved in UCG*

During UCG run, the chemical reactions would occur with specific to the prevailing temperature of the borehole in a coal seam. A list of chemical reactions occurred during coal gasification is shown in Table 2. The reaction zones can be divided into three zones such as oxidation zone, reduction zone, and dry distillation zone. The schematic diagram of UCG process is shown in Fig. 1. In the oxidation zone, multiphase chemical reactions (reaction 2.2, 2.3, and 2.4) take place between the inlet oxygen and coal, and a high-temperature flame front is established for the progress of gasification reaction.

Table 2 Chemical reactions involving in different zones during UCG process

		Heat of reaction	
Oxidation zone			
Oxidation	$C + O_2 \rightarrow CO_2$	+393.8 kJ/mol	(2.2)
Partial oxidation	$2C + O_2 \rightarrow 2CO$	+231.4 kJ/mol	(2.3)
Oxidation	$2CO + O_2 \rightarrow 2CO_2$	+571.2 kJ/mol	(2.4)
Reduction zone			
Boudouard reaction	$C + CO_2 \rightarrow 2CO$	−162.4 kJ/mol	(2.5)
Heterogeneous water gas shift reaction	$C + H_2O_{(g)} \rightarrow CO + H_2$	−131.5 kJ/mol	(2.6)
Hydrogenating gasification	$C + 2H_2 \rightarrow CH_4$	+74.9 kJ/mol	(2.7)
Shift conversion	$CO + H_2O \rightarrow CO_2 + H_2$	+41.0 kJ/mol	(2.8)
Drying and pyrolysis zone			
Coal pyrolysis	$Coal \rightarrow Char + CO_2 + CO + H_2O + CH_4 + H_2 + HC's$		(2.9)

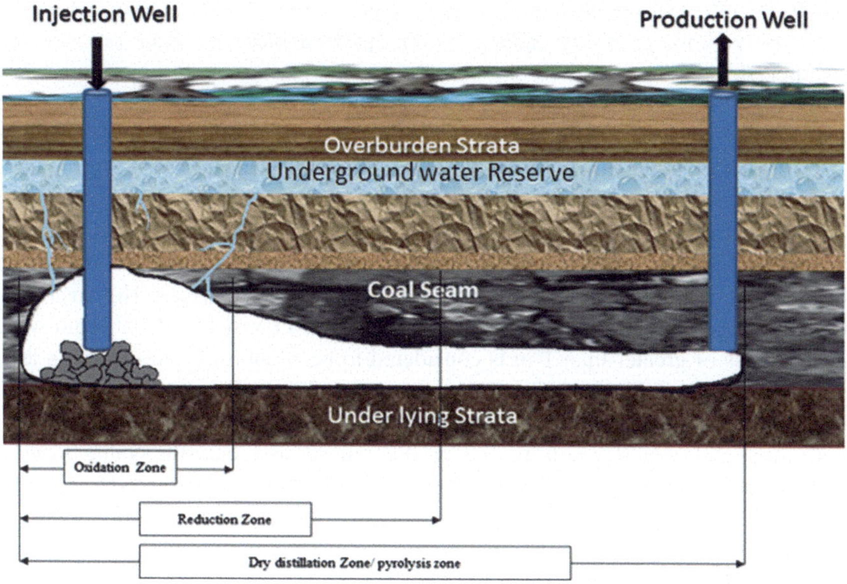

Fig. 1 Reaction zones of underground coal gasification

In the reduction zone, the major chemical reactions (reactions 2.5–2.8) are steam gasification, Boudouard reaction, methanation, and water gas shift reactions, which produce CO, H_2, and CH_4 as the calorific gases. The gasification reactions are endothermic in nature and get progressed with the intake of heat energy from the combustion zone. The residual heat of the gas after gasification reactions and the heat conducted through solid phase between the reaction zones release volatile matters (pyrolysis, reaction 2.9) and moisture (drying) in coal seam. Pyrolysis occurs throughout the coal seam and liberates volatile matter along with product gas. These volatile matters constitute a major proportion of the volume of syngas.

2.4 *Effect of Various Parameters on UCG Process*

As UCG is an in situ gasification technology, the controlling parameters are limited for the occurrence of efficient gasification. The performance of UCG depends on various parameters such as inherent property of coal seam, nature of overburden strata, choice of feed gas, operating pressure, and water seepage. The operating conditions have major effect on cavity growth, temperature, water influx, operating pressure, and gas composition in a UCG process (Perkins and Sahajwalla 2007). It is essential to analyze the geological conditions of a coal seam before selecting it for a UCG process. Several UCG laboratory scale experiments are reported in terms of the effect of operating parameters on product gas composition. In the following section, the effect of parameters such as pressure and coal seam thickness and the choice of various gasifying agents on the performance of UCG process are investigated.

2.4.1 Effect of Permeability and Thickness of Coal Seam

Hard coal seams with less permeability have a tendency to swell and resist the flow of gas inside the coal bed. This would affect the progress of chemical reactions. Thus, it is desirable to conduct UCG in high permeable coal seams. However, the loss of syngas is a major drawback in high permeable coal seams. Coal seam with bed thickness of greater than 1 m is considered to be economic, and thickness less than 0.5 m is unsuitable for UCG (Friedmann et al. 2009). This is due to high loss of thermal energy in thin seams into surrounding strata rather than being utilized for gasification and would result in the production of low calorific value syngas (Stephens et al. 1985).

2.4.2 Effect of Gasifying Agent

The choice of feed gas to UCG depends on the inherent properties and geological conditions of a coal seam. The effect of various combinations of oxidizing and gasifying agents on product gas composition has been reported in several studies. Air/steam, O_2/steam, air/CO_2, O_2/CO_2, and O_2-enriched air/steam are the various injection fluids to UCG.

Air as a gasifying medium produces a less calorific value product gas as it is diluted with nitrogen gas. Further, it reduces the temperature of reaction zones and causes inefficient gasification. Steam/oxygen feed gas produces a medium calorific syngas. However, it is reported that low-temperature steam extinguishes the fire front of a coal seam, and thus, superheated steam is essential for efficient gasification. Syngas with an equimolar proportion of CO and H_2 can be produced using steam as a gasification agent. Daggupati (2011) used steam as a gasifying medium with O_2 and reported that at least 4 h of pre-combustion period is essential for efficient gasification to form sufficient roof surface area for solid–gas reaction. Their studies show that a high oxygen flow rate led to higher combustion rate than gasification, and the gasification efficiency is reduced with excess steam. An optimum steam to O_2 ratio was found as 2.5 for the production of high calorific syngas with 39% hydrogen with a calorific value of 178 kJ/mol. Liu et al. (2009) conducted UCG studies using low-ash lignite coal and concluded that an increase in steam/O_2 ratio resulted in high H_2 production with a sharp decrease in CO concentration. However, an increase in CH_4 and CO_2 concentration is observed due to the progress of methanation and water gas shift reaction. An optimum steam/O_2 ratio of 2:1 is estimated for the production of high calorific value syngas. Yang (2008) conducted two-stage gasification using air and steam mixture as gasifying agents in Jiangsu Province, North China. Their study reported that an average of

65–75% of H_2 (mol%) gas can be produced in two-stage gasification method with H_2O/air ratio of 1:1 (Yang 2008). These trial experiments show that O_2/steam and air/steam produce a high calorific syngas at the optimum molar ratio of feed gas. However, the production and transportation of steam to deep coal seam would lead to operational difficulty for efficient gasification.

Water can also be injected as a gasifying medium along with pure oxygen instead of steam. However, the presence of underground aquifer in the vicinity of coal seams would lead to water influx into the underground cavity. The influx water into a coal seam can act as a natural source of gasification agent. It is essential to control the quantity of water influx/injected into the cavity for better gasification. In case of high rate of water influx, most of the combustion energy can be lost for evaporating water into steam. As a result, a less calorific value syngas is produced with enriched steam content. Thus, water seepage should be controlled during UCG operation. Many real-scale UCG trials conducted in Chinchilla (Australia) and Swan Hills (Canada) (Perkins and Vairakannu 2017) show that water/O_2 is a potential gasifying medium under optimized operating conditions. Another alternative gasifying medium for UCG is CO_2, which is a greenhouse gasification agent. CO_2 can replace steam as a gasifying agent, and thus, the complexity associated with the production and transportation of steam into deep coal seams could be avoided. Further, the interaction of CO_2 gas with volatile matters, tar and char of coal seam, would progress the dry reforming and Boudouard reactions (reaction 2.5) and may enhance the yield of syngas. Although CO_2 gasification has several advantages, the production of pure CO_2 to UCG is expensive. If UCG is integrated with power plants incorporating a carbon capture unit (CCU), the captured CO_2 gas can be recycled to UCG for economic operation.

Table 3 shows proximate analysis of coal, gasifying medium, product gas composition, and calorific value of syngas of UCG trials, which are conducted in laboratory and pilot-scale level. On comparing the data, it was observed that O_2 as a gasifying medium provides a higher calorific value product gas than air. Air gasification resulted in the production of syngas with calorific values in the range of 11–140 kJ/mol, whereas O_2 feed gas provided a higher range of calorific value gas of about 77–195 kJ/mol. The presence of N_2 in air reduces the temperature of flame front in reaction zones and further dilutes the product gas, and as a result, the calorific value of the product gas decreases. As the use of pure oxygen is expensive, O_2-enriched air may be considered as a viable option (Stańczyk 2011) for economic operation. It is observed from Table 3 that with an increase in O_2 to air ratio, the calorific value of product gas decreases because the combustion reactions might get well progressed rather than gasification. A medium calorific value product gas in the range of 92–127 kJ/mol is reported for the case of O_2-enriched gasification. However, the experimental results of steam/O_2 and CO_2/O_2 gasification show the production of high calorific value syngas with hydrogen and CO-enriched syngas, respectively. However, the production of steam and pure CO_2 as a gasifying medium to UCG led to high energy penalty and may reduce the overall efficiency of process.

Table 3 Effect of coal properties and gasifying agent on syngas product composition and its calorific value

Gasifying agent	Coal properties				Product gas composition				Calorific value (kJ/mol)	Reference
	M (%)	A (%)	VM (%)	FC (%)	CH_4	H_2	CO	CO_2		
Air	14.5	8.6	42.8	34.1	0.17	2.5	1.3	12.1	40.1	Stańczyk (2011)
Air	1.5	2.2	32.4	63.9	2.8	11.8	10.3	14.7	80.15	Stańczyk (2011)
Air	31.8	4.4	31	32.4	1.61	17.3	8.9	16.1	90	Cena et al. (1988)
Air	30.1	4.1	32	33.7	2.32	13.9	7.7	18.3	95	Cena et al. (1988)
Air	0.02	16.15	28.68	31.71	5.1	18	11	24	115	Yang et al. (2003)
Air	0.01	6.42	30.37	58.92	8	19	10	21	139	Yang et al. (2003)
O_2	14.5	8.6	42.8	34.1	1.7	19.2	6.2	63.6	77.6	Stańczyk (2011)
O_2	9.3	32.7	31.4	26.7	6	19	8	18	117	Perkins et al. (2016)
O_2	1.4	11	38	49.5	10.1	9.1	10.5	11.4	154	Haines and Mallett (2013)
O_2	5.3	15.4	40.1	39.2	5.5	17.2	27.5	43	163.5	Shu-qin et al. (2009)
O_2	1.5	2.2	32.4	63.9	3.2	31.5	33.2	27.9	195.8	Stańczyk (2011)
O_2-enriched air (4:2)	14.5	8.6	42.8	34.1	2.3	23.1	6.3	49.4	92.14	Stańczyk (2011)
O_2-enriched air (2:3)	1.5	2.2	32.4	63.9	4.2	18.7	17.3	23	127.88	Stańczyk (2011)
O_2:CO_2 (0.5:1)	0	4.5	10.1	85.4	0.794	0.287	41.31	47.33	123.92	Marcourt et al. (1983)
O_2:CO_2 (1.01:1)	0	4.5	10.1	85.4	0.63	0.42	40.7	25.9	177.62	Marcourt et al. (1983)
Steam:O_2 (1:1)	32.83	10.91	24.9	44.5	3.8	29	31.5	36	189.14	Liu et al. (2009)
Steam:O_2 (2.5:1)	32.83	10.91	24.9	44.5	4.7	51.45	11.14	32.7	193.55	Liu et al. (2009)
$H_2O(g)/O_2$ (1:1)	9	13.8	38.9	38.2	9	38	13	39	202	Cena et al. (1988)
Steam:O_2 (3:1)	32.83	10.91	24.9	44.5	7.1	58.11	3.9	30.87	208.5	Liu et al. (2009)
Steam:O_2 (2:1)	32.83	10.91	24.9	44.5	4.17	48.9	24.48	22.4	221.01	Liu et al. (2009)
Steam:O_2 (1.5:1)	32.83	10.91	24.9	44.5	4.3	41.45	31.7	22.5	224.36	Liu et al. (2009)

2.4.3 Effect of Pressure

Pressure has a positive impact on coal gasification process. The increase in operating pressure improves kinetic parameters and could lead to high calorific value product gas (Bhutto et al. 2013). Although high operating pressure can be a benefit for improving the gasification efficiency, it may lead to gas loss in a coal seam during UCG process. If the overburden strata of coal seam are highly permeable and the hydrostatic pressure of the adjacent water body is less than the operating pressure of UCG process, the syngas would easily escape through this high permeable coal seam and leaks into the water body, and thus, it would cause groundwater contamination. The dissolution of contaminants of syngas in water may increase the pH value, which subsequently affects the nearby environment. Also, there is a possibility that the groundwater may invade into high-temperature cavity of the coal seam after the completion of the UCG process. As a result, most of the water may get vaporized and goes back to the surface of the coal seam. However, this water vapor would condense and again get into the coal seam as water. The migration of this contaminated water into the seam after gasification process may lead to leaching of gasification residue (ash, tar, and carbon deposit). The major contaminants of organic and inorganic compounds of the gasification residue are phenolic compounds and total dissolved solids (TDS), respectively (Bicer and Dincer 2015). It is reported that the contamination of water would reduce with decrease in the UCG run period. Also, if the distance between aquifer and coal seam is large, it may diminish the chance of water contamination. The surrounding strata with high thickness of coal seam act as a filter for water intrusion into the cavity (Liu et al. 2009). Low-pressure UCG operation can also cause water seepage into the cavity, which would reduce the temperature of reaction zones. Thus, the operating pressure is a crucial parameter for efficient gasification in a UCG process.

2.5 Worldwide Real-Scale UCG Trials

UCG concept was first introduced by Sir William Siemens in 1868 and later the first experiment was conducted by William Ramsay in the year 1912 in Durham Coalfield, UK (Vyas and Singh 2015). Intensive studies were conducted in the 1930s in the USSR on shallow depth coal seams. And, several industrial scale UCG power plants were started in Russia and Uzbekistan. In 1939, the former Soviet Union had successfully begun a UCG plant at Ukraine, and later, Germany terminated this UCG plant. In the year 1960, a UCG power plant at Angren, Uzbekistan, was constructed, which is still working with 100 MW capacity serving for more than 30 years (Green 2008). This trial demonstrated the gasification of 30,000 tons of shallow depth coals (Klimenko 2009). In South Africa, an UCG power plant with a capacity of 4200 MW was constructed at Majuba coalfield, Johannesburg, in the year 2007. Due to volcanic intrusions, it is currently operating with a capacity of 1200 MW integrated with combined cycle power system (IGCC) (Sajjad and Rasul 2014).

During the 1970s, Department of Energy (DoE), USA, invested billions of dollars to develop efficient UCG system for energy scarcity problems. Lawrence Livermore National Laboratory (LLNL) conducted several UCG pilot-scale plant projects ($\sim$30) in the USA at various locations of Hanna I, II, III, and IV and Rocky Mountain I. Also, they developed two successful UCG test sites in Centralia, Washington, and Hoe Creek, Wyoming. However, due to the decrease in oil price during the 1980s and 1990s, there was a halt in commercializing the UCG technology (Ranathunga et al. 2017). The UK government also commenced a five-year study for the utilization of offshore coal seams using the UCG technology. In Australia, Chinchilla UCG plant was operated during the period 1997–2003. Linc Energy demonstrated the first long-term pilot plant in the world successfully without water contamination problem in Queensland, Australia. Also, Queensland Cougar Energy started another UCG plant with a capacity of 400 MW at Kingaroy, Australia, in 2010. However, it was terminated in a short period due to water contamination issues (Fei 2017; Isa and Hedland 2012).

China also started UCG project in the 1990s, and currently, it became a development leader of UCG technology. In 1991, China conducted 16 pilot plant UCG trials. They have been utilizing UCG syngas for the production of fertilizer and chemicals. Several UCG plants are constructed in China at various locations of Xinhe (Xuzhou), Liuzhuang (Hebei), Hebi (Henan), Feicheng (Shandong), Suncun (Xinwen), and Huating (Shanxi). Most of the UCG process plants are operated using air or oxygen-enriched air as the oxidizing medium. It was reported that the generated syngas of these plants showed a high percentage of calorific value gases in the range of 45–55% H_2, 9–11% CH_4, and 28% CO (Fei 2017).

Figure 2 shows the existing large-scale and pilot-scale UCG plants of various countries in the world. A successful UCG plant was established at Chinchilla, Australia, and it ran for almost 5 years in a lignite coal seam. Also, a UCG-integrated combined cycle power plant was established in 2007 at Majuba, South Africa, and it is the first successful UCG-integrated power plant in the world. LLNL conducted several real-scale shallow deep UCG trials at Hanna, USA. Russia and Spain conducted UCG studies in deep coal seams having a depth of 400 and 600 m, respectively. These worldwide UCG studies influence the commercialization of the UCG technology in India.

2.6 UCG Studies in India

India started UCG studies in the 1980s and conducted a preliminary study for the selection of UCG site in the regions of Mehsana in Gujarat (depth 500–1700 m), Merta road in Rajasthan (depth 100–200 m), Bihar, and Jharkhand. Skochinsky Institute of Mining (SIM) selected coal seams at South Sayal, Medni Rai blocks, and Merta Road block of Rajasthan. The UCG site was selected based on the consideration that there should not be any aquifer nearby surface or underground. In 1984–86, oil and natural gas corporation (ONGC) conducted a pilot-scale UCG trial

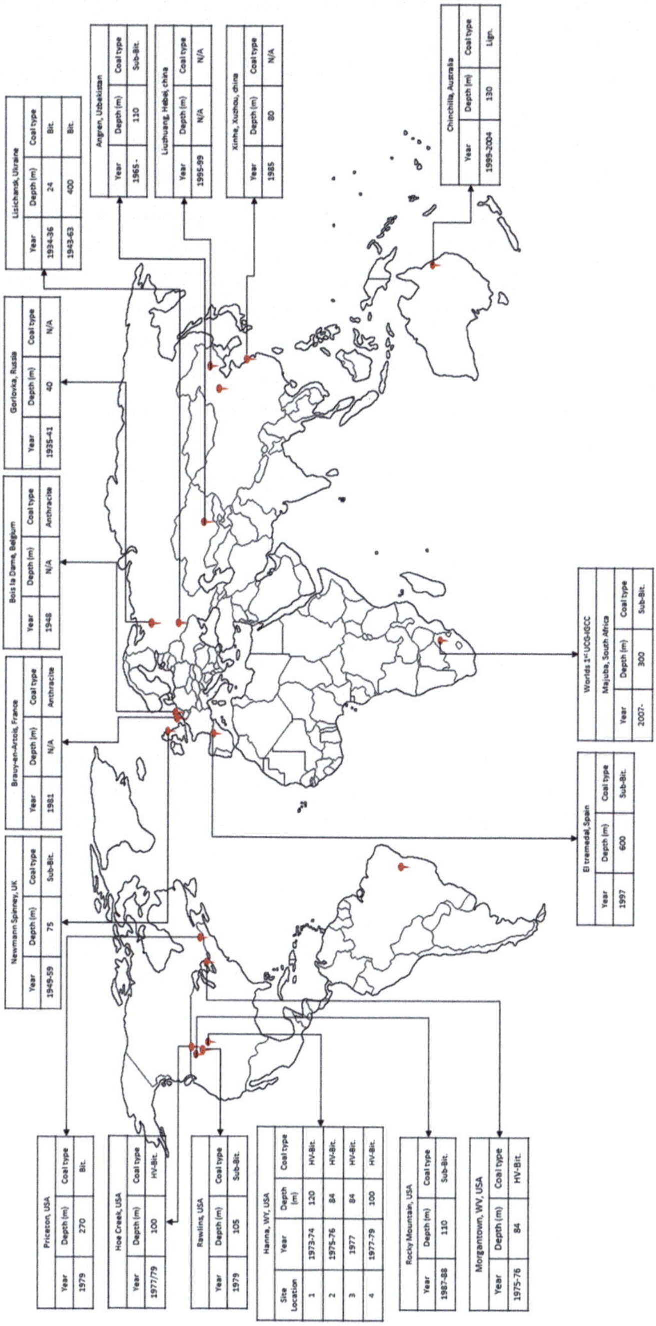

Fig. 2 Location of worldwide UCG sites

in a deep lignite mine (600–1000 m) at Mehsana, Gujarat. Injection and production wells were drilled at a distance of 10 km (Khadse et al. 2007). However, the result of the study has not been reported.

2.7 *Status of UCG in India*

In 2006, a UCG committee was constituted by the Indian government for the implementation of UCG in India. The members of this committee are the representatives of investors such as Essar Oil, ISM, RIL, SCCL, BHEL, CIL, NCL, MECL, CIMFR, GAIL, and CMPDI. The primary objective of this committee was to study the undergoing UCG trials worldwide for establishing a commercial UCG process plant in India. In 2015, Coal and Power Ministry of India framed a draft policy to develop UCG technology in unexplored coal and lignite bearing areas of India. In the financial year 2014–15, Indian government nominated five lignite blocks located in the states of Gujarat and Rajasthan for power/commercial mining/UCG (Black and Aziz 2010).

2.7.1 Oil and Natural Gas Corporation Ltd. (ONGC)

ONGC signed an agreement for collaboration with Skochinsky Institute of Mining (SIM), Russia, in 2004 for the implementation of UCG in India. After several studies, Vastan Mine block of Gujarat Industrial Power Co. Ltd. (GIPCL), Surat (Gujarat), was found suitable for the UCG technology. On the other hand, ONGC and NLC jointly planned to establish UCG plants in Tarkeshwar, Gujarat, and Hodu–Sindhari and East Kurla, Rajasthan. Another jointly identified site by ONGC and GMDC was Surkha, Bhavnagar (Gujarat) (Sahu 2013). Out of all these locations, only a few locations are considered to be suitable for UCG and rest of the mines were rejected. These rejected mines were found to be surrounded by water aquifers or discontinuous in nature. Some of the coal mines were blocked due to the formation of basalt, which would not be suitable for UCG operation.

2.7.2 Reliance Industries Ltd. (RIL)

RIL conducted serious efforts for exploring the viability of UCG implementation in India. It collaborated with Uzbekistan to understand the process conditions of UCG. In Angren, there was a UCG project plant with 45-year-old commercial operation site. RIL signed memorandum of understanding (MoU) with Gujarat Mineral Development Corporation (GMDC) Limited for examining the suitability of coal mines for UCG in Gujarat.

2.7.3 Indian School of Mining (ISM)

During the 1980s, ISM, Dhanbad, constructed a UCG reactor to study the effect of various parameters on the performance of UCG. ISM submitted a multidisciplinary R&D project proposal to the planning commission of India for the grant of 12 crores INR, for setting a "Center of Clean Coal Technology" in its campus for UCG research work (Black and Aziz 2010).

2.8 Application of UCG Syngas

Several studies reported the integration of UCG plant with power generating systems for the utilization of syngas. Net thermal efficiencies of UCG-based power plants such as integrated gas turbine combined cycle (IGCC), integrated steam turbine cycle (ISTC), and integrated solid oxide fuel cell (ISOFC) (Cena et al. 1978; Prabu 2015; Prabu and Geeta 2015; Prabu and Jayanti 2012; Mallick and Prabu 2017) are reported. High-pressure UCG syngas can be effectively utilized in a combined cycle power plant. UCG syngas can also be utilized for production of synthetic diesel, naphtha, and wax or other liquid fuels such as DME and methanol (Khadse et al. 2007).

3 Coal Bed Methane (CBM)

CBM is a clean source of energy as it contains approximately 95–99% pure methane (Mallick and Prabu 2017). Coal seam acts as a source and reservoir for methane gas (Ranathunga et al. 2017). CH_4 is a high calorific value gas having heating value of approximately 55.7 MJ/kg. On the other hand, it is a greenhouse gas and causes global warming. Hence, methane gas from coal bed provides a sustainable energy source for the production of electricity.

3.1 Extraction of CBM Gas

CBM gas can be extracted at different periods of mining process, and it is classified as follows.

(a) Pre-mining methane: capture of methane before coal mining process
(b) Coal mine methane (CMM): capture of the released methane gas during mining process
(c) Post-mining methane: capture of methane that is left in abandoned mining areas.

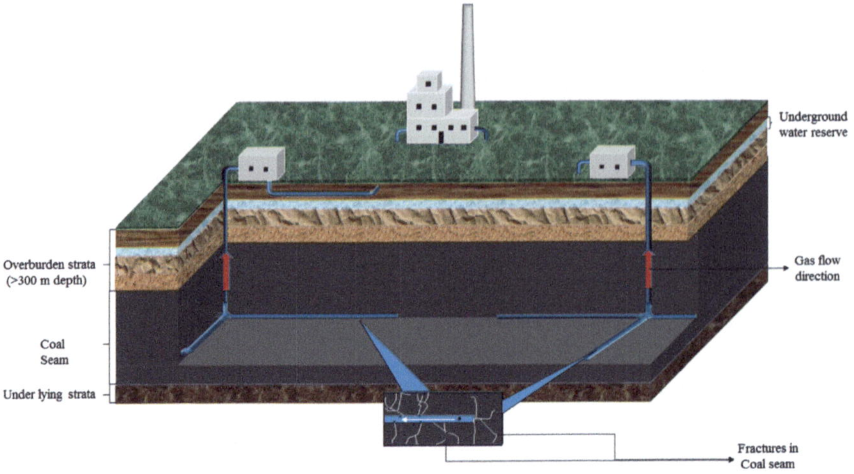

Fig. 3 Schematic diagram of CBM extraction process

Most general methods of CBM extraction are pre-mining and post-mining drainage methods. Pre-mining method requires either high permeable coal seam or continuous cleat system for the capture of methane gas. On the other hand, in post-drainage system, the residual methane gas is captured in underground mining areas, and thus, it avoids the leak of CBM gas into mine ventilation area (Jonathan and Kumar 2006).

Figure 3 shows the schematic representation of CBM process. CBM gases can be extracted from coal seam by drilling a well to a depth in the range of 1000–1500 ft. An annular casing is placed inside the drilled hole, and a high-pressure water stream (facing fluid) is forced into coal seams for creating fractures. The facing fluid and inherent coal seam water are dewatered using a pump. As a result, the pressure of the coal seam gets reduced, and methane gas is detached from the coal seam and flows out to the surface through the annular casing.

3.2 Parameters Affecting CBM Production

The extraction of CBM gas depends on various parameters such as depth, permeability and porosity of coal seam, and the availability of aquifers near the coal seam.

3.2.1 Reserve Quality

It is found that a minimum coal seam thickness of 20 ft is required for economical operation (Doucet and Brown 2015). The quantity of entrapped methane gas in a coal seam depends on the ranking of coal. High-ranking coal accumulates huge amount of methane gas. Coal seam having an aquifer with a sufficient hydrostatic pressure stores a reasonable quantity of CBM gas. CBM process is said to be economical if coal seam contains 50–70 ft^3 of gas/ton of coal.

3.2.2 Permeability of Coal Bed

Coal beds are usually considered as water-saturated bed and low permeable in nature. For the extraction of CBM gases, coal seams should be highly permeable. Therefore, fractures are to be induced artificially through several techniques. It was reported that the natural fracture is high with an increase in the rank of coal (Black and Aziz 2010). However, the pore size of coal matrix decreases with an increase in coal rank (Balan and Gumrah 2009).

3.2.3 Effect of Coal Composition

Moisture and ash play a vital role in methane holding capacity of coal. The quantity of moisture content majorly varies with the rank of coal. Moisture content decreases with an increase in the depth of coal seams and the rank of coal. Low-moisture coals have high adsorption capacity of methane (Bustin and Clarkson 1998). Lignite has high moisture content, and as a consequence, the adsorption capacity of methane is less although it has high pore space. Lamberson and Bustin (1993) reported that with an increase in the inorganic matter (ash content) of coal, the gas holding capacity of coal matrix would become less. The inorganic content of coal acts as a non-adsorbent for methane gas (Crosdale et al. 1998). Thus, high-ash coals have low adsorption capacity as Langmuir volume decreases with an increase in ash yield (Misra et al. 2006).

3.2.4 Environmental Impact

Water pollution is an important issue in the CBM extraction process. Dewatering process of coal seam contaminates the groundwater. As the expelled water from coal seams is acidic in nature, it cannot be used for irrigation without a proper treatment. The removal of water from a coal seam creates void space and would cause land subsidence. Several non-technical challenges are reported by the working industries of the USA and Canada. Some CBM fields are found in dense forest areas, which require road connectivity and other infrastructures (Gruszkiewicz 2009).

3.3 Worldwide CBM Resources

CBM gas exists near to liquid state inside the pores of a coal seam. This gas is also known as sweet gas as it has low sulfur content in ppm level (Hamawand et al. 2013). Approximately, 5000 ft^3 of CBM gas may generate in coal seam per every ton of coal. The worldwide CBM content is estimated as 4944 trillion cubic feet (TCF), and hence, it can be considered as a better choice for sustainable energy (Hamawand et al. 2013). Canada has the highest CBM gas reserves of 2697 TCF in the world. Currently, there are 3000 active CBM wells in Canada. Due to huge availability of CBM resources, Canada plays an important role in the production of fuel gases (Gatens 2001).

The USA has a total production of 49.7 Bm^3 of CBM gas in 2007. This CBM gas accounts for almost 9.1% of the total natural gas production in the USA. The first CBM well was drilled in the 1970s in Black Warrior Basin, USA, and the production was started in 1980. At present, there are approximately 4650 active CBM wells in the USA, and the total gas production capacity is higher than 52 Bm^3 of gases. The largest CBM plant in USA is found in the Powder River Basin, Wyoming, and it is the third largest project worldwide (Shalhevet 2008). The USA has an estimate of total resource of 449–900 TCF of CBM gas (Ernst & Young Pvt. Ltd. 2010).

The exploration of CBM was started in Australia in 1990s at Surat Basin, Queensland. In this coal seam, approximately 2.8 Bm^3 of CBM gas was captured in the year 2007–2008 (Saghafi 2010). A total of CBM resources of 310–510 TCF is estimated in Australia. China possesses the second position in holding the CBM resources approximately 579–1200 TCF. A total of 2500 active CBM wells are reported in China with nine major CBM basins having total gas reserves of 30.9 trillion m^3 during the year 2008. Countries such as Indonesia and UK have less CBM reserves with approximately 3.5–7.1 and 0.2 TCF, respectively (Ernst & Young Pvt. Ltd. 2010).

3.4 CBM Potential in India

According to the report of Director General of Hydrocarbons (DGH), India stands at the sixth position worldwide and contains CBM resources approximately 49.4–

91.8 TCF. In the last few years, India develops technologies for CBM exploration intensively. In 2009, the commercial production of CBM gas in India is estimated as 0.056 Bm^3 (Moore 2012). Due to the increase in energy demand, CBM gas production in India has increased by 456% with an average annual growth rate of 5.2% from 1977 to 2007 (Doucet and Brown 2015).

Table 1 shows the total content of estimated CBM gas resource in India using Langmuir correlation. It shows that a 43.1 Bm^3 of gas reserves is found at a depth greater than 300 m. The potential of gas storage of coal seam depends on prevailing hydrostatic pressure (Prabu and Mallick 2015) and Langmuir volume V_l. Langmuir pressure P_l (Langmuir pressure is the abandon pressure of a coal seam below which CBM extraction is not possible) is calculated using Eqs. (3.1) and (3.2).

$$\log(P_l) = k_3 * \log((FC/VM) + k_4) \tag{3.1}$$

$$\log(V_l) = k_1 * \log((FC/VM) + k_2) \tag{3.2}$$

where FC is fixed carbon content and VM is volatile matter content.

$$V_{\max} = (1 - M\% - A\%) * V_l * P/(P + P_l) \tag{3.3}$$

The volume of CBM gas in a coal seam can be found using Eq. (3.3). *M*, *A*, and *P* are moisture, ash, and pressure of the coal seam, respectively, and V_l, P_l, and *h* are the Langmuir volume, Langmuir pressure, and depth of coal seam, respectively. Table 4 shows the estimated gas volumes of Indian coal mines considering its inherent properties. The capacity of these gas reserves was already estimated in our earlier studies (Prabu and Mallick 2015) and is recalculated in this study using the current estimated coal resources of India. A total of CBM reserves of 43.11 Bm^3 is calculated for the coal seams, which are found at a depth between 300 and 1200 m. The highest CBM gas reserves are estimated in states in the order of Jharia, Jharkhand, and Raniganj, West Bengal. Currently, the CBM sites of these states are operated by GEECL, ONGC, and CIL companies.

3.5 CBM Blocks in India

Presently, CBM gas contributes 5% of total natural gas production in India. In India, CBM blocks are allotted to companies by DGH, Ministry of Coal (MoC), and CMPDI, Ranchi. A total of 33 CBM blocks with covering area of 16.6 km^2 are allotted for the extraction of CBM. Existing CBM plant in Raniganj South block is operated by M/s GEECL since July 2007.

Figure 4 shows the locations of nominated CBM blocks to various companies for the exploration of CBM gas in India (as per Director General of Hydrocarbon). Several CBM seams were awarded through nomination in the year 2001. Three sites were suggested for CBM in Jharia, Raniganj North and Raniganj South coal

Table 4 Estimated CBM resources of India considering coal proximate analysis

S. No.	State wise coal mine	Depth (m)	Total resource (MT)	Coal properties				V (Mm^3)
				M (%)	VM (%)	A (%)	FC (%)	
1. *West Bengal*								
(i)	Raniganj	300–600	8068.59	1.35	21.20	32.4	45.10	3918.89
		600–1200	4409.57	1.35	21.20	32.4	45.10	2141.71
(ii)	Birbhum	300–600	4303.24	3.50	17.55	35.0	43.95	2521.64
		600–1200	1371.43	3.50	17.55	35.0	43.95	803.64
2. *Jharkhand*								
(i)	Raniganj	300–600	525.63	1.35	21.20	32.4	45.05	254.84
(ii)	Jharia	600–1200	5217.64	1.35	13.60	23.5	61.50	9656.76
(iii)	East Bokaro	300–600	1668.89	1.30	26.90	18.2	53.60	886.59
		600–1200	2436.08	1.30	26.90	18.2	53.60	1294.15
(iv)	West Bokaro	300–600	689.97	1.30	26.90	18.2	53.60	366.54
(v)	Ramgarh	300–600	650.25	9.40	27.80	17.6	45.20	225.93
(vi)	North Karanpura	300–600	5031.80	2.50	26.70	29.0	41.80	1543.81
(vii)	South Karanpura	300–600	2095.45	4.70	22.30	33.2	39.80	719.45
		600–1200	918.21	4.70	22.30	33.2	39.80	315.26
(viii)	Auranga	300–600	1290.08	9.40	27.80	17.6	45.20	448.25
		600–1200	70.69	9.40	27.80	17.6	45.20	24.56
(ix)	Hutar	300–600	12.30	9.40	27.80	17.6	45.20	4.27
(x)	Rajmahal	300–600	5041.16	4.70	21.20	33.6	40.50	1918.31
		600–1200	35.56	4.70	21.20	33.6	40.50	13.53
3. *Madhya Pradesh*								
(i)	Pench Kanhan	300–600	1393.68	4.30	24.30	25.5	45.20	578.07
		600–1200	0.86	4.30	24.30	25.5	45.20	0.36
(ii)	Pathakhera	300–600	446.93	1.90	24.30	29.1	44.72	179.11
(iii)	Sohagpur	300–600	2690.73	5.50	33.00	13.3	48.20	875.36
		600–1200	137.25	5.50	33.00	13.3	48.20	44.65
(iv)	Singrauli	300–600	5514.62	7.00	26.50	42.5	24.00	517.64
		600–1200	222.29	7.00	26.50	42.5	24.00	20.87
4. *Chhattisgarh*								
(i)	Sonhat	300–600	971.78	6.20	28.70	12.6	52.50	453.89
		600–1200	568.85	6.20	28.70	12.6	52.50	265.69
(ii)	Hasdeo-Arand	300–600	73.39	6.10	32.41	32.2	29.29	8.39
(iii)	Korba	300–600	2971.23	7.00	26.50	38.0	28.50	400.36
(iv)	Mand-Raigarh	300–600	8542.98	7.30	25.00	31.6	36.10	2053.21
		600–1200	610.88	7.30	25.00	31.6	36.10	146.82
(v)	Tatapani-Ramkola	300–600	1546.07	4.20	26.00	24.0	45.00	584.19
		600–1200	303.36	4.20	26.00	24.0	45.00	114.63

(continued)

Table 4 (continued)

S. No.	State wise coal mine	Depth (m)	Total resource (MT)	Coal properties				V (Mm^3)
				M (%)	VM (%)	A (%)	FC (%)	
5. *Maharashtra*								
(i)	Wardha valley	300–600	1966.57	9.00	26.40	27.1	37.50	481.41
		600–1200	40.36	9.00	26.40	27.1	37.50	9.88
(ii)	Kamptee valley	300–600	1007.32	9.00	26.40	27.1	37.50	246.59
		600–1200	160.44	9.00	26.40	27.1	37.50	39.28
(iii)	Bander	300–600	381.43	7.60	27.30	25.9	39.20	98.88
		600–1200	16.76	7.60	27.30	25.9	39.20	4.34
	Umred	300–600	83.22	9.00	26.40	27.1	37.50	20.37
		600–1200	11.95	9.00	26.40	27.1	37.50	2.93
6. *Orissa*								
(i)	IB-River	300–600	25893.21	6.50	24.00	45.0	24.50	2829.39
		600–1200	2360.36	6.50	24.00	45.0	24.50	257.92
(ii)	Talcher	300–600	16473.45	7.00	24.00	45.0	23.50	1666.13
		600–900	2054.26	7.00	24.00	45.0	23.50	207.77
7. *Andhra Pradesh*								
(i)	Godavari	300–600	692.47	7.50	28.30	21.5	42.70	207.54
		600–900	360.74	7.50	28.30	21.5	42.70	108.11
8. *Telangana*								
(i)	Godavari	300–600	8385.64	7.50	28.30	21.5	42.70	2513.20
		600–1200	3560.18	7.50	28.30	21.5	42.70	1067.00
9. *Assam*								
(i)	Makum	300–600	202.00	2.30	45.20	4.1	48.40	46.00
Total CBM resources below 300 m depth								**43.108 Bm^3**

mines. Presently, these seams are operated in the production phase of CBM gas. Also, a total of 33 CBM sites were allocated for exploration during the years 2003, 2005, and 2008. In the allocated sites, only 12 blocks are under the exploration phase and two blocks of the sites in Orissa have not got approval for exploration. The rest of CBM seams were relinquished or under relinquishment due to the geological conditions of the seams.

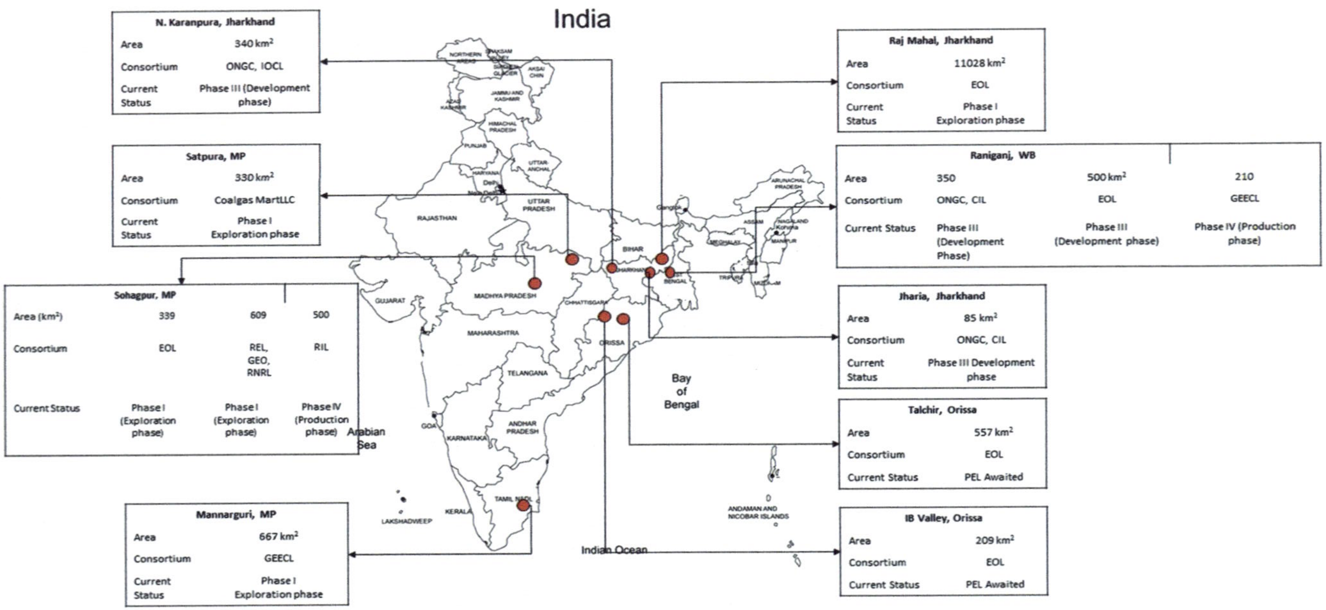

Fig. 4 Location of CBM reserves and coal seams in India

4 Conclusions

The exploration of coal resources located at a depth below 300 m is essential to satisfy our energy demand. The unconventional technologies such as UCG and CBM are suitable for the utilization of such deep coal seams. This study brought out the estimate of existing unutilized coal resources and its associated CBM resources of India for efficient usage. Further, site-specific studies are required to analyze the geological conditions and ecosystems of coal seams.

References

Andrianopoulos E, Korre A, Durucan S (2015) Chemical process modelling of underground coal gasification and evaluation of produced gas quality for end use. Energy 76:444–453

Annual Report (2017) Coal and lignite production. Ministry of Coal (MoC) 2017

Balan HO, Gumrah F (2009) Assessment of shrinkage-swelling influences in coal seams using rank-dependent physical coal properties. Int J Coal Geol 77(2):203–213

Bhutto AW, Bazmi AA, Zahedi G (2013) Underground coal gasification: from fundamentals to applications. Prog Energy Combust Sci 39(1):189–214

Bicer Y, Dincer I (2015) Development of a multigeneration system with underground coal gasification integrated to bitumen extraction applications for oil sands. Energy Convers Manag 106:235–248

Black D, Aziz N (2010) Impact of coal properties and operational factors on mine gas drainage. In: 10th underground coal operators Conference, pp 229–240

Blinderman MS, Saulov DN, Klimenko AY (2008) Forward and reverse combustion linking in underground coal gasification. Energy 33(3):446–454

Bustin RM, Clarkson CR (1998) Geological controls on coal bed methane reservoir capacity and gas content. Int J Coal Geol 38:3–26

Cena RJ, Britten JA, Thorness CB (1988) Resource recovery and cavity growth during the Rocky Mountain I field test. 14th UCG Annual Symposium, Chicago, IL, USA

Cena RJ, Thorsness CB, Aiman WR, Hill RW, Stephens DR (1978) Underground coal gasification experiment with air and O_2/steam injection periods. Lawrence Livermore National Laboratory, SPE-7512

Crosdale PJ, Beamish BB, Valix AM (1998) Coal bed methane sorption related to coal composition. Int J Coal Geol 35:147–158

Daggupati S (2011) Laboratory studies on cavity growth and product gas composition in the context of underground coal gasification. Energy 36(3):1776–1784

Doucet J, Brown C (2015) Coal bed methane the past present and future of Alberta's most abundant non-traditional source of natural gas. Energy Environ 6:48–61

Ernst & Young Pvt. Ltd. (2010) Shale gas and coal bed methane potential sources of sustained energy in future. Annual Report, EYIN1011-125

Fei M (2017) Underground coal gasification (UCG): a new trend of supply-side economics of fossil fuels. Nat Gas Ind 3:312–322

Friedmann JS, Upadhye R, Kong F (2009) Porsprcts for underground coal gasification in carbon constrained world. Energy Proced 1(1):4551–4557

Gatens M (2001) Coal bed methane: practices and progress in Canada. Canada Society for Unconventional Gas, Alberta, p 2001

Geological Survey of India (2016) Inventory of coal resources of India. Central Mine Planning and Design Institute (CMPDI), Report 2016

Green M (2008) Underground coal gasification, state of the art. In: Poland Clean Coal Conference, Badewo

Gruszkiewicz MS (2009) Adsorption kinetics of CO_2, CH_4, and their equimolar mixture on coal from the Black Warrior Basin, West-Central Alabama. Int J Coal Geol 77:23–33

Haines J, Mallett C (2013) Carbon Energy's Blackwood UCG Pilot – Panel 2 2011–2013, Underground Coal Gasification Association Conference, London, UK

Hamawand I, Yusaf T, Hamawand SG (2013) Coal seam gas and associated water: a review paper. Renew Sustain Energy Rev 22:550–560

Hongtao L, Feng C, Xia P, Shuqin L (2011) Method of oxygen-enriched two-stage underground coal gasification. Min Sci Technol 21(2):191–196

Irving W, Tailakov O, Kruger D, Williams D (2001) CH_4 emissions: coal mining and handling. In: IPCC good practice uncertainty management in national greenhouse gas inventories, pp 129–144

Isa M, Hedland P (2012) Australian in situ coal resources. Geosci Aust GeoCat #74097. ISBN # 978-1-922103-56-7

Jonathan PM, Kumar H (2006) An overview of current coal bed methane extraction technologies. Department of Energy & Mineral Engineering & EMS Energy Institute. http://www.rpsea.org/media/files/project. Accessed 31 Aug 2017

Khadse A, Qayyumi M, Mahajani S, Aghalayam P (2007) Underground coal gasification: a new clean coal utilization technique for India. Energy 32(11):2061–2071

Klimenko AY (2009) Early ideas in underground coal gasification and their evolution. Energies 2 (2):456–476

Lamberson MN, Bustin RH (1993) Coal bed methane characteristics of the Gater formation, British Columbia. AAPG Bull., 77(12)

Liu SQ, Wang YY, Zhao K, Yang N (2009) Enhanced-hydrogen gas production through underground gasification of lignite. Min Sci Technol 19(3):389–394

Longwell JP, Rubin ES, Wilson J (1995) Coal: energy for the future. Prog Energy Combust Sci 21:269–360

Mallick N, Prabu V (2017) Energy analysis on coal bed methane (CBM) coupled power systems. J CO_2 Utiliz 19:16–27

Marcourt M, Paquay V, Piel A, Pirard JP (1983) Coal gasification at pressure by mixtures of carbon dioxide and oxygen. Fuel 62:823–828

Misra BK, Singh BD, Singh A (2006) Significance of coal petrological investigations in coal bed methane exploration—Indian context. Curr Sci 91(10):1403–1409

Molenaar J, Bruining J (1995) An integrated 3D model for underground coal gasification. Soc Pet Eng SPE-30790, 929–940

Moore TA (2012) Coal bed methane: a review. Int J Coal Geol 101:36–81

Perkins G, Sahajwalla V (2007) Modelling of heat and mass transport phenomena and chemical reaction in underground coal gasification. Chem Eng Res Des 85(3):229–343

Perkins G, Vairakannu P (2017) Considerations for oxidant and gasifying medium selection in underground coal gasification. Fuel Process Technol 165:145–154

Perkins G, Toit E, Cochrane G, Bollaert G (2016) Overview of underground coal gasification operations at Chinchilla, Australia. Ener Soruc 38(24):3639–3646

Prabu V (2015) Integration of in-situ CO_2-oxy coal gasification with advanced power generating systems performing in a chemical looping approach of clean combustion. Appl Energy 140:1–13

Prabu V, Jayanti S (2012) Laboratory scale studies on simulated underground coal gasification of high ash coals for carbon-neutral power generation. Energy 46(1):351–358

Prabu V, Geeta K (2015) CO_2 enhanced in-situ oxy-coal gasification based carbon-neutral conventional power generating systems. Energy 84:672–683

Prabu V, Mallick N (2015) Coal bed methane with CO_2 sequestration: an emerging clean coal technology in India. Renew Sustain Energy Rev 50:229–244

Ranathunga AS, Perera MSA, Ranjith PG, Wei CH (2017) An experimental investigation of applicability of CO_2 enhanced coal bed methane recovery to low rank coal. Fuel 189:391–399

Remme P, Trudeau U, Gracyk N, Taylor D (2011) Technology development prospects for the Indian power sector. IEA Information Paper, International Energy Agency, Paris
Roddy DJ, Younger PL (2010) Underground coal gasification with CCS: a pathway to de-carbonising industry. Energy Environ Sci 3(4):400–407
Saghafi A (2010) Potential for ECBM and CO_2 storage in mixed gas Australian coals. Int J Coal Geol 82:240–251
Sahu HB (2013) Implementing clean coal technology through gasification and liquefaction—the Indian perspective. Int J Chem Tech Res 5(2):824–830
Sajjad M, Rasul MG (2014) Review on the existing and developing underground coal gasification techniques in abandoned coal seam gas blocks: Australia and global context. In: Proceeding of 1st international e-conference energies, vol 6, pp 14–31
Shalhevet S (2008) Balancing socioeconomic and environmental risks and benefits under multiple stressor conditions: a case study of the powder river basin in wyoming. Real-Time Deliberative Decision Making Applying to Emerging Stress, pp 415–425
Shu-qin L, Yuan-yuan W, Ke Z, Ning Y (2009) Enhanced-hydrogen gas production through underground gasification of lignite. Mining Sci Technol 19:389–394
Stańczyk K (2011) Gasification of lignite and hard coal with air and oxygen enriched air in a pilot scale ex situ reactor for underground gasification. Fuel 90(5):1953–1962
Stephens DR, Hill RW, Borg IY (1985) Underground coal gasification review. Lawrence Livermore National Laboratory. CA, UCRL-92068
Thomas DJ (2015) Direct underground gasification of North Sea coal for future UK energy prosperity. Fuel 185:941–948
Thorsness C, Britten J (1989) Lowrence Livermore National Laboratory (LLNL): UCG project final report. National Technical Infmmutio U.S. Department of Commerce. UCID-21853
Vyas DU, Singh RP (2015) Worldwide developments in UCG and Indian initiative. Procedia Earth Planet Sci 11:29–37
Yang L (2003) Study of the model experiment of blinding-hole UCG. Fuel Process Technol 82(1):11–25
Yang L (2008) Field test of large-scale hydrogen manufacturing from underground coal gasification (UCG). Int J Hydrogen Energy 33(4):1275–1285
Yang L, Liang J, Yu L (2003) Clean coal technology—study on the pilot project experiment of underground coal gasification. Energy 28:1445–1460
Yang L, Liu S, Yu L, Liang J (2007) Experimental study of shaftless underground gasification in thin high-angle coal seams. Energy Fuels 21(4):2390–2397

Part II
Sustainable Energy Policies

Investments in Clean Energy in South Asia: Visiting Barriers and Gaps from the Perspective of Policies and Politics

Anupa Ghosh and Duke Ghosh

Abstract Many countries in South Asia are facing serious challenges concerning energy access. Robust infrastructure for generation, transmission and distribution are still lacking. Simultaneously, many of these countries have enormous potential for generating clean energy that can solve the problem of energy provisioning and transform their energy mix. It can further cushion them from dependence on fuel imports. However, investment in clean energy in most South Asian countries is far below what is expected and desired. In this perspective, it is important to investigate the evidences on existing constraints that are limiting investments in clean energy in South Asia. Transition in energy systems is essentially a long-term process—and involves changes in technology, economy (structure, efficiency), active policies, institutions, behaviour and belief systems. Mere formulation of policies and setting targets (policy goals) has been found to be insufficient for attracting investment in clean energy in most of these countries. There exists a gap in the literature regarding evidences on the barriers to investment in clean energy in South Asian countries. Following the Political Economy Analysis framework, in this article, the authors attempt to build up on evidence for answering the following questions:

(a) What is the nature of energy regimes in South Asia? What are the trajectories of clean energy?
(b) What factors are affecting prospects of investment in clean energy?
(c) What are some of the gaps in policies, regulations, etc. that have impacts on investments in clean energy?
(d) What are the evidences on politics and political mandates of governments affecting investments in clean energy in South Asia?

A. Ghosh (✉)
Department of Economics, The Bhawanipur Education Society College, Kolkata, India
e-mail: anupa.ghosh@gmail.com

D. Ghosh
Science and Policy Research Unit, University of Sussex, Brighton, England
e-mail: duke.ghosh@globalchangeresearch.in

A. Gautam et al. (eds.), *Sustainable Energy and Transportation*, Energy, Environment, and Sustainability,
https://doi.org/10.1007/978-981-10-7509-4_7

(e) How should the risks in investments be approached and managed?
(f) Are there opportunities streaming out of the political economy of clean energy in South Asia? How can these opportunities be leveraged?

Keywords Clean energy · Transition · Political economy framework

1 Introduction

Energy access is essential for poverty reduction, social development and economic growth (AIIB 2017; Bradbrook and Gardam 2006; WCED 1987). In many Asian countries, due to inefficient and insufficient infrastructure for energy generation, transmission and distribution, universal energy access is a serious challenge. Most of these countries can benefit from altering their energy portfolio mix by exploiting their enormous potential for clean energy. This would arrest the drain on their exchequer due to fuel import, increase their self-sufficiency in energy provisioning and reduce their carbon footprint.

South Asia[1] has enormous potential to transit from an energy poor[2] to an energy-sufficient region by harnessing its clean energy resources. Population growth and increased economic activity are driving up energy demand in South Asia. Between 2001 and 2010, the region accounted for 44% of the total global increase in energy demand (CDKN and ODI 2014). In the same period, the global growth in primary energy[3] supply was largely driven by growth in coal consumption in South Asia. Further, during the same period, because of increasing energy demand and substantial dependence on fossil fuels, Asia had the highest global growth rate of 7.89% in greenhouse gas (GHG) emissions from the energy sector (CDKN and ODI 2014). There is, therefore, an urgent need to invest in clean energy resources so as to ensure sustainable energy sector development in the region. This paper attempts to investigate the key barriers/constraints to investment in clean energy in South Asia from the perspective of the political economy assessment (PEA) framework (DFID 2009). The aim is to enquire into the gaps in knowledge and evidence on constraints and opportunities, and how these may be addressed for designing investment plans in the clean energy sector.

---

[1]The South Asian countries are Afghanistan, Bangladesh, Bhutan, India, Maldives, Nepal, Pakistan and Sri Lanka (Intergovernmantal Panel on Climate Change 2007).

[2]Energy poverty is the lack of access to modern energy services like electricity and clean cooking fuel (International Energy Agency 2017).

[3]Primary energy is energy stored in natural resources like coal, crude oil, natural gas, uranium and renewable resources (CDKN and ODI 2014).

Transitions in energy systems are essentially a long-term process—and involves changes in technology, economy (structure, efficiency), institutions, culture, behaviour and belief systems (Patwardhan et al. 2012). In the context of South Asia, this paper tries to build up on evidence on the following:

(a) The nature of energy regimes in South Asia and the trajectories of clean energy.
(b) Factors affecting the prospects of investment in clean energy.
(c) Gaps in policies, regulations, etc. that have impacts on investments in clean energy.
(d) Evidences on politics and political mandates of governments affecting investments in clean energy.
(e) Systems/instruments that will help manage investment risks.
(f) Existing opportunities streaming out of the political economy of clean energy in South Asia and ways to leverage these opportunities.

This paper attempts to collate answers to the above questions on the basis of evidences that are scattered in the literature. A brief discussion on the methodology adopted for the purpose of the present study precedes an analysis of the extant energy regime in South Asia in the context of PEA. This discussions bring to the fore the lacunae in the system that is restricting investments in renewable in the region, while simultaneously highlighting the prospects that may be exploited to effect the scaling up of clean energy.

2 Methodology for Enquiry

The study has been carried out based on desk research. Various secondary sources—Web documents, academic articles, databases, etc. have been consulted for the purpose of the study. Reports prepared by several international think tanks, multilateral and bilateral development financing institutions, non-profit organizations, research groups, etc. have been reviewed. Evidence has also been drawn from policy documents, academic articles and scientific research reports. Further, databases on energy-related issues which are available in the public domain have also been consulted. The focus has been to understand the risk and opportunities for investment in clean energy in South Asia arising out of the political economy. However, there is dearth of literature with regard to some of the countries. This limited the scope for consultation of documents. The framework of analysis adopted for this study is a PEA framework (DFID 2009).

3 Energy Regimes in South Asia

Policy advocating low-carbon development pathways should not just reduce GHG emission, but should simultaneously enhance energy security and energy access (Patwardhan et al. 2012). In most parts of South Asia, both access to electricity and energy security[4] are inadequate (Table 1). Spread over only 3.7% of the total global surface area, South Asia is home to almost a quarter of the world population (The World Bank 2017a). About 20% of the population in this densely populated region has no access to electricity. Energy poverty is particularly pronounced in rural areas where only 72% of the population has access to electricity as against 97% in urban areas (The World Bank 2017a; The World Bank, International Energy Agency 2015). The situation is both socially and environmentally alarming since evidence suggests that the energy poor depends on polluting and less energy-dense fuels (Sovacool 2012) and are, therefore, exposed to the risk of economic losses and indoor air pollution (Sovacool 2013). Further, electricity generation in South Asia is largely dependent on fossil fuels. India, Bangladesh, Maldives are heavily dependent on coal, natural gas and/or oil for electricity generation (Tables 2 and 3). Also, South Asian nations, particularly Maldives, often have to import fossil fuels for energy generation (Maldives Energy Authority 2013). To meet demand shortfalls, energy trade among the nations is also common. This further puts at risk their energy securities (UNDP 2013). Therefore, in order to achieve total energy access, improvements in their balance of payments situation and lower emissions, it is essential that South Asia pursue an alternative development pathway that precipitates a transformative change in its electricity regime where renewable will now play a pivotal role (Howes and Wyrwoll 2015). However, the success of any policy is significantly dependent on an enabling governance structure. Extant politics and political economy would significantly influence the course and outcome of energy reforms in South Asia (Fritz et al. 2009).

3.1 Renewable Potential in South Asia

Investment in renewable can help South Asia attain energy security and curtail the high level of GHG emission that economic growth almost always ensues (Anbumozhi and Kawai 2015; Intergovernmental Panel on Climate Change 2014; United Nations 2015). The geophysical conditions in South Asia endow the region with substantial renewable potential, particularly solar, hydro and wind (Tables 4

[4]The term energy access broadly refers to access to sustainable fuel sources for the purpose of lighting, cooking, etc. (International Energy Agency 2017). Further, the term clean energy is generic and encompasses a large number of sectors like transport, agriculture, industry, etc. Given this rather broad perspective, the focus of this paper is restricted to electricity generation from renewable sources (RE).

Table 1 Electricity access in South Asia in 2014

S. No.	Country	Percentage of total population with access	Percentage of urban population with access	Percentage of rural population with access
1	Afghanistan	89.5	94.3	87.8
2	Bangladesh	62.4	84.2	51.4
3	Bhutan	100.0	100.0	96.0
4	India	79.2	98.3	70.0
5	Maldives	100.0	100.0	100.0
6	Nepal	84.9	99.2	81.7
7	Pakistan	97.5	100.0	95.6
8	Sri Lanka	92.2	97.8	90.9
9	South Asia	80.0	97.1	71.8
10	World	85.3	96.3	73.0

Source The World Bank (2017b)

and 5). Yet fossil fuel dominates the energy mix in the region. Even in India—a significant player in the global renewable energy (RE) market, fossil fuel dominates electricity generation. Though in Bhutan and Nepal, hydropower dominates the electricity generation mix (Table 3), both the countries have tapped only an insignificant proportion of their hydro potential. This is in sync with most Asian nations, where only 10% of the available capacity for RE is leveraged (Sipahutar et al. 2013). Even though the energy policies of most South Asian nations—India, Pakistan, Sri Lanka, Bangladesh, Nepal, Maldives—have committed to mainstream RE (REN21 2016; Azwar and Waheed 2011), it is evident that till date, investment in RE in the region is still lacking, and hence, the full potential of renewable is yet to be harnessed (UNDP 2013).

3.2 *The Political Economy Guiding Energy Regimes in South Asia*

Political and economic factors play an important role in the successful transformation of any regime (Geels 2002; Geels and Schot 2007). The shift from a carbon-intensive energy regime to a low-carbon pathway might require an extensive overhauling of the existing governance systems, institutional arrangements and structural frameworks of an economy (Bhattacharya 2007). Simultaneously, there is also a need for political, social and economic stability for any long-term policy change to be effective (The World Bank 2006). The political scenario in most South Asian countries has been highly volatile in recent years. Since the 1990s, frequent changes in governments, political ideologies and hence governance systems across the region have severely hampered the continuity of policies in many cases.

Table 2 Energy generation in South Asian countries (in billion kWh)

Country	2012				2014			
	Total electricity	Hydro	Non-hydro RE	Share of renewable in total electricity (%)	Total electricity	Hydro	Non-hydro RE	Share of renewable in total electricity (%)
Afghanistan	0.88	0.71	0	80.7	1.05	0.90	0	85.7
Bangladesh	45.69	0.77	0	1.7	52.53	0.58	0.15	1.4
Bhutan	6.75	6.75	0	100.0	7.15	7.15	0	100.0
India	1053.12	112.58	52.75	15.7	1217.71	127.95	67.51	16.1
Maldives	0.29	0	0	0.0	0.33	0	0	0.0
Nepal	3.55	3.50	0.03	99.4	3.8	3.75	0.04	99.7
Pakistan	92.84	29.56	0	31.8	100.38	31.11	0.40	31.4
Sri Lanka	11.36	3.26	0.19	30.4	11.96	4.51	0.33	40.5

Source International Energy Statistics (2014)

Table 3 Electricity generation (as percentage of total generation) by source in 2014

Country	Fossil fuels	Renewable	Hydro power	Non-hydro renewable	Electricity distribution and transmission loss
Afghanistan	14.68	85.32		–	–
Bangladesh	98.7	1.32	1.1	0.3	11.0
Bhutan		99.99		–	–
India	81.7	15.41	10.2	5.2	19.0
Maldives	99.4	0.61		–	–
Nepal		99.97	99.8	0.2	32.0
Pakistan	64.9	30.22	29.8	0.4	17.0
Sri Lanka	60.8	39.20	36.5	2.7	11.0
South Asia	80.0	16.80	11.6	4.6	19.0
World	66.3	22.35	16.2	6.0	8.0

Source The World Bank (2017b)

Table 4 Overview of renewable energy potential in South Asia

Country	Renewable energy potential
Afghanistan	Hydro power: >2300 MW; solar power: Possible in southern Afghanistan; wind power: possible in western Afghanistan (United States Energy Association 2015)
Bhutan	Huge hydro potential of which only 5–6% has been tapped. It is possible to generate an estimated mean annual power of 99,159 GWh that is about 60 times current consumption. Solar, wind and municipal solid waste can also be explored as economically viable alternate RE sources (Asian Development Bank 2012)
Bangladesh	The country has enormous scope for producing power through biogas, biofuels, solar and wind (Islam et al. 2014). This has also been recognized by the Government of Bangladesh (Ministry of Power, GoB 2011)
India	A significant stakeholder in the global renewable market, the country has a commercially exploitable potential of 900 GW of power from solar, hydro and wind. The Government of India has committed to undertake one of the 'largest renewable capacity expansion programmes in the world' (Ministry of New and Renewable Energy, GoI 2017)
Maldives	Largely dependent on imported fossil fuel for energy. Government of Maldives as part of the national carbon neutral plan is committed to generate a minimum of 60% electricity from solar power by 2020 (Azwar and Waheed 2011)
Nepal	Apart from large hydropower, Nepal is endowed with resources conducive to development of micro- and mini-hydropower plants, solar installations and biogas-based power generation system. The Government of Nepal is aiming at harnessing such energy resources and increasing the share of renewables from the present 1% to about 10% of the total primary energy supply by 2030 (Government of Nepal 2011; Climate Investment Funds 2012)

(continued)

Table 4 (continued)

Country	Renewable energy potential
Pakistan	Alternative Energy Development Board (AEDB) identifies solar (in Balochistan, Punjab, Sindh, Cholistan), wind (60,000 MW), bioenergy and small hydro as the dominant potential sources of renewable energy (Alternative Energy Development Board 2015)
Sri Lanka	Substantial hydro, wind and solar potential. While major hydro projects are already in operation, the Government of Sri Lanka is keen on exploiting Small Hydropower projects that have an aggregate technical potential of 800–900 MW. The government is also exploring the other resources in a bid to attain 20% RE share in the energy mix by 2020 (Sri Lanka Sustainable Energy Authority 2016)

Table 5 RE potential in South Asia

Country	Solar power potential (kWh/m^2/day)	Hydropower potential (MW)	Wind power potential (MW)
Afghanistan	6.5	25,000	158,000
Bangladesh	5.0	330	–
Bhutan	–	30,000	–
India	5.0	150,000	102,778
Maldives	4.5–6	–	20
Nepal	4.0	83,000	–
Pakistan	5.3	59,000	131,800
Sri Lanka	5.0	2000	24,000

Source Ershad (2017), Sri Lanka Sustainable Energy Authority (2016), Shukla et. al. (2016), International Renewable Energy Agency (2015), Asian Development Bank (2012), Azwar and Waheed (2011)

Sometimes, the emergence of coalition governments with conflicting interests has hindered the pace of transition leading to policy paralysis (Bhattacharya 2007; Parish 2006). Also, in countries, like India, where multilevel governance prevails, competing priorities (and philosophies) have posed a threat to the transition process even during periods of political stability. While some sub-national governments have been first and fast movers in exploiting the advantages of RE, some others have been slow to effect a policy shift favouring low-carbon pathways (Pal 2013). An intra-governance difference in development priorities and/or pathways can hamper a smooth and uniform transition of regimes across a sovereign state.

In spite of being similar in their general economic and development status, the South Asian nations have a highly diversified political system. Parliamentary democracy, federal republic, constitutional monarchy, etc. coexist in the region. Consequently, prioritization of development issues, policy prescriptions, implementation rules, etc. differs based on the political context, resource base, institutional arrangements and governance system of every individual country in the region. Some of the barriers to the dissemination of RE that such diversity generates have been discussed below.

4 Factors Affecting Investment in Clean Energy

Globally, developing countries are increasingly focusing on RE as a sustainable alternative for addressing their growing energy demand. In 2015, for the first time, total investment in RE (excluding large hydro) in developing economies exceeded that in developed nations by USD 26 billion (FS-UNEP Centre 2016). In South Asia, however, investment in RE has been low, except for India and Pakistan (FS-UNEP Centre 2016; FS-UNEP 2017). In fact, globally, India is among the top investors in renewable technology in the world (REN21 2016; UNEP 2015). According to the Renewable Energy Country Attractive Index (RECAI) 2017, only India and Pakistan in South Asia have been found to be 'attractive' destinations for private investments (Ernst and Young 2017).

4.1 Private Investment

Globally, investment flow in RE has been predominantly domestic (both public and private) or from international public financial institutions[5] (Organisation for Economic Co-operation and Development 2015). The RE financing landscape is dominated by international and bilateral financial institutions and development banks (Table 6) like the World Bank and the Asian Development Bank (Souche 2014) who have introduced new rules of governance in RE (Nakhooda 2011). International private investment in the sector has been historically low (FS-UNEP Centre 2016; Moon et al. 2016). Increase in private investment is essential if a faster transition into a low-carbon pathway is to be achieved (Organisation for Economic Co-operation and Development 2015; Iftikhar et al. 2015; Chakrabarty 2016). However, the ease of doing business in South Asian economies is low because of poorly performing institutional and legal frameworks, slow functioning of administrative and financial machineries, absence of speedy conflict resolution mechanisms, etc. High cost of capital, market barriers and other associated risks deter private investment in areas like Afghanistan, Maldives and Bangladesh (Table 7) (International Renewable Energy Agency 2016).

[5]Public Financial institutions include—International Financial Institutions, Development Finance Institutions, Local Financial Institutions, Export Credit Agencies, Climate Finance Institutions (Organisation for Economic Co-operation and Development 2015).

Table 6 International public finance flow in RE (in USD million)

Region	2009	2010	2011	2012	2013
Afghanistan	2	8	–	3	–
Bangladesh	–	–	–	155	116
Bhutan	–	–	–	–	105
India	240	384	1785	944	277
Maldives	–	–	–	–	–
Nepal	–	–	–	3	414
Pakistan	98	37	305	1017	358
Sri Lanka	–	40	4	29	–
South Asia	340	469	2094	2151	1270
Asia	587	1056	3894	2664	2550
World	16,298	13,704	19,926	25,983	15,094

Source International Renewable Energy Agency (2016)
Note Important financing institutions include Asian Development Bank, European Bank for reconstruction and development, Japan Bank for International Cooperation, Japan International Cooperation Agency, The World Bank

4.2 Barriers to Investment in Clean Energy in South Asia

In South Asia, there are serious impediments to investment in renewable. These barriers are linked to market and policy failures, market conditions, political systems, technical challenges or may be country-specific (Organisation for Economic Co-operation and Development 2016). Some of the specific constraints are.

4.2.1 Actor Constraints

A broad, complex and fragmented network of actors steer the development and deployment process of RE in South Asia (Table 8) (Asian Development Bank 2015a; United Nations Development Programme 2013; Ölz and Beerepoot 2010; The World Bank 2011; Krishna et al. 2015). Within each class of actors, there are groups of actors with competing priorities and constraints (United Nations Development Programme 2013). RE investors therefore have to coordinate with multiple agencies who have their own goals, priorities and targets (World Wind Energy Association 2014). Network dynamics, divergence of actor perspectives, power struggle, etc. can impede the scaling up of RE (International Renewable Energy Agency 2012). Malfunctioning of networks can generate risks and uncertainties that escalate the transaction cost of doing business and delays project implementation (Asian Development Bank 2015a; Ölz and Beerepoot 2010; Sovacool 2010). Again, capacity deficiency in terms of knowledge and training on RE and RE potentials and solutions can negatively impact the returns on investments as knowledge acquisition and aligning them with the goals and aspirations of

Table 7 Ease of doing business: ranking of South Asian nations according to parameters

Economy	Ease of doing business	Starting a business	Dealing with construction permits	Getting electricity	Registering property	Getting credit	Protecting minority investors	Paying taxes	Trading across borders	Enforcing contracts	Resolving insolvency
Afghanistan	183	42	186	159	186	101	189	163	175	180	159
Bangladesh	176	122	138	187	185	157	70	151	173	189	151
Bhutan	73	94	97	54	51	82	114	19	26	47	169
India	130	155	185	26	138	44	13	172	143	172	136
Maldives	135	65	62	145	172	133	123	134	147	135	105
Nepal	107	109	123	131	72	139	63	142	69	152	89
Pakistan	144	141	150	170	169	82	27	156	172	157	85
Sri Lanka	110	74	88	86	155	118	42	158	90	163	75

Source The World Bank (2017)

Table 8 Network of actors for developing and deploying RE in Asia

Class of actor	Entities	Roles
International	United Nations framework convention on climate change, intergovernmental panel on climate change, etc.	Sets the global climate goals; influences policies and actions of various countries
National government	Various ministries, agencies, etc.	Sets national level goals, policies and regulations concerning developing and deploying RE. Various line ministries may be crucial actors as RE has linkages with and consequence upon various other sectors—conventional power generation, grid management, transport, industry, rural development, trade and commerce, agriculture, finance, geo-political considerations, etc.
Sub-national governments	Various ministries, agencies, etc.	Wherever a federal structure exists, provision of energy and deployment of RE can be a responsibility of both national and sub-national governments. Hence, the sub-national government entities come to play a role in framing sub-national goals, policies, programmes, regulations, etc. concerning RE and sectors linked with RE
Technology developers	Government or private entities; domestic or foreign entities	Experimentation, developing pilots and large-scale feasible solutions
Technology implementers/ project developers	Government or private entities; domestic or foreign entities	Deploy RE technologies, implement projects and be responsible for operation and maintenance. May operate individually or in groups (consortia, joint ventures, PPP, etc.)
Investors	Corporate and retail investors, venture capitalists, financial intermediaries, financial regulators, multilateral and bilateral finance institutions	Channelizes and/or facilitates flow of investments in the RE sector. May finance through a variety of modes—grant, equity, debt, etc. May operate individually or in groups
Activists	Civil society organizations/ non-governmental organizations	Generation of awareness; ensuring last mile delivery of RE
Consumers	Households, businesses	Accept/oppose RE as a solution; provide feedback on RE programmes and applications

Source Ghosh and Ghosh (2016)

actors is time-consuming (The World Bank 2011; International Renewable Energy Agency 2012).

Equity financing in RE is constrained by the absence of international private investors. Domestically also, national banks often refrain from participating in RE projects due to associated risks.[6] Considerable dependence on debt therefore exposes RE financing to the risk perception of the lender. Consequently, small-scale projects are more attractive to investors leading to the loss of scale economies that large projects generate (Souche 2014; The World Bank 2011).

In energy-starved South Asia, consumers generally demand reliable and affordable electricity, and not necessarily sustainable energy. This is challenging for niche RE technologies that have to compete with the often subsidized and already mainstreamed conventional fossil fuel-based energy systems (United Nations Economic Commission for Europe 2011). Articulation by actors motivates a country to adopt measures for transition. Socio-economic priorities and constraints, cultural norms, political beliefs, etc. may distort the views of some key actors on the delivery of global public good (Roy et al. 2013). The divergence over local priorities and global climate agenda may therefore undermine the intent of actors to work together to scale up RE (Howes and Wyrwoll 2015).

It is sometimes anticipated that the emphasis on RE may lead to job losses in energy-intensive sectors. Simultaneously, there is also uncertainty regarding the possibility of creating green jobs (Ghosh and Ghosh 2016). This is alarming for developing economies where poverty removal through livelihood generation is a policy priority. The potential trade-off between short-term social costs and long-term climate goals therefore influence the decisions by governments, and hence, RE projects get compromised (Howes and Wyrwoll 2015).

The existing infrastructure for transmission and distribution of fossil fuel-based energy system makes integrating RE into this infrastructure a major trial for technology developers and investors (The World Bank 2011). Further, large-scale RE deployment requires land. In South Asian nations, actors' perceptions, policies, laws, etc. concerning land acquisition—particularly agricultural land, are a sensitive issue (Asian Development Bank 2015a). RE investor's are therefore wary due to the risks involved.

Civil societies are influential players in the actor network. They often facilitate the vital last mile delivery of energy—particularly in rural areas. However, sometimes they have acted as deterrents in the deployment of hydropower projects in South Asia. Activist groups have opposed projects because dams often destroy riparian ecosystems, human habitats, impact catchment area agriculture and livelihood, and threaten food security (Howes and Wyrwoll 2015). Hence, civil society activities are also perceived as major entry barriers by RE investors.

---

[6]In April 2015, the Reserve Bank of India has accorded RE the status of priority sector for lending by commercial banks. Thus all benefits associated with priority sector finance are now available to this sector in India.

4.2.2 Policies, Politics and RE Investment

Policies can play a crucial role in 'protecting' niche innovations from the vagaries of market selection (Geels 2002). However, such 'investment grade policy regimes'[7] can also be offset by policies that prevent investment in RE. The type of policy regime to be adopted by an economy is determined by political mandates and government ideologies. Consequently, energy regimes are steered by actors whose constraints and priorities are guided by the socio-political contexts in which they operate (Asian Development Bank 2015b; Sen et al. 2016). Based on their priorities, the actors can be broadly categorized into—(a) those who believe in the supremacy of the market and (b) those who advocate public benefit above all else (Ghosh and Ghosh 2016). Most South Asian nations have generally prioritized socio-political objectives above market efficiencies. Apprehending welfare loss and rising unemployment arising out of corporatization and unbundling of the energy sector, the political process in these countries has often refused to advocate laissez-faire (Dubash and Rajan 2001). In countries like India, Pakistan, Sri Lanka and Nepal, there have been instances when public dissatisfaction has stalled the privatization of the power sector (Nepal and Jamasb 2011). Also, in South Asia, it has been observed that welfare objectives have resulted in the development policy support in the form of energy subsidies, input subsidies, etc. (Rasul 2016). This has often triggered resource misuse and unsustainable development. The situation has been worsened by fragmented institutional structures and uncoordinated actor networks.

Thus, policies and politics simultaneously impact the trajectory of energy transition. A flawed policy space often puts at risk the scaling up of RE in South Asia (UNDP 2013). Varied in nature, these flaws distort the political economy of RE and restrict investment flows to the sector.

The high cost of RE technologies has often restricted the large-scale deployment of RE, particularly in countries where affordability is an issue (The World Bank 2011; Sovacool 2010; Salim 2015). In South Asia, where economic- and energy-poverty eradication is a development priority, energy has been traditionally subsidized. Economic rationality dictates that as long as energy prices do not 'internalize externalities and fail to take into account the wider global and local environmental impacts of different technologies, as well as their contributions to reducing the price volatility of energy and increasing energy security', renewable energy technologies will be costlier than their conventional counterparts (The World Bank 2011). Also, policies on RE tariff determination fail to account for the multiple benefits that RE deliver—diversification of the energy mix, reduction in fuel imports, hedge against price volatility of fossil fuels and supply interruptions, off-grid systems that save the costs of transmission and distribution, etc. (The World

[7]Systems where policies promote the harnessing of financial resources in building clean energy systems (Hamilton 2009).

Bank 2011). Thus, in most South Asian countries, RE remains a costly option—deterring demand and hindering investments (Ghosh and Ghosh 2016).

Sometimes, RE deployment policies may have a faulty design, thereby rendering it ineffective in terms of achieving intended goals, providing support to target population groups, etc. Studies indicate that RE policies have been found lacking in provisioning adequate incentives for investors (The World Bank 2011; International Renewable Energy Agency 2012); bridging the capacity gaps in the network of actors (United Nations Development Programme 2012); focusing on the entire range of RE options available so as to ensure last mile delivery of energy (United Nations Economic Commission for Europe 2011), etc. These lacunae discourage investments in RE projects (Souche 2014; International Renewable Energy Agency 2012; Hamilton 2009).

Sometime, policy makers overlook the necessity to explicitly enunciate necessary definitions, laws, regulations, etc. thereby rendering the RE policy incomplete (Krishna et al. 2015). This generates ambiguities, and investors hesitate to enter a market that they fear is inefficient (International Renewable Energy Agency 2012; Liu et al. 2013).

Most RE technologies are characterized by high upfront costs and relatively low operation and maintenance expenditures. The returns are delayed due to a set of risks—implementation risk, technology risk, risk of grid connectivity, etc. Hence, the planning horizon for a RE project viability is relatively long (The World Bank 2011). Contractual financial obligations like power purchase agreements (PPAs), structure of feed-in tariffs and other relevant contracts (subsidy, incentives, etc.) also have a direct bearing on the financial viability of RE projects. The contracts have to be unambiguous and have to be supported by long-term guarantees through policies and regulations (International Renewable Energy Agency 2012; Cambridge Economic Policy Associates Limited 2014). Some South Asian countries are still in an energy regime defined by ambiguous commitments. Hence, the risks of RE investment are high (Liu et al. 2013; The Asia Foundation 2013).

Debts extended to RE developers typically need to be of long tenure. However, credit policies in the selected nations sometime restrict financial institutions from extending long-tenure debts, thereby discouraging investments (The World Bank 2011).

Political volatility in South Asian nations generates an atmosphere of insecurity that restricts the entry of potential RE investors. With the change of governments, policies change. The unpredictability in policies creates uncertainty regarding future policy support for RE projects (Jacobsson and Bergek 2004). This phenomenon is more complicated when the country operates under a federal structure of governance—as changes in government and policies can be at both national and sub-national levels (UNDP 2013). Policy volatility adversely impacts the predictability of cash flows and hence affects investments (Nakhooda 2011; The World Bank 2011; Hamilton 2009).

In many countries, it is found that policies and institutions for governing the future direction of energy systems are guided by political mandates and not by a systems approach. Short term gains for the political class scores over long-term social impacts and consequences. Therefore, motivating the political class in policy-framing and decision-making processes assume importance (Krishna et al. 2015). Policies supporting renewable energy are indeed about lobbying over policy goals and design of a favourable institutional structure (Jacobsson and Bergek 2004). If the renewable energy lobby of a country is weak, then investor interest in RE in the region wanes.

5 Managing Investment Risks

For business as usual scenario, efficient management of investment risk is essential for the successful scaling up of RE technologies in South Asia. Both financial and non-financial actions can be deployed to mitigate the different risks to investment. Financial instruments (Table 9) alone may not be enough to hedge RE investment uncertainties. Investors can further abate the associated risks through investments in programme designing, negotiating and articulating promises with governments and other actors, convincing actors about goals and benefits of RE projects, etc. In the developed world, renewable power producers generally undertake the following risk management strategies: (i) geographic and technological diversification so as to mitigate political, regulatory/policy and weather-related volume risks; (ii) broad basing of financing partners so as to reduce financing risks, in response to macroeconomic and political changes; (iii) internal capacity generation in technology, legal matters, lobbying, etc. so as to mitigate capacity and technology-related risk; and (iv) employment of risk transfer instruments like hedging instruments, derivatives, customized insurance, catastrophe bonds, etc. (The Economist 2011). Developing South Asia can gain by adopting these measures in combinations that suit their socio-economic-political contexts.

Table 9 Selected instruments for managing financial risks in RE projects

Type of risk	Instrument for managing risk
Political risks	• Country credit default swaps (CDS) • Risk-sharing schemes between developers and investors • Political risk insurance
Economic risks	• Joint ventures and strategic agreements • Insurance schemes • Guarantees (dismantling guarantee, weather guarantee, etc.)
Credit risk	• CDS • Alternative risk transfer (securitization, insurance linked securities, etc.)
Technical risk	• Guarantees • Insurance

Source Altran and Little (2011)

6 Leveraging Opportunities for RE in South Asia

The distinctive geo-political socio-economic scenario in South Asia provides an environment, which if properly leveraged, can transform the carbon-intensive energy space in the region. Investors in RE can benefit from a thorough understanding of the local context—the existing energy architecture, the RE potential, the framework of the indigenous energy market, the extant policies and regulations, the governance structure, etc. (International Renewable Energy Agency 2012). Knowledge of the local environment in the context of the indigenous energy narrative makes the buy-in from the stakeholders a relatively easy task. A meticulous assessment of the market not only can help ascertain the appropriate technology, solutions and target groups, but can also facilitate the identification of business opportunities that match the expectations on financial returns. For governments and development financing institutions, it is important to identify markets and technologies where returns are below market expectations but above financial losses (International Renewable Energy Agency 2012).

An understanding of the local context can further facilitate the identification of suitable business models for the successful deployment of RE projects (Marro and Bertsch 2015). Investors can then: (i) effectively identify and bring on board actors who not only have technical expertise in RE but also possess lobbying and negotiating power with the government to orient policies and institutions in favour of RE deployment; (ii) build strong partnerships across the value chain; (iii) achieve effective partnership with financial institutions; (iv) produce competitive renewable solutions and hence gain consumer support and market confidence essential for scaling up RE.

Further, in South Asia, there appears to be an urgent need for aligning market opportunities with the national energy targets and policies. This can help trigger a more effective communication between consumer groups, investors, governments, policymakers, civil societies, etc.

7 Concluding Discussions

South Asia has the potential to become a key stakeholder in the international renewable energy generation space. A wealth of untapped renewable energy sources, global environmental concerns, rising energy demand due to population growth and economic development, fuel market volatilities, adverse trade balances are some of the major drivers for the upscaling of RE in the region. However, incomplete policies, political mandates and governance instabilities, inefficient institutional and regulatory arrangements, dearth of infrastructure and knowledge, and an indifferent consumer group are impeding the flow of investment into the sector. Absence of investor confidence, escalating project costs and delays are continuing to lock the region into a carbon-intensive energy regime.

Even though South Asian energy policies are keen on a renewable regime, under the business as usual scenario, RE investors have to be careful to embed risk mitigation mechanisms right from the stage of project design. There are mechanisms to manage risks. However, their efficacy in the South Asian context is yet uncertain. Regional unions like the South Asian Association for Regional Cooperation (SAARC) helps in creating an enabling environment to attract investment in the region. However, there remains a dearth of knowledge and evidence on the exact national context of each country and its distinctive nuances that can generate a more effective environment for the scaling up of RE in South Asia.

References

AIIB (2017) AIIB energy sector strategy: sustainable energy for Asia. Discussion draft for consultation. Asian Infrastructure Investment Bank, Beijing, China

Altran, Little AD (2011) Risk quantification and risk management in renewable energy projects. International Energy Agency, Hamburg

Anbumozhi V, Kawai M (2015) Towards a low-carbon Asia: challenges of economic development. In: Anbumozhi V, Kawai M, Lohani BN (eds) Making the transition to a low-carbon economy: perspectives, policies and practices from Asia. Asian Development Bank Institute, Tokyo, Chap 2, pp 11–44

Alternative Energy Development Board (2015) (Online). Cited: 13 May 2017. http://www.aedb.org/index.php#

Asian Development Bank (2012) Sustainable energy for all—Bhutan: rapid assessment and gap analysis. Asian Development Bank. http://www.se4all.org/sites/default/files/Bhutan_RAGA_EN_Released.pdf

Asian Development Bank (2015a) Renewable energy developments and potential in the greater Mekong subregion. ADB, Manila

Asian Development Bank (2015b) Assessment of power sector reforms in Sri Lanka. ADB, Manila

Azwar A, Waheed A (2011) Renewable energy status in Maldives (Online). Cited: 10 June 2017. http://www.globalislands.net/userfiles/Maldives5.pdf

Bhattacharya SC (2007) Power sector reform in South Asia: why slow and limited so far? Energ Policy 35(1):317–332

Bradbrook AJ, Gardam JG (2006) Placing access to energy services within a human rights framework. Hum Rights Q 28:389–415

Cambridge Economic Policy Associates Limited (2014) Policy risk in renewable energy investments in developing countries. Cambridge, UK

CDKN, ODI (2014) The IPCC's fifth assessment report: what's in it for South Asia. s.l.: Climate & Development Knowledge Network and Overseas Development Institute

Chakrabarty M (2016) Energy security in South Asia. Observer Research Foundation. [Online] 16 Feb 2016. Cited: 9 June 2017. http://www.orfonline.org/research/energy-security-in-south-asia/

Climate Investment Funds (2012) Scaling-up renewable energy in low income countries programme: Nepal. Climate Investment Funds, Kathmandu

Dubash NK, Rajan SC (2001) Power politics: process of power sector reform in India. Econ Polit Wkly, 3367–3390

DFID (2009) Political economy analysis: how to note. Department for International Development, Government of UK, London

Ernst & Young (2017) RECAI—batteries: leading the charge. Ernst & Young Global Limited, UK

Ershad AM (2017) Institutional and policy assessment of renewable energy sector in Afghanistan. J Renew Energ. https://doi.org/10.1155/2017/5723152
Frankfurt School-UNEP Centre (2015) Bloomberg new energy finance: global trends in renewable energy investment 2015. Frankfurt School of Finance and Management, Frankfurt
Fritz V, Kaiser K, Levy B (2009) Problem-driven governance and political economy analysis: good practice framework. The World Bank, Washington, D.C.
FS-UNEP Centre/ BNEF (2016) Global trends in renewable energy investment 2016. Frankfurt: Frankfurt School-UNEO Collaborating Centre for Climate and Sustainable Energy Finance
FS-UNEP (2017) Global trends in renewable energy investment 2017: key findings. Frankfurt: FS-UNEP Collaborating Centre for Climate and Sustainable Energy Finance, Bloomberg New Enkfurtnergy Finance
Geels FW (2002) Technological transitions as evolutionary reconfiguration processes: a multi-level analysis and a case study. Res Policy 1257–1274
Geels FW, Schot J (2007) Typology of sociotechnical transition pathways. Res Policy, 399–417
Ghosh D, Ghosh A (2016) Evidence and gaps in evidence on the principle political economy constraints and opportunities to successful investment in clean energy in Asia. Evidence on Demand, UK [Online]. doi:https://doi.org/10.12774/eod_hd.january2016.ghoshdetal
Government of Nepal (2011) Scaling up renewable energy programme: investment plan for Nepal. Government of Nepal, Kathmandu
Hamilton K (2009) Unlocking finance for clean energy: the need for investment grade policy. Energy, Environment and Development Programme Paper: 09/04. Chatham House, London
Howes S, Wyrwoll P (2015) Evaluation of current pledges, actions and strategies. In: Anbumozhi V, Kawai M, Lohani BN (eds) Managing the transition to a low-carbon economy: perspectives, policies and practices from Asia. Asian Development Bank Institute, Manila, Chap 4, pp 85–146
Iftikhar MN, Najeeb F, Mohazzam S, Khan SA (2015) Sustainable energy for all in South Asia: potentials, challenges and solutions. Working Paper #151. Sustainable Development Policy Institute, Islamabad, Pakistan
Intergovernmantal Panel on Climate Change (2007) Contribution of working group I to the fourth assessment report of the intergovernmental panel on climate change. In: Solomon S, Qin D, Manning M, Chen Z, Marquis M, Averyt KB, Tignor M, Miller HL (eds) Cambridge University Press, Cambridge, United Kingdom, New York, NY, USA
Intergovernmental Panel on Climate Change (2014) Mitigation of climate change. In: Contribution of working group III to the fifth assessment report of the intergovernmental panel on climate change. Cambridge University Press, Cambridge, United Kingdom and New York, USA
International Energy Agency (2014) International energy statistics. US Energy Information Administration (Online). Cited: 31 May 2017. https://www.eia.gov/beta/international/data/browser/
International Renewable Energy Agency (2012) Financial mechanisms and investment frameworks for renewables in developing countries. IRENA, Abu Dhabi
International Renewable Energy Agency (2015) Renewable energy roadmap: the republic of Maldives—background report. International Renewable Energy Agency. http://www.irena.org/EventDocs/Maldives/Maldivesroadmapbackgroundreport.pdf
International Renewable Energy Agency (2016a) Unlocking renewable energy investment: the role of risk mitigation and structured finance. IRENA, Abu Dhabi
International Renewable Energy Agency (2016b) Renewable energy statistics 2016. IRENA, Abu Dhabi
International Energy Agency (2017) Energy poverty. International Energy Agency [Online]. Cited: 27 May 2017. https://www.iea.org/topics/energypoverty/.i
Islam MdT, Shahir SA, Iftakhar Uddin TM, Saifullah AZ (2014) Current energy scenario and future prospect of renewable energy in Bangladesh. Renew Sustain Energ Rev 39:1074–1088
Jacobsson S, Bergek A (2004) Transforming the energy sector: the evolution of technological systems in renewable energy technology. Ind Corp Change 13(5):815–849

Krishna C, Sagar AD, Spratt S (2015) The political economy of low-carbon investments: insights from the wind and solar power sectors in India. Institute of Development Studies, Brighton

Liu H, Masera D, Esser L (eds) (2013) World Small Hydropower Development Report 2013. United Nations Industrial Development Organization; International Center on Small Hydro Power, Vienna, Hangzhou

Maldives Energy Authority (2013) Maldives energy supply and demand survey 2010–2012. Maldives Energy Authority, Government of Maldives, Male

Marro P, Bertsch N (2015) Making renewable energy success in Bangladesh: getting the business model right. ADB South Asia Working Paper Series: No 41, pp 1–27

Ministry of New and Renewable Energy, GoI (2017) Annual Report 2016–2017. Ministry of New and Renewable Energy, Government of India, New Delhi. http://mnre.gov.in/file-manager/annual-report/2016-2017/EN/index.html

Ministry of Power, GoB (2011) Power system master plan, 2010. Ministry of Power, Government of Bangladesh, Dhaka

Moon J-Y, Song J, Lee S (2016) Catalyzing investment for renewable energy in developing countries. World Economy Update 6:1–6

Nakhooda S (2011) Asia, the multilateral development banks and energy governance. Glob Policy 2(Special Issue):120–132

Nepal R, Jamasb T (2011) Reforming small power systems under political volatility: the case of Nepal. EPRG Working Paper 1114, Cambridge Working Paper in Economics 1133. s.l. http://www.econ.cam.ac.uk/research-files/repec/cam/pdf/cwpe1133.pdf

Ölz S, Beerepoot M (2010) Developing renewables in southeast Asia: trends and potentials. International Energy Agency Working Paper, pp 7–159

Organisation for Economic Co-operation and Development (2015) OECD policy perspectives: green investment banks. OECD, Paris

Organisation for Economic Co-operation and Development (2016) OECD business and finance outlook 2016. OECD, Paris

Pal A (2013) Power sector in India: growth. Policies Challenges 3(3):527–536

Patwardhan A, Ines A, Foran T, Patankar M, Rao A, Raven R, Samaras C, Smith A, Verbong G, Walawalkar R (2012) Chapter 16—transitions in energy systems. Global energy assessment—toward a sustainable future. Cambridge University Press & International Institute for Applied Systems Analysis, Cambridge, UK and New York, NY, USA & Laxenburg, Austria, pp 1173–1202

Parish D (2006) Evaluation of power sector operations in Pakistan. Asian Development Bank, Manila

Rasul G (2016) Managing the food, water and energy nexus for achieving the sustainable development goals in South Asia. Environ Dev 18:14–25

REN21 (2016) Renewables 2016 global status report. Renewable Energy Policy Network for the 21st Century, Paris, France

Roy J, Ghosh D, Ghosh A, Dasgupta S (2013) Fiscal instruments: crucial role in financing low carbon transition in energy systems. Curr Opin Environ Sustain 5:26–269 (Elsevier)

Salim E (2015) Pro growth, pro job, pro poor, pro environment. In: Anbumozhi V, Kawai M, Lohani BN (eds) Managing the transition to a low-carbon economy: perspectives, policies and practices from Asia. Asian Development Bank Institute, Tokyo, Chap 1, pp 3–10

Sen A, Nepal R, Jamasb T (2016) Reforming electricity reforms? Empirical evidence from Asian economies. The Oxford Institute of Energy Studies, University of Oxford, Oxford

Shukla AK, Sudhakar K, Baredar P (2016) Renewable energy resources in South Asian countries: challenges, policy and recommendations. Resour Efficient Technol. https://doi.org/10.1016/j.reffit.2016.12.003

Sipahutar R, Bernas SM, Imanuddin MS (2013) Renewable energy and hydropower utilization tendency worldwide. Renew Sustain Energ Rev 17:2013–2015

Souche A (2014) Financing of clean energy projects in South-East Asia—challenges and opportunities. DFDL, Bangkok

Sovacool BK (2010) A comparative analysis of renewable electricity support mechanisms for Southeast Asia. Energy 35(4):1779–1793

Sovacool BK (2012) The political economy of energy poverty: a review of key challenge. Energy Sustain Dev 16(3):272–282

Sovacool BK (2013) Energy access and energy security in Asia and the Pacific. Asian Development Bank, Manila

Sri Lanka Sustainable Energy Authority (2016) Renewables (Online). Cited: 12 June 2017. http://www.energy.gov.lk/renewables/renewable-energy-resources

The Asia Foundation (2013) A political economy analysis of electricity tariff restructuring in Nepal. The Asia Foundation, Kathmandu

The Economist (2011) Managing the risk in renewable energy. The Economic Intelligence Unit Limited, Geneva

The World Bank (2006) Understanding socio-economic and political factors to impact policy change. Social Development Department, Report No. 36442-GLB, Washington, D.C.

The World Bank (2011) Financing renewable energy: options for developing financing instruments using public funds. The World Bank, Cape Town

The World Bank, International Energy Agency (2015) Progress toward sustainable energy 2015: global tracking framework report. The World Bank, International Energy Agency, Washington D.C., USA

The World Bank (2017a) Data: access to electricity. The World Bank (Online). Cited: 28 May 2017. http://data.worldbank.org/indicator/EG.ELC.ACCS.ZS?view=map

The World Bank (2017b) Doing business. [Online]. Cited: 12 June 2017. http://www.doingbusiness.org/rankings

UNDP (2013) Achieving sustainable energy for all in the Asia Pacific. United Nations Development Programme—Asia Pacific Regional Centre, Bangkok

United Nations Development Programme (2012) Renewable energy for rural livelihoods. UNDP, Kathmandu

United Nations Development Programme (2013) Accelerating energy access for all in Myanmar. UNDP, Yangon

United Nations (2015) Sustainable development goals: 17 goals to transform our world (Online). Cited: 14 May 2017. http://www.un.org/sustainabledevelopment/energy/

United Nations Economic Commission for Europe (2011) Summary on reports from the national experts on development of renewable energy in The Russian Federation and CIS countries. UNECE, Geneva

United States Energy Association (2015) Energy consultation and available energy resources in Afghanistan. United States Energy Association (Online). Cited: 13 May 2017. https://www.usea.org/sites/default/files/event-file/522/Afghan_Power_Sector_Briefing_June_2011.pdf

World Wind Energy Association (2014) Scaling up wind power deployment in Pakistan: the barriers & the way forward. World Wind Energy Association, Bonn, Germany

WCED (1987) Our common future. World Commission on Environment and Development, Melbourne

A Narrative Analysis of State-Level Renewable Energy Policies in India

Shyamasree Dasgupta and Pooja Sankhyayan

Abstract In the lead up to the Paris Agreement (2015), the Nationally Determined Contribution that India has announced included a reduction in emissions intensity of Gross Domestic Product (GDP) by 33–35% below 2005 levels and an increased share of non-fossil fuels to 40% with respect to power generation (as compared to 11% in 2010) by 2030. In this respect, solar, wind, biomass and small hydro-based electricity generation have emerged as the most important sources of grid-interactive renewable energy generation in India. However, given the diversity in state-level resource availability and policy implementation, it is difficult to formulate a uniform renewable policy in the country. This paper presents the general policy framework applicable including command and control, price and quantity instruments to boost renewable energy and a review of state-level policies in Tamil Nadu, Maharashtra, Gujarat and Rajasthan. These policies include solar policy, policy of repowering wind power projects and wind-solar hybrid policy. The major challenges towards the implementation of these policies are identified in terms of demand-supply mismatch, optimum financial strategy to pay for the high-end initial costs in the off-grid applications, high risk perception of using Renewable Energy Certificates, lack of credit available for developers, the lack of coherence between national renewable energy targets set by National Action Plan on Climate Change and state Renewable Purchase Obligation targets, etc.

Keywords Renewable energy policy · India · State · Solar · Wind

---

S. Dasgupta (✉) · P. Sankhyayan
School of Humanities and Social Sciences, Indian Institute of Technology, Mandi 175005, Himachal Pradesh, India
e-mail: shyamasree.dasgupta@gmail.com; shyamasree@iitmandi.ac.in

A. Gautam et al. (eds.), *Sustainable Energy and Transportation*,
Energy, Environment, and Sustainability,
https://doi.org/10.1007/978-981-10-7509-4_8

1 Introduction

In the recent past, India has set a number of domestic and international targets to be achieved with respect to energy production, distribution and access. It includes electrification of all inhabited villages by 2018, universal 24 × 7 electricity access by 2022, reduction of energy intensity of GDP by 33–35% by 2030 over 2005, increase in the share of non-fossil fuel-based energy up to 40% in the energy mix by 2030 (GOI 2016a; UNFCCC 2016). Clearly, there is an increasing emphasis on renewable energy sources with an ambitious target of 175 GW of installed capacity including 100 GW solar and 60 GW wind power by 2022 (GOI 2016b). Prior to this, the Jawaharlal Nehru National Solar Mission under the National Action Plan on Climate Change (2010) set the target of deploying 20 GW of grid-connected solar power by 2022. It emphasized on the reduction of the cost of solar power generation through long-term policy, large-scale deployment goals, aggressive R&D and domestic production of critical raw materials, etc. Irrespective of these frameworks, renewable energy is often conceptualized to be ideal only for thinly populated areas in the country (GOI 2016a). This reservation is attributed to the difficulty in achieving 24 × 7 power supply due to variability in the supply of the natural resources. This variability in fact poses certain challenges unique to renewable power supply. This is augmented by the fact that renewable energy is still not the existing dominant (incumbent) socio-technical system with respect to energy generation. Appropriate government interventions and public policies are therefore required to protect, upscale and embed these experiments (Roy et al. 2013). Globally, there are ample examples that show how such enabling policy instruments can lead to successful innovation, upscaling and mainstreaming of renewable energy. Countries like Germany, Denmark, Sweden and Spain showed appreciable progress in scaling up innovations in renewable energy harnessing a transition to sustainable energy systems. In these countries, appropriately designed fiscal instruments incentivized entrepreneurs/firms to increase investments in renewable energy. On the other hand, in countries like the Netherlands, in spite of the technology being in place, the absence of policy with right incentives to renewable sources like solar, wind and biomass posed difficulties to overcome the strong resistance from the incumbent regime (Verbong and Geels 2007; Raven 2012). Given the federal structure of India, not only the national framework, but states also play important roles in such policy intervention and shaping the scenario. Not only the production of renewable energy depends on state-specific availability of natural resources, but at the same time, in India, state-level institutions are responsible for transmission and distribution, demand-supply parity, tariff determination, etc., under the broad national framework.

Against this background, it is important to understand the policy frameworks implemented at the state level, the opportunities created by the same and the challenges faced. This paper presents the general policy instruments for renewable energy and then provides a narrative analysis of state-level solar and wind policies in Tamil Nadu, Maharashtra, Gujarat and Rajasthan. These are the four states majorly contributing to renewable energy generation in the country. The paper concludes with a discussion on challenges and opportunities with respect to these policies.

2 Renewable Power Generation in Tamil Nadu, Maharashtra, Gujarat and Rajasthan

Tamil Nadu, Maharashtra, Gujarat and Rajasthan are major renewable energy-producing states in India. About 35% of solar and more than 60% of wind energy of 175 GW of renewable target is expected to be met by these four states (Table 1). Not only they have higher contributions in the national renewable energy pool (Table 2), but at the same time, a large proportion of power requirement of the states are met by renewable sources. More than one-third (8054 MW) of the power generation in Tamil Nadu comes from renewable sources with major shares of solar and wind (TEDA 2014). Solar-powered greenhouse schemes, rooftop capital

Table 1 Renewable power target (in MW) to be achieved by 2022 in Tamil Nadu, Maharashtra, Gujarat and Rajasthan

State	Solar power	Wind power	Small hydel projects	Biomass power
Tamil Nadu	8884	11,900	75	649
Maharashtra	11,926	7600	50	2469
Gujarat	8020	8800	25	288
Rajasthan	5762	8600	–	–
All India target	99,533	60,000	5000	10,000

Source Ministry of New and Renewable Resources, Government of India (http://mnre.gov.in/file-manager/UserFiles/Tentative-State-wise-break-up-of-Renewable-Power-by-2022.pdf)

Table 2 Installed solar and wind power in Tamil Nadu, Maharashtra, Gujarat and Rajasthan

State	Cumulative capacity of grid-connected solar power till 31-03-16 (MW)	Cumulative capacity of installed wind power operational till 31-03-16 (MW)
Tamil Nadu	1061.82	7613.86
Maharashtra	385.76	4653.83
Gujarat	1119.17	3948.61
Rajasthan	1269.93	3993.95
All India	6762.85	26,777.45

Source Ministry of New and Renewable Resources, Government of India (http://mnre.gov.in/file-manager/UserFiles/State-wise-wind-power-potential-utilized.pdf and http://mnre.gov.in/file-manager/UserFiles/grid-connected-solar-power-project-installed-capacity.pdf)

incentive schemes, wind-solar hybrid systems, small hydro and micro hydel projects are identified as priority programmes in the state. In Maharashtra, 11% of the installed capacity is based on wind power and an additional 6% on other renewable energy sources (Thombare et al. 2016). This is the second highest installed capacity of renewable power generation among Indian states. In fact, Maharashtra has a large potential for biomass-based power generation, which is beyond the scope of this paper. Several policies and methodologies for grid-connected and off-grid projects on renewable energy sources including grid-connected rooftop system are prioritized in the state (MEDA 2017). The state of Gujarat also has high potential in renewable energy generation. Rann of Kachchh receives abundant sunshine for ~300 days a year and good winds along its 1600-km-long shoreline (GEDA 2017). Out of ~25,000 MW of installed capacity, ~5400 MW (21.5%) comes from renewable sources in the state. This accounts for 9% of the total installed capacity of renewable energy in the country (JUVNL 2017). Solar, wind, bio-energy and small hydel projects are thrust areas identified in the state of Gujarat (GEDA 2017). The situation is quite similar in Rajasthan, the neighbouring state of Gujarat. With ~300–330 clear days with sunshine and high wind flow rate, 26% of the total power generation capacity in Rajasthan comes from renewable energy sources. Mega solar power projects, rooftop solar PV, incentivization of decentralized and off-grid solar applications, including hybrid systems, wind power and biomass-based energy generation, received much attention in the state (RRECL 2017).

3 Policy Instruments to Promote Renewable Energy

Given the method of implementation and execution, renewable energy policies can be either command and control-based or market-based. While command and control policies are prone to set up mandates for increased production and use penetration of renewable energies, market-based mechanisms work towards the removal of market barriers through price- or quantity-based instruments (Table 3). This section provides a basic understanding of the working of these policies, specifically in India.

3.1 Regulation and Standard

Under command and control policies, certain mandates are fixed and it is legally binding for the obligated stakeholders to comply with those mandates. Setting up regulations and standards under the Energy Conservation and Building Codes (ECBC) is one such command and control instrument. ECBC sets minimum energy standards for new commercial buildings with a grid-connected load of 100 kW or greater or a contract demand of 120 kVA. The latest ECBC, launched in June 2017,

Table 3 Policy instruments to promote renewable energy

Types of policy	Regulation and standard	Quantity instruments	Price instrument
Principle	A command and control mechanism where stakeholders are bound to act according to mandates	A market-based mechanism that targets absolute quantity for renewable energy production	A market-based mechanism that creates a favourable price regime for renewable energy and let market determine quantity
Types	• Renewable energy mandates such as building codes • Flexible grid access through net metering	• Renewable Portfolio Standards (RPS)/ Renewable Purchase Obligation (RPO) • Renewable Energy Certificates (REC) • Renewable Regulatory Fund (RRF)	• Fiscal incentive • Feed-in-tariff

makes provision of installation of renewable energy generation systems mandatory for new buildings (BEE 2017). Under the umbrella framework of ECBC at the national level, state-level polices take shape. In 2016, the Government of Maharashtra decided to make solar water heaters mandatory for all new buildings and installation of solar panels mandatory on government and private buildings to generate 200 MW in the coming five years. This will lead to covering of more than 5 lakh sq. m. and producing about 320 lakh litres of hot water per day. However, such regulation and standard-based polices often face implementation challenges due to cost and other market barriers. In this case, while the builders welcomed it from an environmental point of view, the cost barrier and technological difficulties of installing such water heaters in the lower floors of high-rise buildings were also identified (TOI 2016). Price and quantity instruments play important roles in the removal of such cost and other market barriers.

3.2 *Quantity Instruments*

Quantity instruments are market-based mechanisms that target the quantity of renewable energy to be produced. Renewable Purchase Obligation (RPO) along with Renewable Energy Certificates (REC) is a widespread quantity instrument in India. RPO mandates that obligated entities (primarily electricity distribution companies (discoms), captive consumers and open access users) must ensure that a pre-specified percentage of energy is being generated from renewable capacity installed. If implemented alone, RPO becomes less flexible as for all obligated entities, it may not be cost-effective to install renewable power generation processes. Therefore, RPO is often combined with RECs. The RPOs can be energy- and technology-specific. For example, Tamil Nadu, as a part of their solar policy, made a 5% Solar Purchase Obligation (SPO) mandatory. In India, RPO is established by the State Electricity Regulatory Commission (SERC). If an obligated entity fails to meet the renewable target through direct generation, they have the flexibility to purchase RECs. RECs are non-tangible and tradable, representing two attributes —first, a particular quantity of electricity has been generated, and second, it has been generated from renewable source. These two attributes can be bought and sold separately or together. When both the attributes, i.e. electricity and the renewable attribute embodied in it, are transacted together, RECs are said to be 'bundled'. When they are transacted separately, it is called 'unbundled'. An REC represents one MWh of electricity generated from renewable sources. They are issued to companies that produce power from renewable sources but do not find the tariff to be preferable to sell it to distribution companies. The flexibility of implementation of this policy reduces the overall cost of compliance. Another penalty and incentive structure created to ensure stability of power supply through renewable sources in India is Renewable Regulatory Fund (RRF) for wind and solar projects. Under RRF, wind/ solar generators are mandated to undertake forecasting and give schedule of their generation on a day ahead. They are responsible for forecasting with an accuracy of at least 70%. If the actual generation deviates beyond ±30% of the schedule, they are liable to pay deviation charge which is linked to grid frequency. This policy provides stability to achieve the quantity targets.

3.3 Price Instruments

Although the cost of production of renewable energy is declining worldwide, it is still quite high as compared to sources based on fossil fuel. This leads to significant price disincentive for the investors and producers. Fiscal incentives and Feed-in-Tariff (FiT) are two important market-based price instruments to reduce such disincentives. They are devised to create a favourable price regime for renewable energy and let the market determine the optimum quantity to produce. There are several direct fiscal incentives such as accelerated/enhanced depreciation, capital buy down, investment tax credits, production tax credits along with some indirect fiscal incentives such as increase in taxes on fossil fuel, differential taxation for renewable and non-renewable sources. The primary objective of fiscal incentives is to lower the cost of production of renewable energy in order to make them competitive in the market. Another price instrument to promote active investment in renewable energy is FIT. It typically reduces inherent risk of producers by offering long-term contracts and guaranteed price of power generated from renewable sources and thus encourages diversification of energy technologies. Along with FiTs, Indian Renewable Energy Development Agency (IREDA) also provides Generation Based Incentive (GBI) to power producers using renewable sources. This is in addition to the tariff approved by the SERC. However, this is accessible only to producers who do not use the accelerated/enhanced depreciation benefits under the Income Tax Act.

4 State-Level Solar and Wind Energy Policies

4.1 Tamil Nadu: Solar Energy Policy—2012 and Wind Energy Policy

The Solar Energy Policy of 2012 in Tamil Nadu set the target of 3000 MW of solar energy generation by 2015 (MNRE 2012). The collective figure of 3000 MW of solar energy is to be achieved by producing 1000 MW each year for 3 years. The mechanism to generate solar energy is through utility-scale projects (1500 MW), rooftops (350 MW) and REC (1150 MW). The state made Solar Purchase Obligation (SPO) as mandatory at 6% (3% till December 2013 and 6% January 2014 onwards). The SPO mandates heavy energy users such as Special Economic Zones (SEZs), information technology parks, telecom towers, educational institutions, industries guaranteed with 24 × 7 power supply and buildings with build-up area of 20,000 sq. feet or more. The SPO is administered by Tamil Nadu Generation and Distribution Corporation Limited (TANGEDCO). A single-window agency Tamil Nadu Energy Development Agency (TEDA) has been set up for clearance of solar energy projects in 30 days. But the order of the government under the SPO was set aside by Tamil Nadu Electricity Tribunal which mandates SPO for heavy power consumers. The Renewable Energy Programme in Tamil Nadu falls under the domain of Tamil Nadu Electricity Board which is now TANGEDCO. TANGEDCO deals with non-conventional energy sources such as wind energy, biomass energy.

TEDA also has significant focus on the generation of electricity from wind power. With the declared tentative gross potential of 5500 MW, various central and state government incentives are offered to the investors (IREDA 2015). The incentives include accelerated depreciation, import of wind electric generator under concessional customs duty, tax exemptions for 10 years in respect of profits from the private wind electric generators. Government of Tamil Nadu additionally offered incentives to buy surplus energy at a stipulated price, concessional wheeling charges and banking facility subject to 5% charges for using power any time of the year up to 31st March of the respective financial year.

4.2 Rajasthan: Solar Energy Policy 2014 and Policy for Promoting Generation of Electricity from Wind, 2012

With a view on exploring the vast potential of solar energy due to the situational advantage, the Rajasthan Government came up with the solar energy policy in 2014 (MNRE 2014). The main aim of this solar policy is to create an enabling environment for the installation of 25,000 MW of solar power through state/private enterprises or through public–private partnership (PPP) or individual efforts. The tariff is to be determined by Rajasthan Electricity Regulatory Commission (RERC) through a competitive bidding process to the extent of RPO target fixed by RERC. Other initiatives taken under the policy are the promotion of development of Photo Voltaic (PV) solar power plants connected to Low Tension (LT) under Net Metering Scheme as per guidelines of RERC. The policy also talks about decentralized and off-grid solar applications so as to benefit individual manufacturers as well as individual consumers. It includes hybrid systems to meet electrical and thermal energy requirements for domestic and commercial use and setting up of solar power plants for sale of power to individuals through its own distribution system. For remote areas, the policy promotes the establishment of local solar grid and stand-alone solar systems. Rajasthan Renewable Energy Corporation (RREC) has been set up as the nodal agency for the promotion of solar parks in Rajasthan.

Under the Policy for Promoting Generation of Electricity from Wind, 2012 (Government of Rajasthan 2012), the wind power plants can go for direct sale of power to discoms of Rajasthan from 2013 onwards as follows: 2013–14: 300 MW, 2014–15: 400 MW, 2015–16: 500 MW. As per the policy, Rajasthan will promote wind power plants of unlimited capacity for captive use/sale to third parties located within the state of Rajasthan at mutually agreed rates. The RPO requirement for the discoms of Rajasthan will be governed by the RERC. It will act as the nodal agency for wind power projects.

4.3 Gujarat: Solar Power Policy 2015 and Wind Power Policy—2016

The Gujarat Solar Power Policy 2015 (GEDA 2015) aims to promote large-scale addition of solar power generation capacity in Gujarat. It plans to set up MegaWatt (MW) scale of solar power projects with the help of solar PV and solar thermal technologies. As for the KiloWatt (kW) scale of solar projects, the set up will be in the form of rooftop systems. The installation capacity targets will be as per the RPO defined by Gujarat Electricity Regulatory Commission (GERC). The minimum size for MW scale project should be 1 MW and for kW project would be 1 kW. 'Any company or body corporate or association or body of individuals whether incorporated or not or artificial judicial person shall be eligible for setting up Solar Power Generator (SPG) either for the purpose of captive use and/or for selling of electricity to distribution licensee or third party whether or not under the REC mechanism in accordance with Electricity Act 2003'. The tariff is determined differently for different stakeholders.

Gujarat has adopted its Wind Power Policy in 2016 (GEDA 2016). This policy will remain in operation up to 30 June 2021. The eligible unit includes any individual, company or body corporate or association or body of individuals whether incorporated or not or artificial judicial person for setting up of Wind Turbine Generators (WTG) either for the purpose of captive use or for selling of electricity to obligated entity(ies). The eligible site for WTGs will be notified by the Gujarat

Energy Development Agency (GEDA) or any other sites identified as potential site within the state by nodal agency or developer. GEDA will be the state nodal agency for facilitation and implementation of Gujarat Wind Power Policy 2016. Regarding the sale of electricity to obligated entities, they may purchase power from wind power project to fulfil their RPO at tariff determined by GERC or rate determined through competitive bidding.

4.3.1 Maharashtra: Renewable Energy Policy 2015

Maharashtra has launched renewable policies for various sectors under the umbrella name of Maharashtra Renewable Energy Policy 2015 (MEDA 2015). This policy envisages setting up of grid-connected renewable power projects. Maharashtra Energy Development Agency (MEDA) will be responsible for the implementation of this policy. Maharashtra Electricity Regulatory Commission (MERC) will set the tariff for the projects commissioned under the policy. Under the policy, solar power project of 7500 MW capacity will be developed. Out of this, 2500 MW capacity will be developed by the Maharashtra State Power Generation Co. Limited (MAHAGENCO) in PPP mode to fulfil the Renewable Generation Obligation (RGO). Rest 5000 MW will be developed by other developers. About 10% of solar power project through MAHAGENCO on the PPP model will be implemented along canals, lakes, water bodies through agreement with the Water Resource Department. Electricity generated from the projects under the PPP mode will be used to fulfil RPO requirements. The minimum capacity to be developed under this policy will be 1 MW. These projects can also be developed by way of solar parks. Under the wind energy policy, the target is to commission 5000 MW capacity of wind energy projects. Out of the total, 1500 MW will be developed to meet the procurement requirement of distribution licensees under the RPO regime. The remaining 3500 MW capacity is free to be used for interstate/intrastate, open access/captive consumption/REC, etc.

5 Discussion

While national-level policies posit the umbrella framework, given the federal structure of the country, states play important role to shape the entire process of overcoming challenges related to renewable energy in the country. The most triggering fact in this regard is that the state utility companies are responsible to achieve a demand-supply balance on their own grid. It requires access to a mix of reliable conventional generator and new experiments with different levels of flexibility to internalize the variation in load and supply from renewable sources on the grid. The demand-supply mismatch arises due to fluctuation in either of them. RRFs are introduced to incentivize solar and wind power producers to forecast their production with a certain degree of accuracy. This is expected to neutralize a part of unforeseen supply fluctuations. However, the introduction of greater system flexibility can further support the process. For example, in Tamil Nadu, the uncertainty associated with wind power generation is one of the major challenges to meet the demand-supply parity. This can be partly neutralized by the available reservoir-based hydro capacity in the state. But this is contingent upon irrigation release schedules and can be practically used only during high inflows into reservoirs when hydropower generation cannot be curtailed (The WIRE 2016). The second layer of complexity arises when the renewable power generation is greater than expected. The state utilities with lack of flexible balancing resources force the renewable power producers to back down. Forced shutdown of windmills to adjust the power supply, when there is a demand dip, is a frequently reported problem in all these states. This leads to instances of state power distribution utilities (discoms) unplugging their generating capacity from the grid, delayed payment and signing of Power Purchase Agreements (PPAs). This remained a cause of significant worry in three of the four states that we discuss here: Maharashtra, Tamil Nadu and Rajasthan (Livemint 2017; Green Clean Guide 2013; The Economic Times 2017). This also leads to a situation where the RPO targets set for the discoms become difficult to be met. This shows that even if the appropriate quantity and price policies are in place, lack of transmission infrastructure, absence of demand or inability of the grid to buy power may lead to a stalemate and create disincentives for renewable power generation. Sometimes, there is lack of consistency between the national-level targets and the state-level policies. For example, National Action Plan on Climate Change set a national target of 7% (with 5% as national RE target in 2010 and 1% increase annually till 2020), while the cumulative state targets add up to 5.44%. This leads to a deficit of 1.56%, which is translated into nearly 14,268 million units (www.indianpowersector.com).

System integration is also emerging as a solution to some of these challenges. Several European countries such as Denmark, Germany and Spain have integrated a large amount of solar and wind power. However, the efficacy with which individual sources of renewable power generation as well as the system integration will work is highly contingent upon the technological and institutional policies in place. Interestingly, the states with higher potential of wind power generation are also the states with higher potential to generate solar power in the country. This gives a unique opportunity to provide impetus to solar and wind energy simultaneously. The forthcoming National Wind-Solar Hybrid Policy is a move towards such integration.

6 Concluding Remarks

Tamil Nadu, Maharashtra, Gujarat and Rajasthan—these four states in India—are expected to play major roles in production of renewable energy in the country. The policy framework that has emerged to boost renewable energy generation in these four Indian states shows that there is a mix of regulatory policies with market-based instruments. The incentives provided are both from the central and state governments. However, there are challenges towards the implementation of these policies, including demand-supply mismatch, optimum financial strategy to pay for the high-end initial costs in the off-grid applications, lack of generation of storage capacity, high risk perception of using RECs, lack of credit available for developers. In fact, the ambitious targets to have renewable power generation capacity may prove to be unfruitful without proper utilization of the power generated. It calls for two important areas of intervention—cost-effective storage with flexible generation at the state level and a well-developed, widespread transmission mechanism. Therefore, the development of interstate High Capacity Power Transmission Corridors along with green energy corridors and intrastate transmission mechanisms can prove to be an essential complementary policy of renewable energy targets in the country.

References

BEE (2017) Energy Conservation Building Code 2017. Bureau of Energy Efficiency, Ministry of Power, Government of India, New Delhi

GEDA (2015) Gujarat Energy Development Agency. http://geda.gujarat.gov.in/policy_files/gujarat_solar_power_policy_2015.pdf. Accessed on 26 June 2017

GEDA (2016) Gujarat Energy Development Agency. https://geda.gujarat.gov.in/policy_files/Gujarat%20Wind%20Power%20Policy-2016.pdf. Accessed on 26 June 2017

GEDA (2017). http://geda.gujarat.gov.in. Accessed on 26 June 2017

GOI (2016a) Energy access in India—review of current status and role of renewable energy (Twenty Second Report submitted by Standing Committee on Energy, Ministry of Power/Ministry of New and Renewable Energy). Lok Sabha Secretariat, New Delhi

GOI (2016b). http://mnre.gov.in/file-manager/UserFiles/Draft-Wind-Solar-Hybrid-Policy.pdf

Government of Rajasthan (2012). http://energy.rajasthan.gov.in/content/dam/raj/energy/common/Notification-Wind%20Policy,%202012.pdf. Accessed on 26 June 2017

Green Clean Guide (2013) Wind Power Producers forced to back down. http://greencleanguide.com (11 Aug 2013). Accessed on 26 June 2017

http://indianpowersector.com/home/2013/05/powering-ahead-with-renewables-leaders-and-laggards-indian-states-status/. (2017). Retrieved from http://indianpowersector.com: http://indianpowersector.com/home/2013/05/powering-ahead-with-renewables-leaders-and-laggards-indian-states-status/. Accessed on 26 June 2017

IREDA (2015) Compendium of State Government Policies in Renewable Energy Sector. http://www.ireda.gov.in/writereaddata/CompendiumStatePolicyRE/Data/TN2.pdf Accessed on 05 Sept 2017

JUVNL (2017). http://www.gseb.com/guvnl/index.aspx. Accessed on 26 June 2017

Livemint (2017). Renewable energy firms worry about back-down by power discoms. Livemint (21 Oct 2016). Accessed on 26 June 2017

MEDA (2015) Maharashtra Energy Development Agency. https://www.mahaurja.com/meda/data/grid_wind_power/state_policy/Policy%202015_2.pdf. Accessed on 26 June 2017

MEDA (2017). https://www.mahaurja.com/meda/. Accessed on 26 June 2017

MNRE (2012) Ministry of New and Renewable Resources, Government of India. http://mnre.gov.in/filemanager/UserFiles/guidelines_sbd_tariff_gridconnected_res/Tamilnadu%20Solar%20Energy%20Policy%202012.pdf. Accessed on 26 June 2017

MNRE (2014) Ministry of New and Renewable Resources, Government of India, New Delhi. http://mnre.gov.in/file-manager/UserFiles/Grid-Connected-Solar-Rooftop-policy/Rajasthan-Solar-Energy-Policy-2014.pdf. Accessed on 26 June 2017

Raven R (2012) Analysing emerging sustainable energy niches in Europe: a strategic niche management perspective. In: Verbong G, Loorbach D (eds) Governing the energy transition. Routhledge, New York, pp 125–151

Roy J, Ghosh D, Ghosh A, Dasgupta S (2013) Fiscal instruments: crucial role in financing low carbon transition in energy systems. Curr Opin Environ Sustain 5(2013):261–269

RRECL (2017). http://energy.rajasthan.gov.in. Accessed on 26 June 2017

TEDA (2014). http://teda.in/. Accessed on 26 June 2017

The Economic Times (2017) Developers reel under losses as Rajasthan companies shut off wind power supply. http://economictimes.indiatimes.com/industry/energy/power/developers-reel-under-losses-as-rajasthan-companies-shut-off-wind-power-supply/articleshow/59327646.cms. Accessed on 26 June 2017

The WIRE (2016) Is the government trying to pull a fast one with its solar power promises? 1 Mar 2016. https://thewire.in/23104/is-the-government-trying-to-pull-a-fast-one-with-its-solar-power-promises/. Accessed on 26 June 2017

Thombare SP, Gunjal V, Bhandarkar S (2016) Overview of renewable energy in Maharashtra. Int J Adv Res Comput Commun Eng 5(8):200–203

TOI (2016) Maharashtra makes solar panels must for all new buildings. Times of India, Nagpur (26 Jan 2016). Accessed on 26 June 2017

UNFCCC (2016). http://newsroom.unfccc.int/paris-agreement/india-deposits-instrument-of-ratification-of-paris-agreement/. Accessed on 26 June 2017

Verbong G, Geels F (2007) The ongoing energy transition: lessons from a socio-technical, multi-level analysis of the Dutch electricity system (1960–2004). Energ Policy 35:1025–1037

Improving the Sustainability of Wastewater Treatment Through Solar-Assisted Multiple Effect Evaporators

Soundaram Ramanathan and Dibakar Rakshit

Abstract India's current water use is around 600 BCM of which 60% is sourced from surface water bodies and 40% from ground. Consumption wise 89% is used by the agricultural sector, 7% by domestic, and 4% by industries. While agricultural sector water use can be considered to be a part of the water cycle, establishing techno-economic treatment systems and wastewater reuse facilities is essential for domestic and industrial sectors. Three stages of wastewater treatment systems are available—primary, secondary, and tertiary. Each stage provides varying qualities of treated water, with the tertiary systems providing the best water quality with total dissolved solids less than 50 parts per million (ppm) permitting its reuse in agriculture and industrial sectors. In India, only primary and secondary treatment systems are widely available. Tertiary systems are retrofitted now, to reuse water. Treated water reuse increases water security by preventing water pollution and decreasing stress on water bodies due to excessive water withdrawals. Preventing water pollution, treating water, and its reuse is vital; similarly, deployment of low-carbon sustainable technologies for water treatment is imperative. Else water pollution prevention efforts would reshape as air pollution issue. Energy consumption by a treatment system correlates to the air pollution potential. Primary, secondary, and tertiary treatment systems consume 2, 3, and 6 kWh/m^3 electrical energy, respectively, in addition, the tertiary systems also require thermal energy which is about a kg steam for every 3 kg effluent to concentrate the waste (Tamil Nadu Water Investment Company Ltd. (TWIC), 2014). Environmental policies are therefore necessary to ensure the use of only sustainable treatment systems. Multiple effect evaporators are the widely used tertiary treatment system. It uses steam as a source of heating medium to evaporate wastewater. Steam used in MEE is of process heat quality with temperatures 150–200 °C and 10 bar pressure. The present chapter would discuss how conventional wastewater treatment systems

S. Ramanathan
Centre for Science and Environment, Delhi, India

D. Rakshit (✉)
Indian Institute of Technology, Delhi, India
e-mail: dibakar@ces.iitd.ac.in

A. Gautam et al. (eds.), *Sustainable Energy and Transportation*,
Energy, Environment, and Sustainability,
https://doi.org/10.1007/978-981-10-7509-4_9

could source their energy requirements sustainably and the policy implications surrounding such a promotion.

Keywords Multiple effect evaporator · Sustainability · Water · Pollution
Energy · Low-carbon technology · Solar

1 Introduction

Water is one of the important resources supporting life. India's current water use is around 600 BCM of which 60% is sourced from surface water bodies and 40% from ground. Total water resource available for extraction and use in India is around 1,123 BCM (690 BCM surface water and 433 BCM ground water) (WRIS). It is forecasted that the water consumption would grow from current 600 to 1,093 BCM by the year 2025 to serve the demands of increasing population and growing economy (Ministry of water resources 2000). This would limit the resource availability in the forthcoming decade. Hence policy measures are essential to streamline use of water and promote its reuse. Implementation of a policy measure requires a span of at least a decade. Therefore, concerted and timely measures should be taken now to avoid water crisis.

Consumption wise 89% of water in India is used by the agricultural sector (534 BCM), 7% by domestic (42 BCM), and 4% (24 BCM) by industries (WRIS). Water used by the agricultural sector is a non-negotiable requirement, however, can be reduced by improving efficiency of irrigation systems.

Food and agriculture are globally the largest consumer of water—an essentiality of life. Nil wastewater generation by the sector indicates an absence of scope for recycling. Therefore, policy measures are needed to reduce consumption of water by these sectors. Reduction of 30–65% water use by the sector is possible when compared to traditional systems by deploying micro-irrigation measures. Micro-irrigation measures include installation of micro-sprayers, micro-bubblers, drippers, and drip tubing. Apart, common system measures like agricultural canal lining, improving canal structures, tail water return systems can also help in reducing water consumption. While domestic and industrial segments generate wastewater (Clarke 2013). Mature technologies are available to reuse both domestic and industrial wastewater.

Domestic wastewater reuse: Reuse of domestic wastewater after secondary treatment is more economical than treating it further for reuse in agricultural and industrial sectors. Techno-economic studies indicate reuse of domestic wastewater for agriculture after secondary treatment is more efficient. However, untreated domestic wastewater polluting rivers is becoming a common occurrence. Inadequate infrastructure due to lack of funds is a major cause. Only 37% of domestic wastewater in urban areas has facilities for secondary treatment (Central Pollution Control Board 2015–2016). The existing policies and laws governing domestic wastewater treatment in India are: the National Environment Policy, 2006;

National Sanitation Policy, 2008; Hazardous waste (Management and Handling) Rules, 1989; Municipalities Act; District Municipalities Act, etc. According to existing policies, sewerage infrastructure creation is the responsibility of the state government. With central government schemes such as National River Conservation Plan, National Lake Conservation Plan, Jawaharlal Nehru National Urban Renewal Mission, and Urban Infrastructure Scheme for Small and Medium Town secondary treatment structures are being built. Concerted efforts of both the state and central government will be required to achieve sufficient infrastructure (Banerjee).

2 Industrial Wastewater Treatment System

Treatment technology of wastewater depends upon the quality and quantity of wastewater. Physical, chemical (Gupta et al. 2012), biological, or a combination of the processes (Sonune and Ghate 2004) are deployed to remove solids, organic matter, and sometimes nutrient from wastewater. The treatments systems are categorized into—preliminary, primary, secondary, tertiary, and advanced wastewater treatment systems depending upon the degree of treatment, in order of increasing treatment level.

Preliminary treatment: Preliminary treatment uses coarse filters, grit comminutors or sewage grinders, chemical mixing systems to control pH, pre-aerators for odor control. The wastewater is stabilized in preliminary treatment system. After preliminary treatment depending upon the biological oxygen demand and chemical oxygen demand ratio, the wastewater is subjected to aerobic or anaerobic treatment (Lettinga et al. 1997).

Primary treatment: In primary treatment, sedimentation chambers are typically used to physically separate suspended solids from wastewater in the case of aerobic treatment systems. While in anaerobic treatment systems, digesters are used which generate methane through anaerobic digestion of wastewater.

Typically wastewaters from domestic, paper and pulp, food processing industry, etc., are subjected to anaerobic treatment because of high BOD–COD ratio and presence of carbonaceous materials (see Table 1).

Secondary treatment: Biofilters, aeration tanks/oxidation ponds are used to allow degradation of carbonaceous materials in wastewater. Further after degradation the wastewater is allowed to settle in clarifiers to separate sludge.

Tertiary treatment: Physical separation of impurities by the use of filters such as sand filters, activated carbon filters, reverse osmosis units, evaporators is tertiary treatment. The filters are chosen depending upon the salt concentration or TDS to be removed from the wastewater. Industrial wastewater with high TDS (over 2,000 mg/L) prefers installation of reverse osmosis units for separation of salts and reuse of wastewater. While those with high BOD–COD ratio prefers to install activated carbon filters to remove odors.

Table 1 Characteristics of different wastewater (Ranganathan et al. 2007; Lokhande et al. 2011; Singare et al. 2010; Lettinga et al. 1993)

Parameter	Textile	Iron and steel	Food processing	Fertilizers	Pharmaceuticals	Paper and pulp	Domestic wastewater
pH	8.6	9.1	5.8–11	6.6	5.1	8.2	6–9
Suspended solids (SS)	675	600	70	487	654	517	720
Chemical oxygen demand (COD)	950	2,400		698	2,654	253	500
Biological oxygen demand (BOD)	342	1,200	500	227	1,011	86	220
Total dissolved solids (TDS)	6,644	Less than 500	Less than 500	2,078	3,412	3,067	500

Reverse osmosis unit's separates only 60–80% of the water from wastewater, the remaining 20–40% of the sludge is then subjected to evaporation through devices such as multiple effect evaporators for final separation of water and inorganic materials from reverse osmosis rejects (Ranganathan et al. 2007).

3 Energy Requirements

Preventing water pollution, treating water, and its reuse is vital; similarly, deployment of low-carbon sustainable technologies for water treatment is imperative. Else water pollution prevention efforts would reshape as air pollution issue. Energy consumption by a treatment system correlates to the air pollution potential. Typically, if all three levels of treatments are carried out for industrial and domestic wastewater, about India's half the current energy consumption would be required. Primary, secondary, and tertiary treatment systems consume 2, 3, and 6 kWh/m^3 electrical energy, respectively, in addition, the tertiary systems also requires thermal energy, about a kg steam for every 3 kg effluent to concentrate the waste (Tamil Nadu Water Investment Company Ltd. (TWIC), 2014). Environmental policies are therefore necessary to ensure the use of only sustainable treatment systems (Ramanathan and Rakshit 2015).

Rapid transition to sustainable systems would be a challenge. So in the present section, promoting the use of multiple effect evaporator—the most energy consuming unit of the tertiary treatment system is discussed. Application of renewable thermal energy system could be a low-hanging fruit to initiate the transition to sustainable wastewater treatment technology.

4 Solar-Assisted Multiple Effect Evaporators

Multiple effect evaporation process is *a thin film evaporation process*, where the vapor in one chamber/effect, condenses in the next one, providing a heat source for further evaporation (see Fig. 1).

Multiple effect evaporators are essentially low-pressure steam generators. They are classified by the number of effects. Unlike traditional boilers where heat is supplied to vaporize products, in single-effect evaporator systems, steam provides the necessary energy for vaporization of the product and the vapor product is condensed and removed from the system, while in the *double-effect* evaporator, the vapor product from the first effect is used to provide energy for a second vaporization unit. This cascading of effects continues for many stages. Vapor from Effect I will be used as a source of heat in Effect II, which will operate at lower pressure consequently. These continue through as a chain: pressure drops through the sequence so that the hot vapor will travel from one effect to the next. The advantage of the process is the steam economy. It is estimated that for a single-effect evaporator nearly 0.8 kg of water can be evaporated with 1 kg steam, similarly with '*n*' number of effects 0.8n kg of water can be evaporated with 1 kg steam. The quantity of steam consumed by multiple effect evaporators is inversely proportional to the number of effects. In wastewater treatment systems, multiple effect evaporators are used to treat the reverse osmosis rejects to avoid damage due to scaling by direct heating of the rejects.

Effects are numbered beginning with the one heated by steam which will have the lowest pressure. Normally, all effects in an evaporator will be physically the

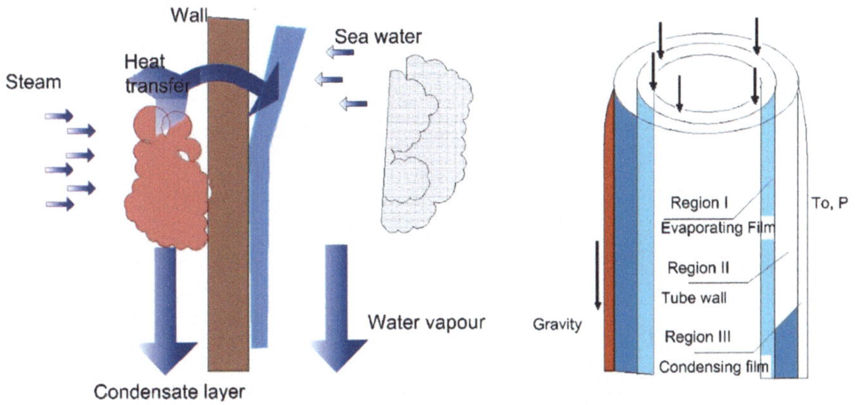

Fig. 1 Multiple effect evaporator concept

same in terms of size, construction, and heat transfer area. Unless thermal losses are significant, they will all have the same capacity as well. Multiple effect evaporator though discussed as a separate type of evaporator, physically is the coupling of individual evaporators (see Fig. 2) (Costa and Lima 2003).

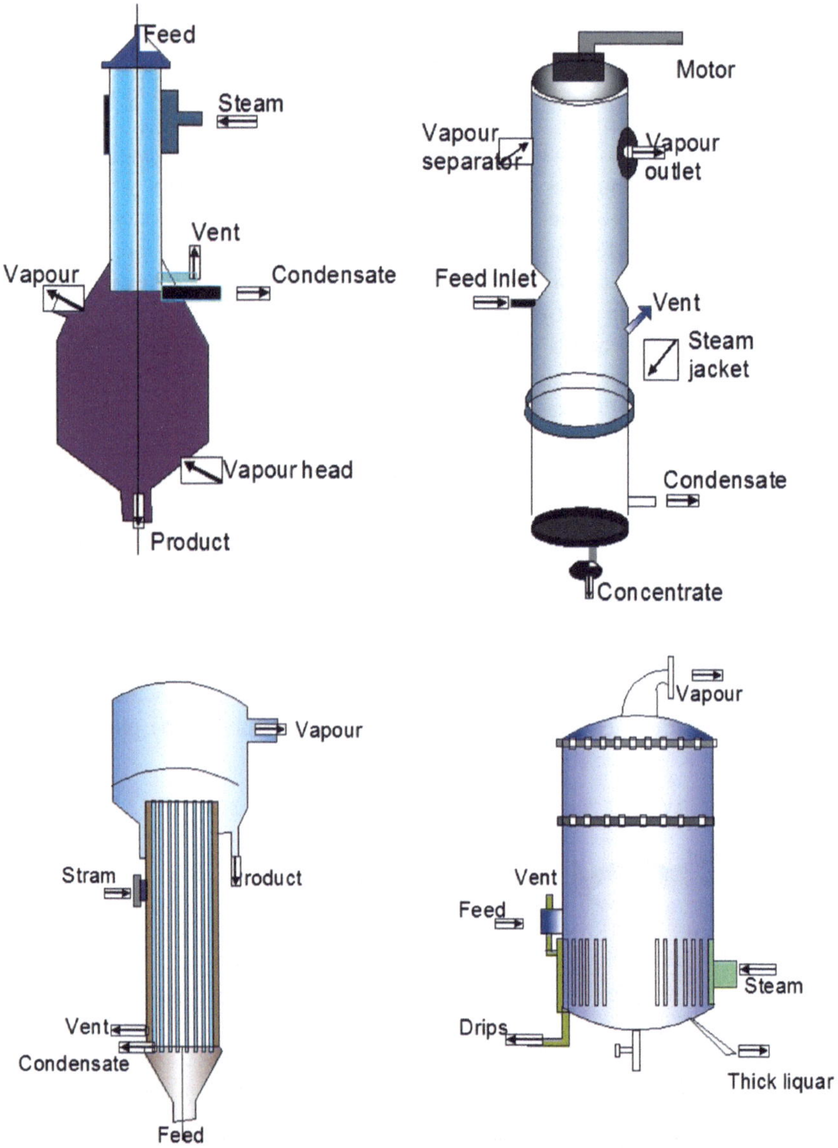

Fig. 2 Different types of individual evaporators which are coupled to form the MEE

Table 2 Solar concentrators in Indian markets (United Nations Development Program 2014–2015)

Tracking type	Collector type	Concentration ratio	Temperature (°C)	Cost (INR/m^2)
Non-tracking	Compound parabolic concentrator	1.5	60–200	12,000
Single axis	Linear Fresnel	10–50	60–250	18,000
	Parabolic trough	10–50	60–300	18,000
Two axis	Parabolic dish	100–500	100–150	20,000
	Arun	150–300	Up to 350	20,000

The industrial units must be advised through 'consents' to adopt sustainable power solutions like solar concentrators to be attached with these evaporators rather than conventional diesel-powered boilers (Blanco et al. 2009). Solar concentrators —Compound Parabolic Concentrator, Linear Fresnel, Parabolic Trough, Scheffler and Arun are available in the Indian market (see Table 2) (United Nations Development Program 2014–2015).

Various institutes, The Cyprus Institute (Georgiou), Turkey and Plataforma Solar De Almeria, Spain (García-Rodríguez and Gómez-Camacho 2001) have coupled multiple effect evaporators with solar concentrators. K G D S, Renewable Private Limited has installed solar Fresnel collectors coupled multiple effect evaporator units in Ramanathapuram district, Tamil Nadu (KGDS). The performance of solar-assisted multiple effect evaporator system is given by

$$Nu = 0.183\, Re^{0.513}\, Pr^{1.692}\, \eta_0^{0.283}\ \text{concentration ratio}^{0.180} \tag{1}$$

where Nu is the Nusselt number, Re is the Renoylds number, Pr is the Prandtl number, η is the optical efficiency of the solar collector and cc concentration ratio (Ramanathan and Dibakar 2016).

Technical concern expressed about deployment of sustainable technology is variability in operation. To alleviate which the present study has also done a Trnsys modeling of solar concentrator with multiple effect evaporator to assess the operational efficacy of the device in varied climatic zones—composite as in Kanpur, Uttar Pradesh—tannery industry cluster, hot and dry as in Pali, Rajasthan—textile industry cluster, moderate as in Gulbarga, Karnataka—cement industry cluster, and Murshidabad, West Bengal—arsenic-affected area (Ramanathan and Rakshit 2015).

Primarily during the monsoon days in June, July, August the concentrator did not produce the desired temperature of steam, hence fluctuations of temperatures were observed on isolated days of monsoon. It was observed that the system performed optimally releasing steam at desired temperature of 150 °C as desired for 88% of the time in hot and dry climatic zone (Fig. 3), 86% in warm and humid climatic zone (Fig. 4), 85% in composite climatic zone (Fig. 5), and 83% in moderate climatic zone (Fig. 6).

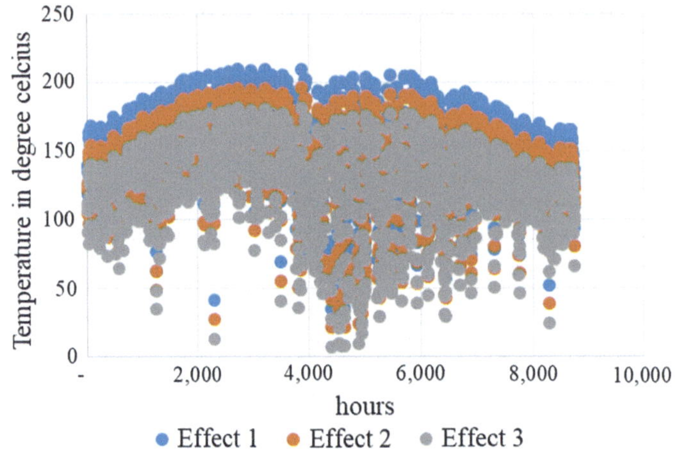

Fig. 3 Temperature profile of different effects/chamber/evaporator of the multiple effect evaporator at hot and dry climatic zone

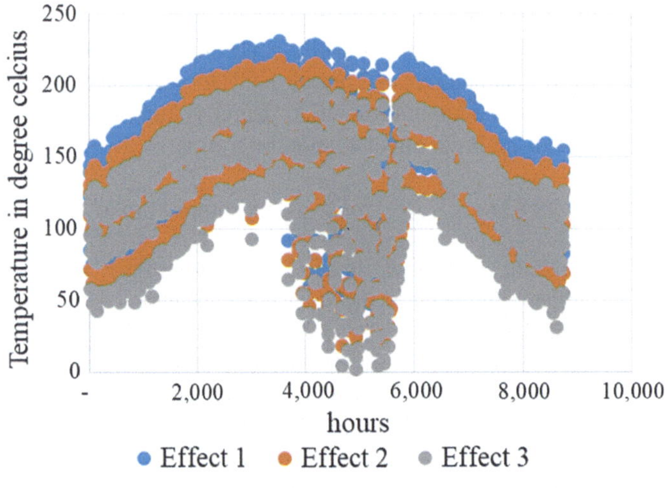

Fig. 4 Temperature profile of different effects/chamber/evaporator of the multiple effect evaporator at warm and humid climatic zone

To manage the fluctuations two options were available—(1) adding supplementary systems to raise the temperature of steam from 100 to 150 °C; (2) fine tuning the system to increase the mass of steam generation according to the temperature—the specific mass ratio of water needed to evaporate wastewater in solar-assisted multiple effect evaporator system is given by

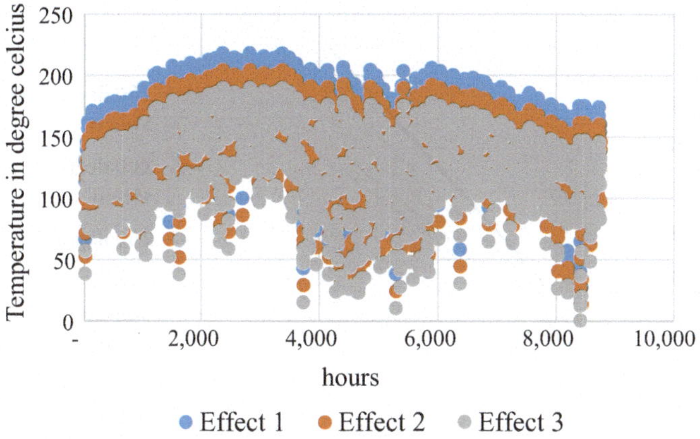

Fig. 5 Temperature profile of different effects/chamber/evaporator of the multiple effect evaporator at composite climatic zone

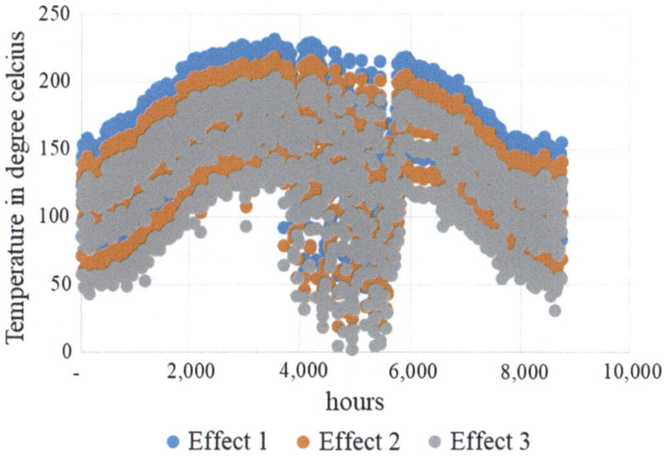

Fig. 6 Temperature profile of different effects/chamber/evaporator of the multiple effect evaporator at moderate climatic zone

$$m_{\text{wwe}} = \frac{h}{H_{fg\_\text{w}}} = \frac{\left(0.183 * (Re)^{0.513} * (Pr)^{1.692} * \eta_0^{0.283} * (\text{cc}\ )^{0.180}\right) * k}{l} \tag{2}$$

Assuming unit dimensions for k and l based on the Nusselt number, the mass of wastewater which can be evaporated with a kg steam was found ranging from 0.4–8 kg/kg depending on the temperature of the steam sent to the first effect of

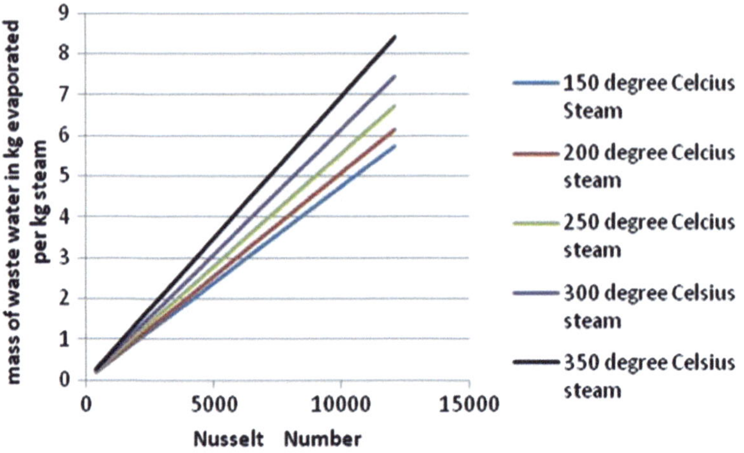

Fig. 7 Variation of specific mass ratio of evaporation with Nusselt number

the MEE. '*k*' is the thermal conductivity (W/m K) of the wastewater and '*l*' is the height (*m*) of the multiple effect evaporator (see Fig. 7). (Ramanathan and Dibakar 2015).

5 Discussion

Promoting sustainable energy in wastewater treatment system can increase its uptake and reduce the illegal discharge of effluents in the river, as the operating cost of treatment system squares down to zero. This also benefits any agricultural units in the vicinity that use river water for crop irrigation. Reducing the pollution from industrial units will significantly reduce the impact of human populations becoming ill through drinking contaminated water. Similar benefits will accrue for aquatic life-forms.

Industrial wastewater reuse: Currently, industrial water consumption in India excluding power sector is roughly 24 BCM (Table 3).

Zero liquid discharge is mandated in industries under Water Act, 1974. Deployment of multiple effect evaporator is essential. However, use of sustainable technology powered evaporators is not emphasized by the governments through their consents. Largely, these systems are being powered by diesel. By switching to solar concentrators, a foreign exchange transaction of about INR 127,392 crores (per BCM wastewater treated) every year on oil purchase can be avoided, and 88 million tonnes of CO_2 emission reduction is possible.

Table 3 Water consumption projection by different industrial sectors in India

Industry	Specific water in m^3/ unit of product	Production in 1000' tones unit of product	Total water consumption in MCM
Textile and jute	50	95,093.9	4,755
Integrated iron and steel	22	265,350	5,838
Food processing	6.8	506,150	3,442
Inorganic chemicals	200	8,000	1,600
Cement	4.5	219,000	986
Fertilizer	16.7	37,782	631
Pharmaceutical	25	8,370	209
Paper and pulps	200	10,350	207
Sugars	2.2	32,330	71
Distillery	22	3,059.6	66
Leather products	30	2,191.3	66
Petrochemicals and refinery	17	1,800.6	31
Smelters	82.5	292.6	24
General engineering	2.2	10,754	24
Chemicals	5.5	1,854.6	10
Rubber	6.6	651.5	4
Pesticides	6.5	306.2	2
Total			17,965

Source Ministry of drinking water and sanitation, 2002

6 Conclusions

Water scarcity is discussed worldwide and is forecasted to become a major challenge by 2025, at this juncture treating and reuse of water by the domestic and industrial segments are unavoidable. Treating wastewater to desirable levels up to tertiary treatment would require roughly half the energy consumption of the country. Vast energy consumption of the sector translates to energy-water nexus (Cho 2010). In addition, with coal energy being largely used in India, wastewater treatment power consumption would also mean increase in air pollution. Hence, uptake of low-carbon or sustainable technology in wastewater is important for meeting the dual objectives of prevention of water scarcity and reducing pollution.

The advancement of low-carbon technologies in India's wastewater treatment sector has been insufficient and has not yet maximized the potential energy and sustainable policy opportunities (Tiwari 2002). There exists opportunity to build sustainable systems. Low hanging immediate implementable attempt that could pave way could be solar-assisted multiple effect evaporators.

These multilevel patterns in India's wastewater treatment sector are shaped by country-specific institutional characteristics, capacity constraints, and interactions between domestic energy requirements and resulting pollution (Sosbey et al. 2008). Notably, at times these factors weaken strategies and initiatives. At other times they open up institutional spaces for new mature technologies to engage in and advance efforts. A successful attempt at advancing technologies through multiple governance levels—and in fact, making effective use of the opportunities afforded by such linkages—will require a careful consideration of system characteristics. Reflecting on the case of multiple effect evaporator, this paper has attempted to showcase how low-carbon technology deployment is possible.

References

Banerjee A, Legal and policy framework for wastewater treatment and reuse in India: a background

Blanco J, Malato S, Fernández-Ibañez P, Alarcón D, Gernjak W, Maldonado MI (2009) Review of feasible solar energy applications to water processes. Renew Sustain Energy Rev 13(6):1437–1445

Central Pollution Control Board (2015–16)

Cho A (2010) Energy's tricky trade offs. Science 329(5993):786–787

Clarke R (2013) Water: the international crisis. Taylor and Francis group. https://books.google.co.in/books?id=DbxcAgAAQBAJ&printsec=frontcover&source=gbs_ge_summary_r&cad=0#v=onepage&q&f=false

Costa AO, Lima EL (2003) Modelling and control of an industrial multiple-effect evaporator system. Can J Chem Eng 81(5):1032–1040

García-Rodríguez L, Gómez-Camacho C (2001) Exergy analysis of the SOL-14 plant (Plataforma Solar de Almeria, Spain). Desalination 137(1):251–258

Gupta VK, Ali I, Saleh TA, Nayak A, Agarwal S (2012) Chemical treatment technologies for waste-water recycling—an overview. Rsc Adv 2(16):6380–6388

India Web Enabled Water Resources Information System (WRIS), National Remote Sensing Centre

KGDS Renewable Energy Private Limited—KGiSL. www.solar.kgisl.com

Lettinga G, De Man A, Van der Last AR, Wiegant W, Van Knippenberg K, Frijns J, Van Buuren JC (1993) Anaerobic treatment of domestic sewage and wastewater. Water Sci Technol 27(9):67–73

Lettinga G, Field J, Van Lier J, Zeeman G, Pol LH (1997) Advanced anaerobic wastewater treatment in the near future. Water Sci Technol 35(10):5–12

Lokhande RS, Singare PU, Pimple DS (2011) Study on physico-chemical parameters of waste water effluents from Taloja industrial area of Mumbai, India. Int J Ecosyst 1(1):1–9

Marios C Georgiou, Thermal Desalination performance analysis, The Cyprus Institute

Ministry of water resources (2000)

Ramanathan S, Dibakar R (2015) Design and simulation of a tool to evaluate the performance of solar assisted multiple effect evaporator system, India. Int Sci Festival, Ministry Sci Tech

Ramanathan S and Dibakar R (2016) Thesis: study of solar assisted multi effect evaporator, IIT Delhi

Ramanathan S and Rakshit D (2015) Multi-criteria optimization of solar assisted multiple effect evaporator. Global Energy Summit
Ranganathan K, Karunagaran K, Sharma DC (2007) Recycling of wastewaters of textile dyeing industries using advanced treatment technology and cost analysis—case studies. Resour Conserv Recycl 50(3):306–318
Singare PU, Lokhande RS, Jagtap AG (2010) Study of physico-chemical quality of the industrial waste water effluent from Gove industrial area of Bhiwandi City of Maharashtra India. Interdisc Environ Rev 11(4):263–273
Sonune A, Ghate R (2004) Developments in wastewater treatment methods. Desalination 167: 55–63
Sosbey et al (2008) Point of use household drinking water filtration: a practical, effective solution for providing sustained access to safe drinking water in the developing world. Environ Sci Technol. doi:https://doi.org/10.1021/es702746n
Tiwari GN (2002) Solar energy—fundamentals, design, modelling and applications. Narosa Publishing House, New Delhi
United Nations Development Program (2014–15)

Making Indian Power Sector Low Carbon: Opportunities and Policy Challenges

Nandini Das and Joyashree Roy

Abstract It is both a challenge and an opportunity for the Indian power sector to strike a strategic balance between fast-paced economic growth, structural shift and social development without being on high greenhouse gas (GHG) emission pathway. The goal of this paper is to develop alternative scenarios for low GHG emission trajectory for next three decades. It needs to be reiterated that since all the growth will happen in coming decade or two, there are opportunities for India for making the right choices in technology, policy and investment decisions so that faster economic growth does not lock the economy into a high emission trajectory. This has become even more important with India's commitment as party to Paris Agreement. In this paper, quantitative methods of scenario and trajectory development are followed and assessment is based on review of power sector policy reforms and visions as expressed in various official documents.

Keywords Low-carbon growth · Power sector · India · Paris agreement

1 Introduction

Current per capita electricity consumption and per capita carbon emissions in India are both low in comparison to other countries (Fig. 1). India's per capita electricity consumption was 805 KWh in 2014 against world average of 3144 KWh (World Bank 2015) with total installed capacity of 263 GW (INDC 2015). However, the reality is, there is a serious need for energy supply to expand in order to achieve universal energy access and also to power the economic growth process as in India, a considerable part of the population is deprived of modern energy supply. Decoupling economic growth from carbon emissions is a major challenge for current as well as future development strategies in India given that energy demand is going to increase. One of the major sectors identified for transition towards a

N. Das · J. Roy (✉)
Department of Economics, Jadavpur University, Kolkata, India
e-mail: joyashreeju@gmail.com

A. Gautam et al. (eds.), *Sustainable Energy and Transportation*,
Energy, Environment, and Sustainability,
https://doi.org/10.1007/978-981-10-7509-4_10

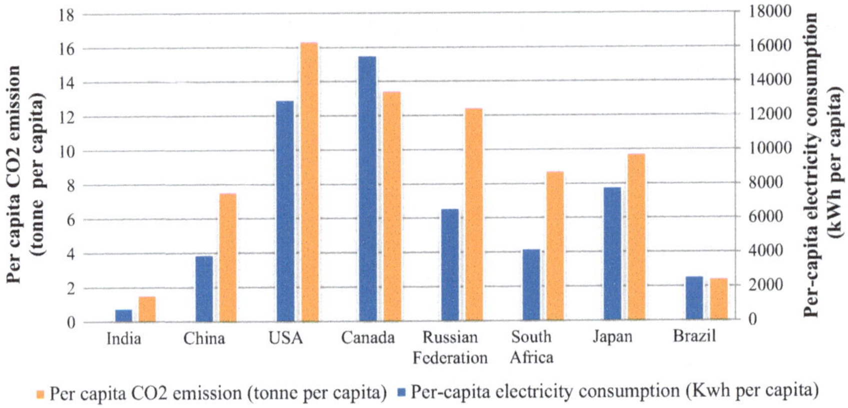

Fig. 1 Per capita CO_2 emission and per capita electricity consumption (2014). *Source* Compiled from World Bank Data

low-carbon future in India is the energy supply sector (INCCA 2010; NAPCC 2008). As emissions from the coal-fired power plants are responsible for a large burden of global as well as local environmental pollution in India (Sethi 2015), the need for shifting from coal-fired power plant has become an imperative.

India is the third largest producer of power in the world with 1304.8 TWh generation in 2015 only after China and the USA, 5810.6 and 4303 TWh, respectively, (BP 2016). However, more than 300 million of Indian population still are deprived of access to electricity (INDC 2015). The main objective of India's energy policy is to achieve three goals, energy access, energy security and lower environmental impact (INDC 2015). Indian power sector is one of the most diversified sectors in the world (MNRE 2014–15). Sources of power generation range from conventional sources such as coal, lignite, natural gas, oil, hydro and nuclear power to viable non-conventional sources such as wind, solar, small hydropower (SHP), bio-power, waste to power (Fig. 2). Being a non-Annex I country, India has voluntarily committed to reduce its emissions intensity by 20–25 percent of its 2005 levels by 2020 as recognition towards growing problem of climate change. The Emission Gap Report (2014) of United Nations Environment Programme (UNEP) has recognized that India is among the countries that have the ability to achieve its voluntary goal.

In October 2015, prior to Paris Agreement, India has submitted its Intended Nationally Determined Contribution (INDC) to UNFCCC. In India power sector being the highest polluting sector got most thrust by several policy framework and mitigation initiatives. In post-Paris phase, the NDC of India clearly reflects its desire for economic growth with reduced climate risk. According to NDC, by 2030, India will reduce the emission intensity of its GDP by 33–35 percent from 2005 levels and 40% of its cumulative installed capacity will be comprised of non-fossil sources. To achieve these targets, importance is given to energy mix mainly through solar, wind and efficient generation process with introduction of supercritical

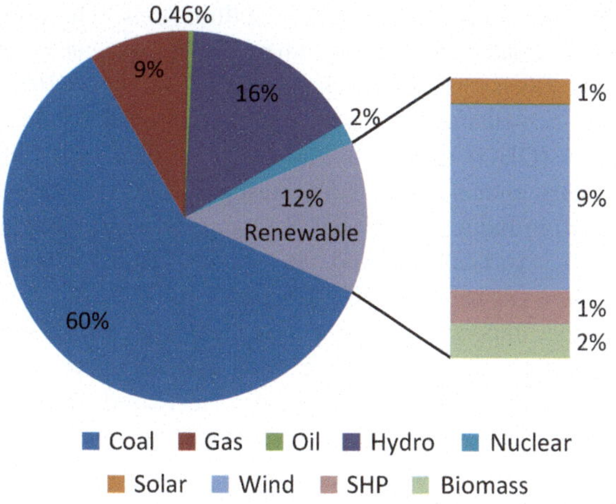

Fig. 2 Total installed capacity according to source in India as on 31.12.2014. *Source* MNRE (2014–15)

technology, ultra-supercritical technology, integrated gasification-combined cycle (IGCC) and carbon capture storage (CCS) in coal-based power plants (INDC 2015; NITI Aayog 2016). It is emphasized that technology and finance are going to play an important role in achieving these goals.

In a recent report of UNFCCC (2016) on assessment of the NDC submitted by the member countries, it has been shown that even with full implementation of all NDCs it is difficult to reach global climate goal of stabilizing global temperature rise to 1.5 °C from pre-industrial level by turn of the century. This indicates that there is a need for additional mitigation measures with respect to our national circumstances and ambitions.

Given this background, this paper aims to develop evidence based low-carbon pathway for Indian power sector by considering different challenges and opportunities. This will help in understanding whether India can deliver NDCs and beyond, given its growth objectives and emission targets. First to set the scene, Sect. 2 provides an overview of the current status of the Indian power sector, followed by Sect. 3 which develops alternative scenarios for low-carbon pathways for the Indian power sector, Sect. 4 presents the various policy challenges that can help in achieving the future global scenario and Sect. 5 summarises the findings and scope for future research.

2 Overview of Present Status of the Indian Power Sector

Generation of thermal power is more polluting than any other single industry (IPCC 2014). 60.8% of power generated in India comes from coal-based power plants (INDC 2015). According to the report of Indian Network for Climate Change Assessment (INCCA 2010), CO_2 emissions from electricity generation for the year

2007 are estimated as 715.83 million tons. The power sector in India consumes approximately 43% of commercial primary energy and generates 40% of India's CO_2 emissions (Bhattacharya and Cropper 2010). Apart from carbon dioxide (CO_2), methane (CH_4) and nitrous oxide (N_2O) are the other harmful greenhouse gases (GHG) from coal-based power plant. CO_2 emissions are increasing at an average annual rate of 5.6% during 2001–02 to 2009–10 from power plants (Mittal 2010) in India. The annual N_2O emissions from energy sector in India for the year 2010 was 12.06 Gg and out of that 11.68 Gg are emitted from electricity production (Abrol et al. 2017). Sulphur dioxide (SO_2), though not a listed GHG, it is another important polluting agent which is emitted from coal plant. Sulphur content in Indian coal varies between 0.3 and 0.55 percent (Chandra and Chandra 2004), but annual SO_2 emissions have increased from 2.5 million tons in 2001 to 3.8 million tons in 2009 at an average annual rate of 169.39 thousand tons per year (Mittal 2010). India imports most of its thermal coal from Indonesia, Australia and South Africa. Although it needs more evidence, but some studies say that the increasing dependency on imported coal (Chandra and Chandra 2004) can be one of the reasons for this rise in emission of SO_2, as most of the Indian thermal power plant do not use more than 30% of domestic coal (Ministry of Power 2017).

On the efficiency scale as well Indian power sector has scope for improvement. Plant load factor (PLF), which is the ratio of actual output to optimum output of thermal power plants, is an important parameter of operational efficiency of thermal power plant. In Indian thermal power plants, PLF is declining over time (Table 1). A decline in PLF leads to a higher cost of electricity generation due to more fuel usage which results in much higher carbon footprints. One of the reasons behind this falling PLF of thermal power plant is increasing capacity of renewables and hydro (CEA 2016). A study of Central Electricity Authority has shown that any increase in renewable electricity generation is mostly replacing the thermal generation. Therefore, with the increased infusion of renewable generation into the grid, the PLF of the coal-based plants decreases. It needs to be mentioned that the PLF for Indian thermal power plants has increased from 55.3% in 1991–92 to 69% in 2000–01 (Shanmugam and Kulshreshtha 2002) which resulted in a higher growth rate for the energy generation from same installed capacity (Beerbaum and Weinrebe 2000).

Though coal-based power as of now accounts for about 60.8% (167.2 GW) of India's installed capacity, India aims to achieve a significant capacity expansion

Table 1 PLF in India during 2009–10 to 2014–15

Year	Target (%)	Actual (%)
2009–10	77.2	77.5
2010–11	72.1	75.1
2011–12	68.7	73.3
2012–13	70	69.9
2013–14	69.6	65.6
2014–15	65.52	64.46

Source CEA (2016)

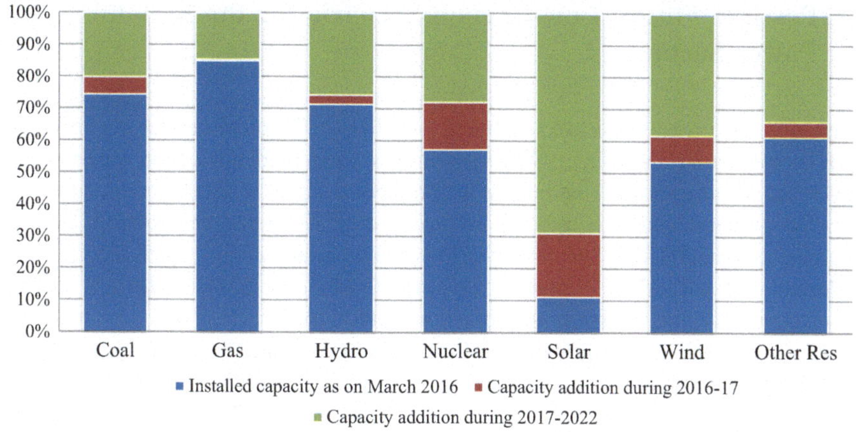

Fig. 3 Installed capacity and capacity addition of different energy sources under 12th and 13th plan. *Source* CEA (2016)

Table 2 Promotion of clean energy

Non-fossil energy sources	Status (CEA 2016)	Target (INDC 2015)
Wind	23.76 GW installed capacity	60 GW installed capacity by 2022
Solar	4.06 GW installed capacity	100 GW by 2022
Biomass	4.4 GW current capacity	10 GW by 2022
Hydro	46.1 GW current installed capacity out of 4.1 GW small hydro and 41.99 GW large hydro	
Nuclear	5.78 GW current installed capacity	63 GW by 2032

Source INDC (2015), CEA (2016)

programs in renewables (Fig. 3). The share of renewable grid capacity has increased over 13% between 2002 and 2015, and India aims to achieve 175 GW of renewable energy capacities by 2022 (INDC 2015) from 42.85 GW as on March, 2016 (CEA 2016).

Renewable clean energy sources contribute almost 12.3% of total installed capacity in India (CEA 2014). The installed capacity of grid-connected renewable energy has touched 38.1 GW in December 2015 with an increase of 18% from last year. Installed capacity of off-grid captive power from renewable sources has crossed 1.23 GW. Though wind continues to dominate grid-connected power in India, solar power has shown the highest expansion of more than 100% increase. Untapped potential for overall non-fossil energy in India is more than 200 GW which shows a huge growth potential for renewable energy in India (Table 2).

3 Possible Future Low-Carbon Scenario for the Indian Power Sector

Scenarios are broadly constructed based on UNFCCC guidelines for reporting of GHGs projection (UNFCCC 1999). According to these guidelines, there are basically two kinds of projections. First is baseline projection which is developed on currently adopted policies, and it is referred as projection "with measures" often called as business as usual (BAU). Second scenario is aspirational scenario which incorporates "additional measures" and policies beyond (O'Mahony 2014; UNFCCC 2013). Environmental and emission scenarios are interlinked across the sectors. The approach to develop a scenario merges both "bottom-up" and "top-down" approaches as it integrates technologies with national or global climate goal. Within this interaction, complex combination of driving forces of energy efficiency and emission reduction can be explored. Several methodologies are discussed in the literature to construct scenarios, both mathematical and analytical. Here we are adopting analytical methodology to draw some future possible scenario based on assessment of announced policies, vision documents and global commitments.

In this section, we present three alternative scenarios and pathways which are developed for Indian power sector by taking 2015 as base year and 2050 as the target year. The year 2015 is BAU scenario, but named in this article as "Reference Scenario" which projects energy supply scenario based on "Report of The Working Group on Power for Twelfth Plan" by the erstwhile Planning Commission of India. Second and third scenarios are two aspirational scenarios, referred as NDC scenario and Global Action Scenario, respectively. NDC scenario is developed based on the India's official submission to UNFCCC as Intended Nationally Developed Contribution (INDC) on October 2015 which can be viewed as a transitional phase to achieve bigger climate goals. As we already know that all the initiatives declared in NDCs are not sufficient to deliver stabilization goal of global temperature rise within 1.5 °C from pre-industrial level by turn of the century, so we have developed as "Global Action Scenario" where we consider deployment of newer technologies going beyond NDCs and fuel mix reflecting a transformative change in installed capacity in the Indian power sector.

3.1 Scenario Design and Assumptions

Taking into consideration the end-use demand growth leading to GDP growth projections, following scenarios for future-installed capacity are developed mainly by changing energy mix, introducing new technology and improving energy efficiency to understand the CO_2e emission scenarios and pathways up to 2050 which is summarised in Table 3.

Table 3 Scenario design and assumptions

	Coal	Gas	Nuclear	Hydro	Wind	Solar	Biomass	SHP
Reference scenario	As projected under 13th plan (2017–2022) and assumption is that this share will continue till 2050	As projected under 13th plan (2017–2022) and assumption is that this share will continue till 2050	As projected under 13th plan (2017–2022) and assumption is that this share will continue till 2050	As projected under 13th plan (2017–2022) and assumption is that this share will continue till 2050	Up to 2022 share of renewables are the same as projection of 13th plan. It has been assumed that same trend will continue till 2050, and when each fuel source reaches its maximum potential, we have shifted the excess capacity to solar			
INDC scenario	Introduction of supercritical technology. Capacity of supercritical technology will be 100 GW by 2022. Assumption is that annual growth rate will continue till 2050	Assumption is that capacity will be same as Reference Scenario due to supply constraint	Capacity is going to be 63 GW in 2032. Assumption is that annual growth rate will continue till 2050	Same as Reference Scenario due to sociopolitical issue on land-related dispute regarding large hydro	Capacity is going to be 60 GW in 2022. Assumption is that annual growth rate will continue till 2050	Capacity is going to be 100 GW in 2022. Assumption is that annual growth rate will continue till 2050	Capacity is going to be 10 GW in 2022. Assumption is that annual growth rate will continue till 2050	Capacity is going to be 5 GW in 2022. Assumption is that annual growth rate will continue till 2050
Global action scenario	CCS in subcritical plant from 2025 and its capacity will be 8 GW in 2047. Ultra-supercritical technology and IGCC will be operational after	Assumption is that capacity will be same as Reference Scenario due to supply constraint	Capacity is going to be 40 GW in 2030. Assumption is that annual growth rate will	Capacity addition to large hydro 12000 MW during 13th plan (2018–2022). Total installed	Assuming that the capacity addition would follow the 12th and 13th plan trajectory. By 2017, capacity would reach close to 32.25	Following IESS level 2 scenario by 2017, capacity would reach close to 8 GW in line with the 12th plan projections, while by 2022 it	Assumption is that potential of biomass is 17.5 GW which is achievable by 2050	Assuming that the capacity addition would follow the 12th and 13th plan trajectory,and following that SHP will achieve its

(continued)

Table 3 (continued)

	Coal	Gas	Nuclear	Hydro	Wind	Solar	Biomass	SHP
	2027 and 2037, respectively. At 2050, share of this two technology in total coal capacity will be 5% each		continue till 2050. At the end of 2050, installed capacity of nuclear will be 85.5 GW	capacity at the end of 2047 is going to be 75 GW	GW in line with the 12th plan projections, while by 2022 it would reach 54.35 GW. Capacity addition increases strongly, thereafter culminating in a cumulative capacity of 202 GW by 2047	would reach 17.9 GW. Capacity addition increases strongly thereafter culminating in a cumulative capacity of 150 GW by 2047		20 GW potential in 2050

Source Planning Commission of India (2012), MNRE (2014–15), INDC (2015), GOI (2016), Aayog (2016), CEA (2016)

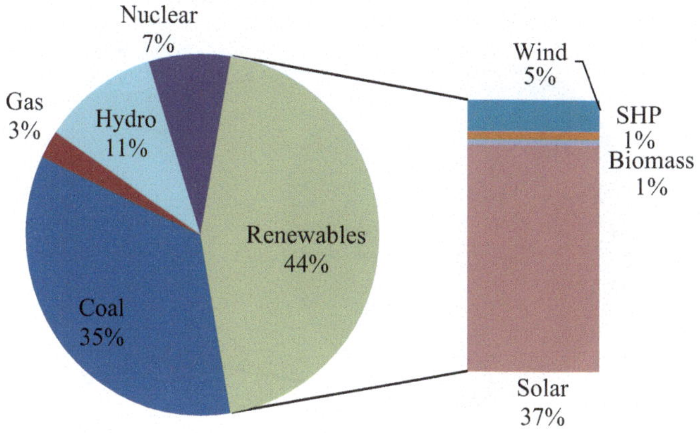

Fig. 4 Capacity mix projection in 2050—reference scenario

Reference Scenario: Reference Scenario has been constructed mainly based on the capacity addition projections under 12th five-year plan document (2012–2017) and projection for 13th plan (2017–2022) by erstwhile Planning Commission of India. Here, it has been assumed that coal with subcritical technology will continue to dominate power generation in India, with increasing thrust on gas-based power generation along with increasing share of renewable energy. Hence in this scenario, there is no change in existing technology (Fig. 4).

NDC Scenario: In NDC scenario, capacity of respective fuel sources is projected based on India's submission to UNFCCC. In this scenario, it has been assumed that 40% of installed capacity will be from non-fossil source by 2030. Efficiency improvement of coal-fired power generation with supercritical technology is taken into consideration (Fig. 5).

Global Action Scenario: In this scenario, it has been assumed that India will try to enhance its actions to increase its voluntary contribution in global goal. More stringent policies are going to be adopted to improve electricity generation with improved carbon intensity by introducing more efficient technology such as ultra-supercritical technology and IGCC in coal-fired power plant along with vigorous promotion of non-fossil energy. It has been further assumed that carbon capture storage (CCS) will be introduced in subcritical coal-fired power plant from 2025. This will help in scaling up low-carbon transition of Indian power sector and to achieve a transformed state of low carbon by 2050 (Fig. 6).

In total, installed capacity is calculated using linear trend of previous periods. In reference scenario for renewable, first we have assumed that the share of 2029 will continue till 2050. Then capacity has been adjusted based on maximum potential projected by government of India. For wind, small hydroprojects (SHP) and biomass, once the maximum potential is reached, for rest of the period, we have kept it to that level only by shifting the excess capacity to solar. The capacity of solar has

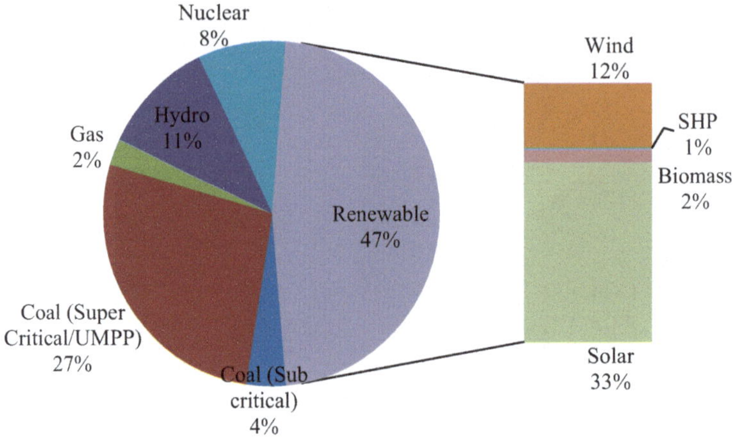

Fig. 5 Capacity mix projection in 2050—INDC scenario

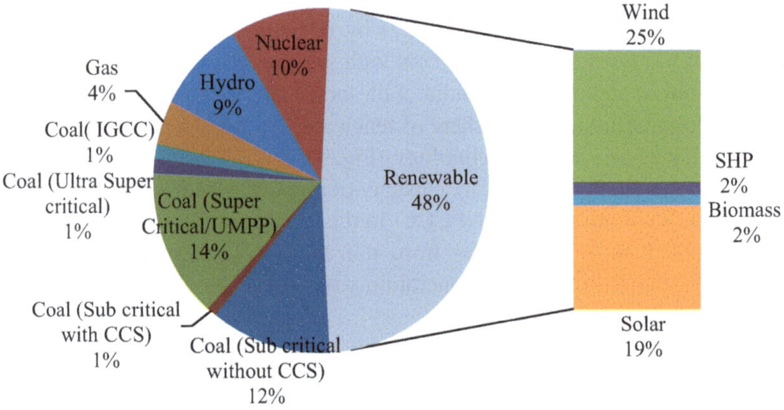

Fig. 6 Capacity mix projection in 2050—Global action scenario

been calculated by adding up its actual capacity and excess capacity from wind, SHP.

Macro-Socio-economic Parameters

Macro-socio-economic parameters such as GDP growth rate, population growth rate used in scenario development are based on various national and international vision documents (Table 4).

We assume that in supply side, capacity expansion is needed to meet the growing demand for electricity and this demand expansion is related to these socio-macro parameters (Table 4). We consider population growth, resultant

Table 4 Macro-socio-economic parameters

	2015	2020	2025	2030	2035	2040	2045	2050
GDP growth rate	7.5	7.5	7.5	7.5	7.5	7.5	7.5	7.5
Population growth rate	1.26	1.15	1.02	0.88	0.74	0.6	0.48	0.38
Macro economic structure								
Agriculture (%)	17	16.5	16	15.5	15	14.5	14	13.5
Industry (%)	29	28.5	27	26.5	25	24.5	23	22.5
Services (%)	53	54.5	56	57.5	59	60.5	62	63.5

Source UNDP (2015), MoSPI (2016), World Bank (2017)

activity growth and the structure of the economy reflected in sector shares in GDP that drive the demand for power.

3.2 *Fuel Mix in Installed Capacity Under Different Scenarios*

Under all three scenarios (Figs. 4, 5 and 6) we have found that thermal-installed capacity using coal and natural gas will continue to dominate till middle of the century. However, the share of coal capacity declines in all the three scenarios. For India, additional mitigation effort in the global scenario beyond NDC for India can come from efficiency enhancement of the thermal power sector. Therefore, in Global Action Scenario which represents India's additional effort, we introduce higher penetration of new technology like ultra-supercritical technology, IGCC and CCS in subcritical power plants, but these technologies will not be operational before 2025.

If we look at the overall fuel mix in all the three scenarios (Figs. 4, 5 and 6), we can see that there is an unambiguous increase in share of non-fossil energy sources in power generation in forms of hydro, nuclear, wind, solar, biomass and SHP. Large hydro is an important source of power generation in India. The capacity of large hydro under Reference Scenario was 48 GW in 2015 which is 18% of total installed capacity. The capacity of large hydro remains same in Reference Scenario and NDC Scenario. However, in Global Action Scenario, capacity of hydro will become 77 GW in 2050. For India, nuclear is an economically viable as well environmentally safe source in power generation (INDC 2015). In all the three scenarios, steady increase in share of nuclear in total installed capacity is assumed as per assessment based on official documents. Capacity of renewables comprising of wind, solar, biomass and small hydro has increased in all the three scenarios. India's expansion of renewable capacity is targeted mainly on solar photovoltaic. However, wind is at present the dominant source in renewable capacity structure. Biomass and small hydro are also an important part of renewable power capacity.

Total installed capacity in 2050 is not very different across three scenarios, as total demand is not varying across scenarios. Installed capacity as projected are 1227, 1473 and 1100 GW under Reference, NDC and Global Action scenario, respectively. The difference is due to the difference in the projection of respective vision documents which are used in developing that particular scenario.

3.3 Power Generation Scenarios and Share of Different Fuels

Power generation scenario (Figs. 7, 8 and 9) is not the same as the installed capacity scenario for many primary sources. Coal is going to be the dominating fuel in terms of generation as well. In Reference Scenario, share of coal-based power generation is going to be 53% (2800 TWh) compared to its capacity of 42% in total installed capacity in 2050. In NDC scenario, share of coal-based power generation will become 58% (3583 TWh) of total generation in 2050, but capacity will be less at 31% (2006 TWh) in 2050. Here contribution in generation increases due to increased share of supercritical technology. In Global Action Scenario, coal-based power generation is expected to be 48% (16% will be from subcritical power plant and 26% will be supercritical power plant) with a capacity of 29%.

In Reference Scenario, gas capacity accounts for 11% in total installed capacity in 2015, but it contributes only 6% in power generation which can be attributed to low PLF of gas-based power station owing to intermittency in supply. Gas-based power generation contributes 7% and 8% in total generation in NDC Scenario and Global Action Scenario, respectively. In 2050 in all the three scenarios, both gas capacity and its generation share account for nearly 3% of total capacity and generation, respectively. This decline in gas capacity is mainly attributed to low and irregular supply of gas.

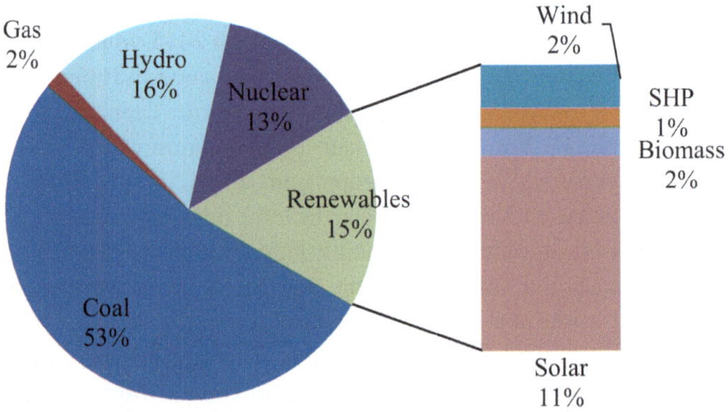

Fig. 7 Annual generation projection 2050 reference scenario

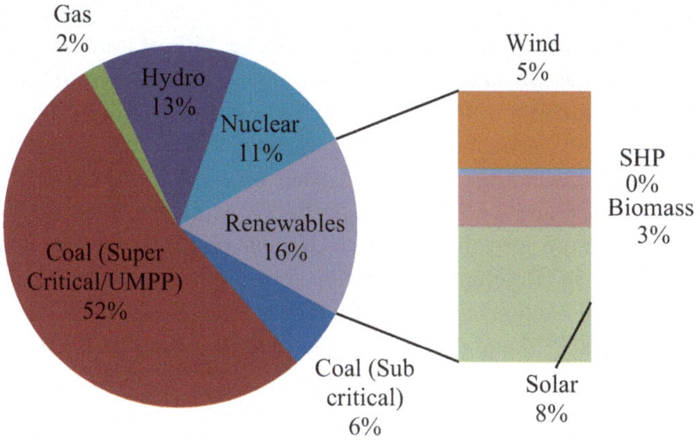

Fig. 8 Annual generation projection 2050 INDC scenario

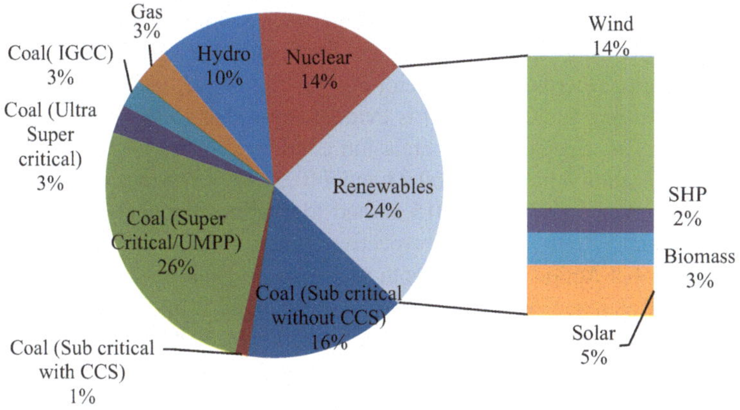

Fig. 9 Annual generation projection 2050 global action scenario

Hydro comes second in terms of power generation, after coal. In 2015, hydro 256 TWh of electricity which is 21% of total generation. In all the three scenarios, hydro has shown a sluggish growth with a decline in share in total generation which will become 16, 13 and 10%, respectively, in Reference Scenario, INDC scenario and Global Action Scenario, in 2050. In Reference Scenario and INDC scenario, total generation from hydro will stand at 839 TWh, as it is assumed that gas capacity will remain the same in these two scenarios and in Global Action Scenario it will decline to 406 TWh.

In all the three scenarios, nuclear power generation has shown significant progress in terms of generation comparing to its capacity. In Reference Scenario, nuclear-based power generation accounts for 687 TWh of generation in 2050 which is 13% of total generation. In NDC scenario, this will become 768 TWh and share in total generation becomes 12%. In Global Action Scenario, power generation from nuclear will be 601 TWh in 2050 which will be 14% of total generation. This increasing share of nuclear-based power generation is coming from improved PLF of nuclear power plant from 50% in 2008 to 78% in 2012 (NITI Aayog 2016).

In 2015, electricity generation from renewable source was around 86 TWh which is 7% of total generation. Out of this, share of wind, solar, biomass and small hydro was 41, 4.5, 28 and 12.5 TWh, respectively. In all the three scenarios solar has exhibited maximum growth. The share of solar power in total generation will become 604 TWh in Reference Scenario, 530 TWh in INDC Scenario and 191 TWh in Global Action Scenario. Significant push for solar power in the country under the JNNSM and decreasing price in installation are the major reasons behind this growth.

3.4 Emission Scenarios

Corresponding to projected capacity and generation scenarios, emission scenarios are developed by using emission factors and methods (IPCC 2006) (Fig. 10). In all the three scenarios, thermal power sector comprising of coal and gas and biomass are the only emitting fuel sources. It is evident from Fig. 10 that under Reference Scenario and INDC Scenario, total emission continues to rise. However, emission per unit of generation which was 769.23 and 740.68 kg/KWh under Reference and INDC scenario, respectively, in 2015 started to decline (Fig. 11) up to 2028 and become 710 and 576.07 kg/KWh, respectively, and after that it becomes almost constant. In Global Action Scenario, both total emission and per capita emission are declining. In Reference Scenario, total emission rises from 1169 million tonne CO_2e in 2015 to 3987 million tonne CO_2e in 2050. In NDC scenario, this rise is from 1157 to 3490 million tonne CO_2e. In Global Action Scenario, total emission

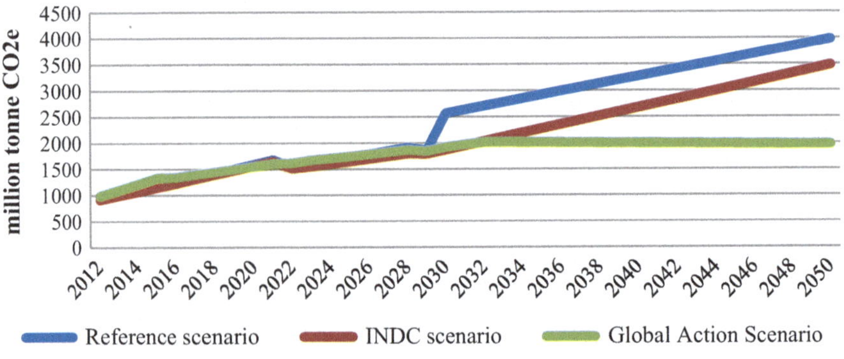

Fig. 10 Emission scenario

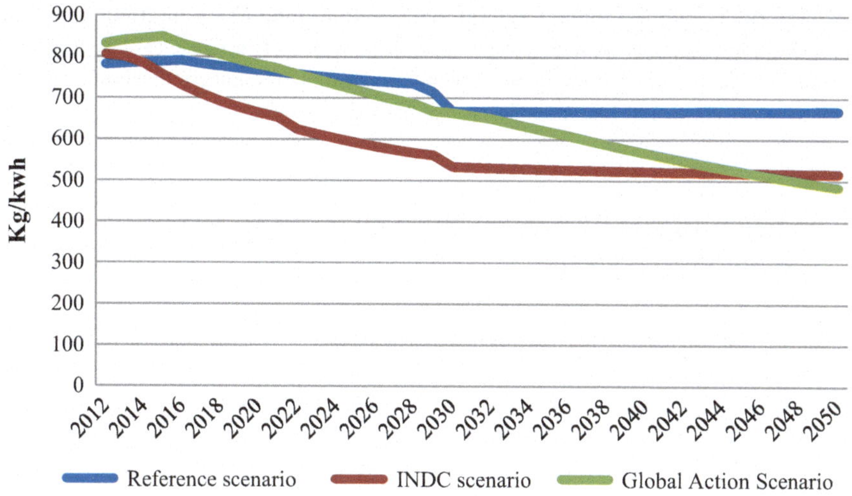

Fig. 11 CO_2e emissions per unit of generation

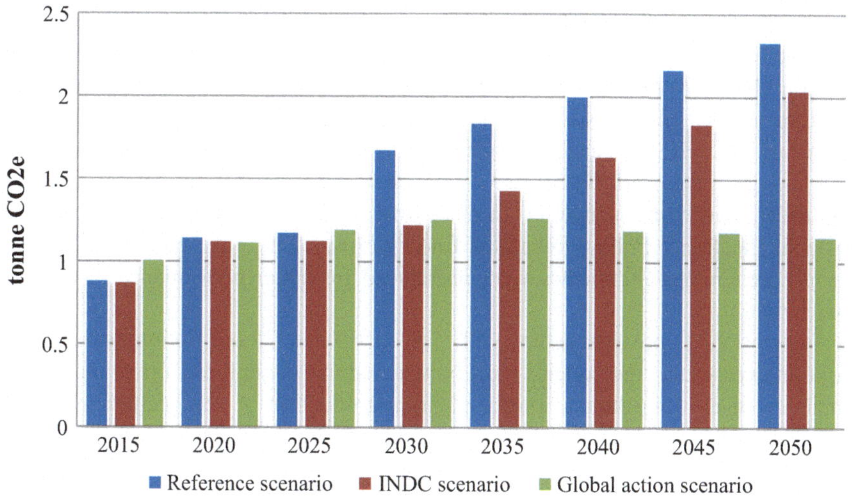

Fig. 12 Per capita emission from power generation

was 1333 million tonne CO_2e in 2015 and it continues to rise until 2032. After 2032, it starts to decline, and in 2050 it becomes 1980 million tonne CO_2e. Emission per KWh of generation in Reference Scenario is going to be 625 kg CO_2e in 2050. This will decline to 516 and 477 kg CO_2e in NDC Scenario and Global Action Scenario, respectively.

Per capita emission from power generation under Reference Scenario and INDC Scenario continues to rise till 2050 (Fig. 12). In Reference Scenario, per capita emission rises from 0.9 tonne CO_2e in 2015 to 2.3 tonne CO_2e in 2050, which is much higher when compared to NDC scenario, from 0.9 tonne CO_2e in 2015 to 2 tonne CO_2e in 2050. In Global Action Scenario per capita emission starts to decline from 2035 and becomes 1.1 tonne CO_2e in 2050.

4 Policy Support to Achieve the Low-Carbon Future for the Power Sector

Preceding sections show the opportunities for power sector in India to keep 2050 emissions level (1979.65 million tonne CO_2e) almost at 2030 level (1929.30 million tonne CO_2e). The analysis shows that a low-carbon future for Indian power sector is technically achievable, but electricity being a highly capital intensive sector, financing for the transition assumed in NDC scenario and transformative shift in Global Action Scenario are going to be the major challenge. Additional investment requirement supported by proper institutional structure is important to work out.

India's National Action Plan on Climate Change (NAPCC) provides information on proposed interventions to address transition of power sector. NAPCC is implemented through eight National Missions out of those National Missions for Enhanced Energy Efficiency (NMEEE) and Jawaharlal Nehru National Solar Mission (JNNSM) outline the priorities of more efficient power sector. These two missions are designed in such a manner that both demand and supply side can be taken care of. JNNSM is a supply-side effort to increase the share of solar energy in total energy mix, and NMEEE is a demand-side effort with energy-efficient appliances and various market mechanisms. Reference Scenario does include National Electricity Policy, Integrated Energy Policy, Perform Achieve and Trade. Renewable Energy Certificates were introduced for better performance of thermal power sector.

As mentioned above, to achieve NDC Scenario and Global Action Scenario for Indian power sector, the policies must be targeted mainly in three forms, increasing efficiency of existing thermal power plants, introduction of new technologies and improving fuel mix with increasing share of non-fossil energy sources, but additionally for Global Action Scenario we need introduction of CCS.

Efficiency improvement

Several policy initiatives are taken to improve the efficiency of thermal power plant to continue with domestic reliance on electricity generation. As a part of clean coal policies, all existing coal-based generating units are mandated to undergo Renovation and Modernization (R&M) and life extension (LE) in a phased manner (INDC 2015). Also, development of Ultra Mega Power Projects (UMPP) has been identified as thrust area. These UMPPs are coal-based but will use supercritical technology with a higher efficiency which results in saving of fuel and lower GHG emission. Each of this UMPP has production potential of approximately 4000 MW involving an estimated investment of about Rs. 16,000 crore. This capacity is going to be 124 GW in 2050 under Global Action Scenario. This involves a huge amount of investment. One of the important barriers in achieving this efficiency improvement is financial constraints.

Perform Achieve and Trade (PAT) is the flagship market mechanism under the National Mission of Enhanced Energy Efficiency which is one of the eight missions under NAPCC (Ministry of Power 2012a, b). Thermal power is one of the designated sectors. Total target for mitigation under first cycle of PAT was 6.6 Mtoe and out of that target for thermal power sector was to achieve energy savings of 3.2 Mtoe (Gupta 2015). At the end of first cycle, total 123 designated customers from thermal power sector have been identified to achieve 3.06 Mtoe of energy savings (Ministry of Power 2016).

The government of India created National Clean Energy Fund (NCEF) in 2010 for the purpose of financing and promoting clean energy initiatives and funding research in the area of clean energy in the country. The corpus of the fund is initially built by levying a cess of Rs. 50 (subsequently increased to Rs. 200 in 2015 and Rs. 400 in 2016) per tonne of coal produced domestically or imported. Total NCEF as on 2015–16 is Rs. 34811.87 crore (Ministry of Finance, GoI 2016).

New technology

Majority of the Indian coal is hard coal with high ash content, and the average calorific value is very low compared to the USA and Australian coal. Therefore, increased dependency on imported coal is adding to the import bill. Hence, one of the major policy challenges for Indian power sector is to increase the ratio of domestic coal being used in Indian power plants by improving the technology both for energy security and improving foreign reserve. Supercritical technology in coal-fired power plant is going to be an important tool in achieving this. However, this is associated with huge cost and this can be a barrier in deployment of this technology.

Whether India can go beyond its NDC target will largely depend on these technologies, investments and access to CCS. Major barrier in CCS deployment in India is the associated costs and availability of the commercial scale technology. Powerplant with CCS involves 76% more cost than powerplant without CCS, which is significantly higher (NITI Aayog 2016). Most of the capital cost incurred in implementation of CCS is in the transportation of CO_2 to sinks through pipelines and other means, compression and storage. Thus, the suitable location of the powerplants will reduce this cost and make application of CCS more financially viable (Parikh 2010). There is a need for administrative framework on regulation of safety and protection scheme along with legal provision for land use and technology transfer. Apart from economic viability (Akasha et al. 2016), other major barriers in deployment of CCS at commercial scale is that CCS involves a huge amount of investment risk. India strongly depends on the successful implementation of CCS technology in developed countries (Viebahn et al. 2014; Kapila et al. 2011), which is also another barrier. A further impediment to CCS is the lack of storage capacity assessment, especially in Indian context (Shackley and Verma 2008; Viebahn et al. 2014). More research and demonstration on the technologies like

CCS, IGCC and ultra-supercritical technology should be pursued with targeted efficiency improvement to use Indian coal in thermal power plant. An integrated institutional framework is needed to be developed to encourage deployment of these cutting-edge technologies in a more efficient way.

Integrated energy policy (2006) was formulated to increase the exploitation of renewable energy sources to improve fuel mix in power generation and R&D on alternative energy forms. Hence, this policy targeted both the requirement of improved fuel mix along with efficiency improvement. India's NDC commitment of 100 GW of solar energy by 2022 is an important part of this policy (INDC 2015). Some of the recommendation of this policy is developments of solar parks and Ultra Mega Solar Power Projects, solar PV power plants on canal banks/canal tops (Ministry of New and Renewable Energy 2016).

Improved fuel mix

In all the three scenarios, nuclear power will provide the major contribution to mitigating CO_2 emissions from Indian power sector. Positive policy environment to promote nuclear with Indo–US agreement for cooperation in civilian nuclear power permitted India to import uranium, reactors and technology under international safeguards.

Solar has been identified as a clean power generation source in all the three scenarios. Government of India has adopted several policy initiatives to promote solar as one of the important renewable sources of power generation. To achieve 100 GW solar capacity by 2020, Ministry of New and Renewable Energy (MNRE), GoI has targeted solar rooftop plant and set a target of 40 GW generation from solar rooftop. MNRE has announced a capital subsidy for rooftop solar installations for residential consumers, government buildings, hospitals, educational institutions and other establishments of common use as a part of central financial assistance (CFA). This subsidy is 15% for residential, institutional, government buildings and no CFA for commercial and industrial establishments in the private sector. Initial budgetary allocation for this scheme was Rs. 5000 crore (Ministry of New and Renewable Energy).

Renewable Energy Certificate (REC) mechanism was launched in November 2010. REC is a market-based instrument under NAPCC's Renewable Purchase Obligation (RPO). Aim behind this project was to produce 15% of countries power from renewable source by 2020 (Ministry of New and Renewable Energy 2009). Under RPO, power distribution companies are obligated to buy a certain amount of electricity from renewable sources. RECs are issued to renewable energy generators. This REC can be traded in the market if power distribution companies fail to meet RPO. In this RPO framework, state governments took policy decisions regarding financial incentives, buy-back tariff and other measures targeting investment in renewable energy (Goyal and Jha 2009). Though design of the REC mechanism appears adequate, the performance of the market is not satisfactory. Lack of effective monitoring, enforcement of RPO targets and clarity on compliances (Shrimali and Tirumalachetty 2013) is appearing to be main the main issues in the way of a well-performing REC market in India.

5 Conclusions

Assessment based on India's vision documents shows that India contemplates towards technological upgradation and use of fuel with less or no carbon content to reduce GHG emissions to half by 2050, as compared to Reference Scenario. The emission reduction target can come from deployment of technologies like CCS, IGCC, ultra-supercritical technology and nuclear besides solar and wind. Several policy initiatives come under NAPCC for the Reference Case Scenario. To make further progress, faster additional initiatives are incorporated in NDC through higher penetration of renewable carbon-free sources of energy. However, in Global Action Scenario we assume transformative technologies in the form of CCS for advanced new coal-based installations and also ultra-supercritical coal power generation technologies. However, real challenge now lies in CCS technology and related complex technological advancement and institutional mechanism. It needs to form a major research agenda for India. In this analysis, we have considered only supply side of power generation to realize the Global Action Scenario. However, if we take the energy system as a whole then effort from electricity end-use sectors can help us in avoiding some big investment in high-cost technologies. This is outside the purview of our article but is crucial for understanding the whole range of opportunities and challenges.

References

Aayog NITI (2016) India energy security scenario 2047. Niti Aayog, GOI, New Delhi

Abrol Y, Adhya T, Aneja VP, Raghuram N, Pathak H, Kulshrestha U et al (2017) The Indian nitrogen assessment: sources of reactive nitrogen, environmental and climate effects. Manag Options Policies, Elsevier

Akasha AR, Raoa AB, Chandel MK (2016) Prospects of implementing CO_2 capture and sequestration (CCS) in the proposed supercritical coal power plants in India. Energy Procedia 604–612

Beerbaum S, Weinrebe G (2000) Solar thermal power generation in India—a techno–economic analysis. Renew Energy 2:153–174

Bhattacharya S, Cropper ML (2010) Options for energy efficiency in India and barriers to their adoption. Resources for the Future

BP (2016) BP statistical review of world energy: electricity. British Petroleum

Bureau of Energy Efficiency (N.A.) Venture capital fund for energy efficiency (VCFEE). Retrieved from Bureau of energy efficiency. https://beeindia.gov.in/sites/default/files/VCFEE.PDF

Bureau of Energy Efficiency (N.A.). Retrieved 08 June 2017. From: https://beeindia.gov.in/sites/default/files/PRGFEE_0.pdf

CEA (2016) National electricity plan. Ministry of Power, GOI

Chandra A, Chandra H (2004) Impact of Indian and imported coal on Indian thermal power plants. J Sci Ind Res 63:156–162

GOI (2016) Energy statistics. Ministry of statistics and programme implementation

Goyal M, Jha R (2009) Introduction of renewable energy certificate in the Indian scenario. Introduction Renew Energy Certificate Indian Scenario 13:1395–1405

Gupta P (2015) Perform achieve and trade scheme thermal. Retrieved 08 June 2017, from Excellence Enhancement Centre for Indian Power Sector: http://www.eecpowerindia.com/codelibrary/ckeditor/ckfinder/userfiles/files/PAT.pdf
INCCA (2010) Climate change and India: a 4 * 4 assessment a sectoral and regional analysis for 2030
INDC (2015) Intended Nationally Determined Contribution. GOI
IPCC (2014) Climate change 2014: mitigation of climate change. Cambridge University Press, Cambridge
IPCC. (2006). IPCC guidelines for national greenhouse gas inventories, vol 2. Energy
Kapila RV, Chalmers H, Haszeldinea S (2011) CCS prospects in India: results from an expert stakeholder survey. Energy Procedia 4:6280–6287
Ministry of Finance, GoI (2016) Department of Expenditure. Retrieved 10 Sept 2017, from http://doe.gov.in/sites/default/files/NCEF%20Brief_post_BE_2017-18.pdf
Ministry of New and Renewable Energy (2016, December 01) Integrated energy policy formulated covering all sources of energy including renewable energy sources. Retrieved 07 June 2017, from http://pib.nic.in/newsite/PrintRelease.aspx?relid=154719
Ministry of new and renewable energy (2009) Development of conceptual framework for renewable energy certificate mechanism for India. Retrieved 08 June 2017, from http://mnre.gov.in/file-manager/UserFiles/MNRE_REC_Report.pdf
Ministry of New and Renewable Energy (n.d.) Ministry of new and renewable energy, GOI. Retrieved 09 June 2017, from solar rooftop—grid connected. http://mnre.gov.in/schemes/decentralized-systems/solar-rooftop-grid-connected/
Ministry of Power (2012) Retrieved 06 June 2017, from http://powermin.nic.in/en/content/distribution
Ministry of Power (2017) Retrieved 10 June 2017, from press information Bureau, GOI: http://pib.nic.in/newsite/PrintRelease.aspx?relid=159500
Ministry of Power (n.d.) Integrated power development scheme. Retrieved June 09, 2017, from http://www.ipds.gov.in/Form_IPDS/About_IPDS.aspx
Ministry of power (2016) Overview of PAT scheme: achievements and prospects. Retrieved 20 June 2017, from Bureau of energy efficiency, Ministry of Power, GoI: https://www.beeindia.gov.in/sites/default/files/PAT%20PPT%20%28Overview%29%20Regional%20Workshops.pdf
Ministry of Power (2012) Press information Bureau, GOI. Retrieved 08 June 2017, from http://pib.nic.in/newsite/PrintRelease.aspx?relid=85182
Mittal ML (2010) Estimates of emissions from coal fired thermal power plants in India
MNRE (2014–15) Powering the renewable energy revolution. Government of India
MoSPI (2016) National income statistics. CSO, GOI
NAPCC (2008) NAtional action plan on climate change. GOI
O'Mahony T (2014) Integrated scenarios for energy: a methodology for the short term. Futures 55:41–57
Parikh J (2010) Analysis of carbon capture and storage (CCS) technology in the context of Indian power sector
Planning Commission of India (2012) Twelfth five year plan
Sethi M (2015) Location of greenhouse gases (GHG) emissions from thermal power plants in India along the urban-rural continuum. J Cleaner Prod 103:586–600
Shackley S, Verma P (2008) Tackling CO2 reduction in India through use of CO_2 capture and storage (CCS): prospects and challenges. Energy Policy 9:3554–3561
Shanmugam KR, Kulshreshtha P (2002) Efficiency of thermal power plants in India. Res Art 27:57–68
Shrimali G, Tirumalachetty S (2013) Renewable energy certificate markets in India—a review. Renew Sustain Energy Rev 26:702–716
UNDP (2015) World population prospects: 2015 revision. UNDP

UNFCCC (2016) Aggregate effect of the intended nationally determined contributions: an update. Retrieved 10 June 2017, from UNFCCC: http://unfccc.int/resource/docs/2016/cop22/eng/02.pdf

UNFCCC (2013) Definition and reporting system of climate finance

UNFCCC (1999) Review of the implementation of commitments and of other provisions of conventions: UNFCCC guidelines on reporting and review. Retrieved 10 June 2017, from UNFCCC: http://unfccc.int/resource/docs/cop5/07.pdf

Viebahn P, Vallentin D, Höller S (2014) Prospects of carbon capture and storage (CCS) in India's power sector—an integrated assessment. Applied Energy 117:62–75

World Bank (2017) Global economic prospect

World Bank (2015) The world bank. Retrieved Dec 2016, from http://data.worldbank.org/indicator/EG.USE.ELEC.KH.PC

Waste Heat Recovery Policy

Piyali Sengupta, S. K. Dutta and B. K. Choudhury

Abstract Waste heat recovery (WHR), i.e. the recovery and reuse of by-product heat from a process/facility that would be otherwise rejected to the environment, is often a valuable cost-effective approach to improve overall industrial energy efficiency. With the advent of technologies and evolving demands of energy services, the scope, economics and potential for WHR have become increasingly dynamic in nature, particularly in the context of developing countries where both energy-intensive manufacturing and energy demand are increasing. The growth of global WHR market is expected to be 6.8% compound annual growth rate during 2016–2022. WHR has various benefits, including monetary savings, reduced environmental pollution and emission, energy security of the nation, public health. An appropriate WHR policy could provide a framework to channelize the whole process into a win–win situation for all stakeholders, such as the industry/users (e.g. iron and steel, glass, ceramic, textiles, cement, solar PV and thermal systems), the suppliers, the Government and even the public at large. The overall objective of the policy on WHR in various industry sectors is to optimize the benefits, such as mentioned above, by achieving short-, medium- and long-term goals. Some of the technological solutions already in place include cogeneration through WHR, regenerative burner, low-temperature organic Rankine cycle power generation, heat pump, thermo-photovoltaic cells/systems, thermo-chemical systems, high-efficiency recuperate, etc. The success will also depend on various socio-economic parameters, such as availability of land, project finance, research, manpower. The greatest challenge to prepare the WHR policy, as presented in this chapter in a step-by-step approach to overcome the same will be to take all the stakeholders into confidence and device it in

P. Sengupta · S. K. Dutta · B. K. Choudhury (✉)
Energy Management Department, IISWBM, Kolkata 700073, India
e-mail: binoykchoudhury@gmail.com

P. Sengupta
e-mail: piyali.sengupta@gmail.com

S. K. Dutta
e-mail: enconskd@gmail.com

A. Gautam et al. (eds.), *Sustainable Energy and Transportation*,
Energy, Environment, and Sustainability,
https://doi.org/10.1007/978-981-10-7509-4_11

harmony with intervening policies in place, in order to attain the mission and vision of this policy to utilize its full potential for public good.

Keywords Conservation · Energy efficiency · Policy · Public good
Techno-economic viability · Waste heat recovery

Abbreviations and Acronyms

BCCI	Bengal Chamber of Commerce and Industries
BEE	Bureau of Energy Efficiency
BOOT	Build-own-operate-transfer
CAGR	Compound annual growth rate
CDM	Clean development mechanism
CCM	Continuous casting machine
CEA	Central Electricity Authority
CEM	Certified Energy Manager
CERC	Central Electricity Regulatory Commission
CSR	Corporate social responsibility
DC	Designated consumer
DDG	Decentralized distributed generation
DISCOM	Distribution Company
DPR	Detailed project report
DRI	Directly reduced iron i.e. sponge iron
EEFP	Energy efficiency financing platform
EIF	Electrical induction furnace
EnMS	Energy management system
ESCO	Energy supply company
FEEED	Framework for energy efficient economic development
FIT	Feed-in-tariff
FOSMI	Federation of small and medium industries
FY	Financial year
GBI	Generation based incentive
GEF	Global environment fund
IIT	Indian Institute of Technology
ITI	Industrial Training Institute
JICA	Japan International Co operation Agency
kW	Kilo watt
kWh	Kilo watt hour
MNRE	Ministry of New and Renewable Energy
MoEF	Ministry of Environment and Forests
MSME	Micro, Small and Medium Enterprises
MSW	Municipal solid waste
MW	Mega watt
NMEEE	National mission on enhanced energy efficiency
PAT	Perform achieve and trade

PPA	Power purchase agreement
PV	Photovoltaic
REC	Renewable energy certificate
RPO	Renewable purchase obligation
RFP	Request for proposal
SDA	State Designated Agency
SEC	Specific energy consupmtion
SERC	State Electricity Regulatory Commission
SNA	State Nodal Agency
SIDBI	Small Industries Development Bank of India
SIP	Sponge iron plant
SI	Sponge iron or directly reduced iron (DRI)
SPV	Solar photovoltaic system
SR	Steel re-rolling
SRP	Steel re-rolling plant
TPH	Ton(s) per hour
UNDP	United Nations Development Program
WBERC	West Bengal Electricity Regulatory Commission
WBPCB	West Bengal Pollution Control Board
WBREDA	West Bengal Renewable Energy Development Agency
WBSEDCL	West Bengal State Electricity Distribution Company Limited
WBSIMA	West Bengal Steel Re-rolling Mills Association
WHR	Waste heat recovery
WHRP	Waste heat recovery policy
WHRT	Waste heat recovery technology
WTE	Waste to energy

1 Introduction

Waste heat recovery (WHR) system is the recovery and reuse of by-product heat from a process/facility that would be otherwise rejected to the environment, and is often a valuable cost-effective approach for improving the overall energy efficiency, and surely checks the environmental damage by reducing the impact of thermal pollution to the ecosystem and often reducing the emission as implementation of WHR would reduce the consumption of fossil fuels. Presently, both nationally and internationally, a significant amount of energy resources are lost unutilized as waste heat. According to the global Waste Heat Recovery Market, it is expected that the market will grow at 6.8% compound annual growth rate (CAGR) during 2016–2022 (https://www.psmarketresearch.com/market-analysis/waste-heat-recovery-market).

The market size was USD 44.14 billion in 2015 and is projected to reach USD 65.87 billion by 2021, at a CAGR of 6.90% and considering 2015 as the base year and the forecast period between 2016 and 2021 [14]. Early introduction of the WHR technologies and environmental policies regarding curbing of industrial harmful emissions are the two main propellants of the European WHR market. The Asia-Pacific WHR market is expected to witness the fastest growth at a CAGR of 9.8% during 2016–2022—even higher than the global figure, with a thrust of strong growth in cement industry.

1.1 World Scenario

It is generally accepted that a country's energy supplies are considered of maximum importance at the level of national energy planning. WHR forms a part of the process in many large installations such as power plants, refineries, steel and chemicals. In many industries, WHR is not practised, and its benefits are not tapped. With rising energy costs, the relevance of WHR is gaining importance for most of the medium and smaller installations all over the world.

The estimated US waste heat potential from industrial process heat available at a temperature high enough for power generation, i.e. above 300 °C with present technologies is 6000–8000 MW (Combined Heat and Power Partnership 2012) of electrical generation capacity. Total installed capacity of waste heat power plants in the USA is 557 MW as of 2012 where waste heat power is considered as renewable energy source in 17 the US state's renewable portfolio standards.

The estimated European Union (EU) WHR potential for industrial sector was about 20 TWh of electric energy (Energy Efficiency in Industrial Processes 2017).

1.1.1 WHR Measures in the USA

According to the US Department of Energy [Otis, Paul (2015) CHP Industrial Bottoming and Topping Cycle with Energy Information Administration Survey Data (https://www.eia.gov/workingpapers/pdf/chp-Industrial_81415.pdf accessed on November 15, 2017)], in USA waste heat amounts to approximately 20% to 50% of the energy consumed and is released or radiated from hot equipment surfaces and heated product streams of hot exhaust gas and liquids, as well as through heat conduction, convection, and radiation. Pinch analysis has been said to be successfully used across the full spectrum of chemical process industries. Energy audits has been identified as a tool to deliver 5–10% savings. Further, the bigger energy savings potential has been identified to be on the process side and could be as high as 20–30%.

1.1.2 WHR Measures in China

Currently, after USA, China is the world's second largest energy consumer and expected to become the world's largest energy consumer and the country's industrial sector is responsible for over 70% of the country's energy consumption (Waste Heat Recovery System Market by Application 2016).

To fulfil their energy need, along with other new technologies, waste heat is also a major option to generate electricity. For example, many steel companies in China are using WHR technologies to generate electricity. One of them using captured waste gases and brunt for power and for this they are using specially designed turbines (Waste Heat Recovery System Market by Application 2016).

1.1.3 WHR Measures in Korea

Korea is an energy-intensive nation and is the world's thirteenth largest economy and the seventh largest exporter (China Energy Recovery 2017). As per record, in terms of energy consumption, it is the world's eleventh highest. With their increasing energy consumption situation, from 2012 government focuses on low-carbon and green technologies. The target was to achieve national target of 30% reduction of greenhouse gas emissions by 2020 (China Energy Recovery 2017). With WHR system of combined heat and power operation, both energy-intensive industries and electricity-generating plants showed their potential in Korea. The country also explored electrical application of WHR in direct cooling systems during summer peaks. Continuous development along with implementation of energy remand reduction and mitigation policies, Korea achieved a higher level of green growth and energy independence (www.iea.org/Textbase/npsum/Korea2012SUM.pdf).

1.2 National Scenario

Significant potential exists for power generation through WHR in India, as most of the industrial sectors in India are less energy-efficient compared to other developed nations. In India, WHR is one of the options that industries have to improve their energy efficiency under the Perform Achieve Trade (PAT) scheme under the Bureau of Energy Efficiency. However, PAT is applicable for measures which do not transgress the system boundary, i.e. in this case, the specific industrial unit.

1.2.1 Review of Policies in India

WHR system provides industries a reduction of energy expenditure, reduction of carbon footprint and reduced dependency on non-renewable energy resources. In summary, WHR systems allow a win–win situation of clean energy policy. Therefore, WHR projects prevent greenhouse gases (GHG) emission due to burning of fossil fuels and further eligible to earn carbon credits under the Clean Development Mechanism (CDM). Till August 2017, there are 164 projects registered worldwide under the waste gas/heat utilization scope of CDM framework (International Energy Agency 2012). In India, there are 70 numbers projects registered under waste gas/heat utilization scope. These projects have average annual emission reduction 83,170 (http://suranaind.com/index.php?option=com_content&view=article&id=104&Itemid=161).

Some of the WHR initiatives and projects along with the case studies taken up in recent past are shown in Waste Heat Recovery Manual (WBSEDCL 2014).

1.3 Case Study of Different Industrial WHR Initiatives

1.3.1 Cement Plant-Waste Heat Recovery from Kiln and Cooler

(i) Based on Fig. 1, amount of power which may be generated for the standard sizes of cement plants is: 4500 tpd:5.6 MW

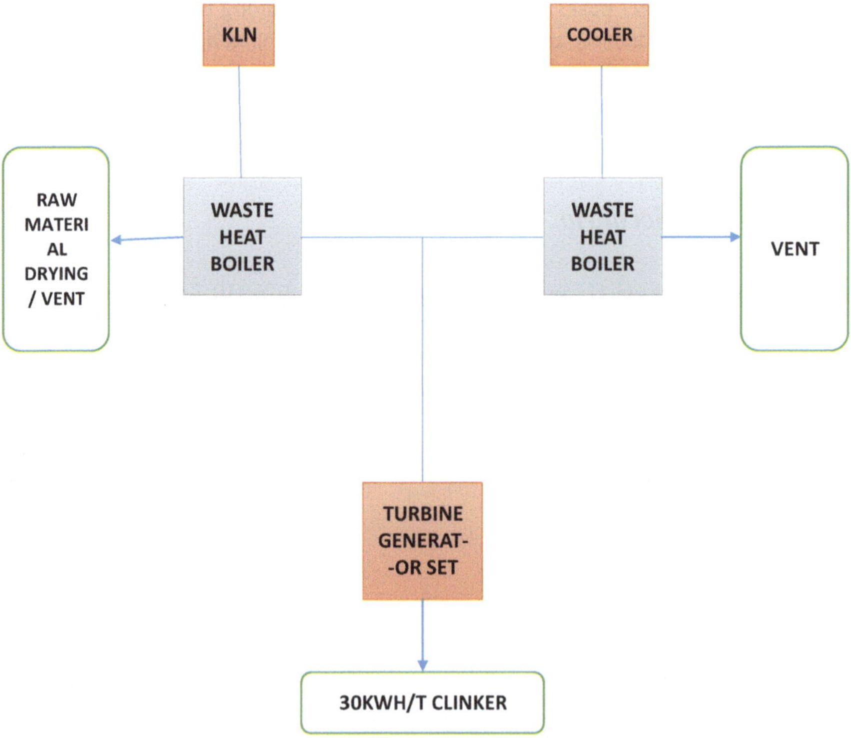

Fig. 1 Schematic of power generation from WHR system of cement plant

(ii) 6000 tpd:7.5 MW
(iii) 10,000 tpd:12.5 MW.

1.3.2 Sponge Iron Plant—WHR in Rotary Kiln Directly Reduced Iron (DRI) Production

Here, directly reduced iron (DRI) is produced in a rotary kiln of 3–6 m diameter and approximately 85 m length installed at an incline. The capacity of the kiln depends on metallization degree and usually does not exceed 225,000–300,000 t/year (Fig. 2). It is operated in counter current flow with solids moving down the incline in opposite direction to the gases. Iron ore and coal are jointly charged to the kiln from the charge end. As the burden progresses down the slope, it is heated up by the gas to a temperature of 1000–1100 °C. A typical retention time of 10–14 h allows the ore to be reduced to DRI by the CO produced by the carbon content of the burden. The pulverized coal/natural gas/oil burner is installed at the discharge end of the kiln. Metallization varies from 83 to 92%, and carbon does not exceed 0.5%. Off-gases produced in the kiln contain considerable sensible and chemical heat, the recovery of which represents the biggest opportunity for improving energy efficiency in this process.

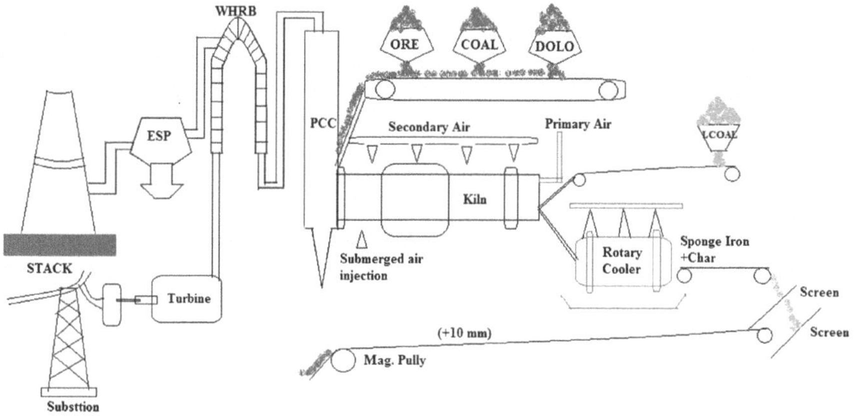

Fig. 2 Schematic of power generation from WHR system of sponge iron plant

2 Benefits of Waste Heat Recovery

Waste heat recovery provides both direct and indirect benefits to the industry.

2.1 *Direct Benefits*

Reduced fuel consumption, which in turn reduces:

(i) To simultaneously meet the demand for electricity and heat in a most cost-effective manner.
(ii) To bring down the specific energy consumption of the plant significantly.
(iii) To bring about best energy-efficient practice in the plant.
(iv) Reduction in pollution: Helps to reduce a number of toxic combustible wastes such as carbon monoxide gas, carbon black off-gases, oil sludge, Acrylonitrile and other plastic chemicals, releasing to atmosphere if/when burnt in the incinerators serves dual purpose, i.e. recovers heat and reduces the environmental pollution levels.
(v) Reduction in equipment sizes: WHR system helps to reduce the fuel consumption, which helps to reduce the flue gas production. So, any industry may gain the benefit of all flue gas handling equipment (like burners, fans, stacks, ducts) sizes reduction.

2.2 *Indirect Benefits*

(i) Mitigates the emission of greenhouse gases, which are affecting the environment adversely.
(ii) Provides economic competitive advantage in the market.
(iii) Reduction in auxiliary energy consumption: reduction in equipment sizes (as mentioned in Sect. 2.1) gives additional benefits of reduction in auxiliary energy consumption (for example electricity for fans, pumps).

3 WHR Opportunities

Sources of waste heat are hot combustion gases discharged to the atmosphere, heated products exiting industrial processes and heat transfer from hot surfaces. Quality of waste heat recovery depends on availability of process parameters, namely—high (around 1000 °C), medium (around 600 °C) and low (around 300 °C). Some typical examples of WHR are placed below.

Combustion Exhaust: Glass-melting furnace (high-quality WHR), cement kiln (high-quality WHR), sponge iron kiln (high-quality WHR), fume incinerator, steel reverberatory furnace (high-quality WHR), boiler (medium-quality WHR), engines (medium-quality WHR).

Process Off-gases: Steel electric arc furnace (high-quality WHR), steel reverberatory furnace (high-quality WHR).

Cooling Water from Thermal Utilities: Furnaces (medium-/low-quality WHR), air compressors (low-quality WHR), internal combustion engines (medium-quality WHR).

Conductive, Convective and Radiative Losses from Thermal Utilities: (medium-/low-quality WHR) from equipment and heated products (medium-quality WHR), hot cokes (medium-quality WHR), blast furnace slags (medium-quality WHR).

Typically, the end use of industrial waste heat is combustion air preheating, boiler feed water preheating, load preheating, power generation, steam generation, water preheating, heat transfer to liquid or gaseous process streams, etc.

4 Types of WHR Technologies

Following are the types of WHR technologies that are working in the sponge iron and steel re-rolling units for recovery of waste heat and power generation from the waste heat thus received in the process itself.

4.1 WHR Boiler with Cogeneration

4.1.1 Technical Description

The WHR boiler is of single drum, water tube natural circulation type and designed to recover heat energy available in the sponge iron rotary kiln exhaust gas. Figure 3 shows a typical WHR system with a boiler iron ore preheating by WHR. Here, the gas cooling temperature is 170 °C and the feed water temperature at the inlet of the economizer is 105 °C. As the mass flow rate of the gas can't be measured accurately, so the steam generation cannot be guaranteed. The boiler is supplemented with all other accessories as conventional water tube boilers, the details of which are given in the DPR. The boiler is controlled by PLC-based SCADA system.

The waste gases generated from the rotary kiln is passed through the boiler to generate the steam. The steam is used for generation of power. The temperature available in flue gases is enough to generate the required steam for power generation through boiler. The boiler is a single drum, water tube natural circulation boiler. The gases leaving the kiln would enter a superheater, and the gases leaving the superheater would enter a set of boiler bank tubes which are expanded into the

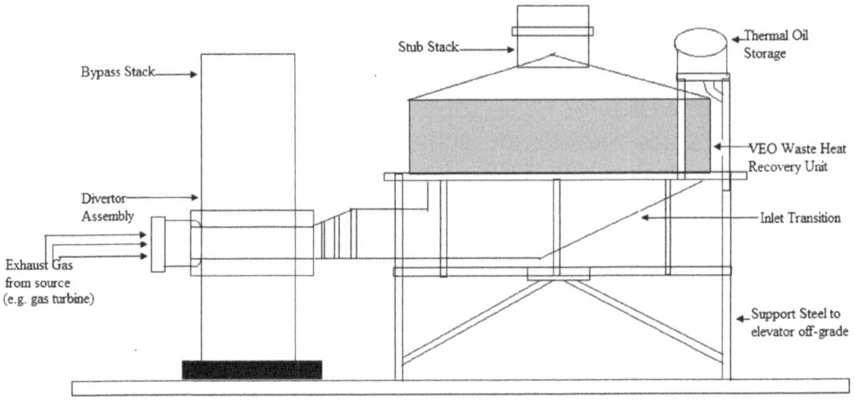

Fig. 3 Typical waste heat recovery system with a boiler iron ore preheating by WHR

Table 1 Cost benefit analysis

S. No.	Particular	Unit	Value
1	Rotary kiln capacity	TPD	100
2	Project cost	INR	91,889,400
3	Electricity saving	kWh/year	14,400,000
4	Monetary benefit	INR	92,160,000
5	Debt equity ratio	Ratio	1.17
6	Simple payback period	Year	1
7	NPV	INR	267,866,375
8	IRR	%age	79.09
9	ROI	%age	28.75
10	DSCR	Ratio	3.8
11	Process downtime	Weeks	8–9

steam and water drums. The gases after passing across the boiler bank would enter a bare tube economizer. This is an inline counter flow economizer and heats up the feed water going to the drum. The gases are reduced to around 152 °C for the economizer. After the economizer, the gases are let into a bag filter which is provided to reduce the dust emission level. The steam turbine is straight condensing turbo generator set type turbine. The generator (alternator) is of 11 kV (Table 1).

4.2 *Iron Ore Preheating*

4.2.1 Technical Description

This technology involves installation of charge preheaters to utilize the waste heat energy content of flue gases released during manufacturing process of the sponge iron in rotary kilns. The charge preheaters will utilize the sensible heat content of flue gases released at 950 °C from the individual kilns to preheat the incoming raw material, i.e. iron ore and dolomite mixture to around 900 °C from ambient temperature of 40 °C. The charge preheaters are of miniature rotary kiln designed to enable adequate mixing of flue gases and raw material mixture for effective heat transfer. Thereafter, the flue gases are released into the atmosphere, complying with the environmental norms. The preheated raw material is then fed to the main rotary kiln for further heating and reduction process to produce sponge iron and hence reduces coal consumption for the same quantity of production. In normal condition, equivalent amount of coal would have been consumed in the main rotary kiln to raise the temperature of the raw material mixture to 900 °C. The proposal thus helps in reduction of coal consumption per tonne of sponge iron produced in the sponge iron kilns, thus leading to greenhouse gas (GHG) emission reductions.

Here, the iron ore passes through the fuel economizer and enters the kiln at about 700 °C where the mixture of coal and dolomite follows the conventional route. The hot gases are blown into the iron ore chamber and the exchange of heat takes place between the iron ore and the hot gases. Thus, the temperature of the iron ore is raised before it enters the kiln. The heat transfer is directly proportional to the surface area which is exposed to the hot gases (Fig. 4; Table 2).

4.3 *1250 kWe WHR ORC Power Project Based on Kiln Exhaust Gas Heat Recovery*

4.3.1 Technical Description

The organic Rankine cycle (ORC) (Fig. 5) is named for its use of an organic, high molecular mass fluid with a liquid–vapour phase change, or boiling point, occurring at a lower temperature than the water–steam phase change. The fluid allows Rankine cycle heat recovery from lower temperature sources such as biomass combustion, industrial waste heat, geothermal heat, solar ponds. The low-temperature heat is converted into useful work that can itself be converted into electricity. A prototype was first developed and exhibited in 1961 by solar engineers Harry Zvi Tabor and Lucien Bronicki. For a 200 TPD sponge iron kiln, the estimated flue gas flow rate would be 28,000 NM^3/hr. Considering average flue gas temperature of 8500 °C, the recoverable heat form this gas would be 5169 kWth

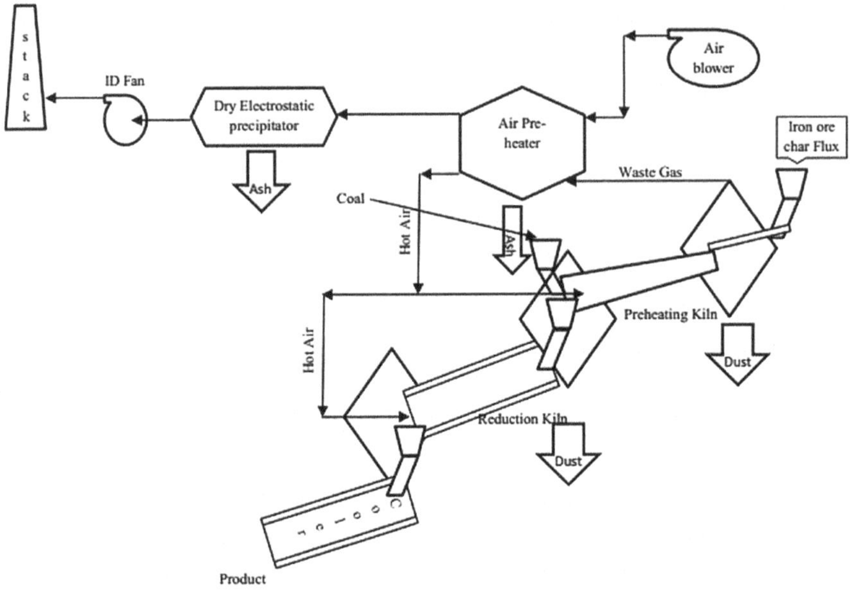

Fig. 4 Schematic of a typical iron ore preheating by WHR of a kiln

Table 2 Cost benefit analysis

S. No.	Particular	Unit	Value
1	Rotary kiln capacity	TPD	100
2	Project cost	INR	20,000,000
3	Fuel saving (coal)	tons/year	408,200
4	Monetary benefit	INR	12,821,482
5	Debt equity ratio	Ratio	2.01
6	Simple payback period	Months	19
7	NPV	%	50.67
8	IRR	%age	50.70
9	ROI	%age	29.21
10	DSCR	Ratio	2.45
11	Process downtime		8

and the gross power recovery using high-efficiency ORC power systems would be 1250 kWe.

Step 1: The working fluid (i.e. water) is pumped from low to high pressure, as the fluid is a liquid at this stage and the pump requires little input energy.

Step 2: The high-pressure liquid enters into AQC and SP boiler where it is heated at constant pressure by using waste hot flue gas from clinker cooler and preheater, respectively, to become a dry saturated vapour.

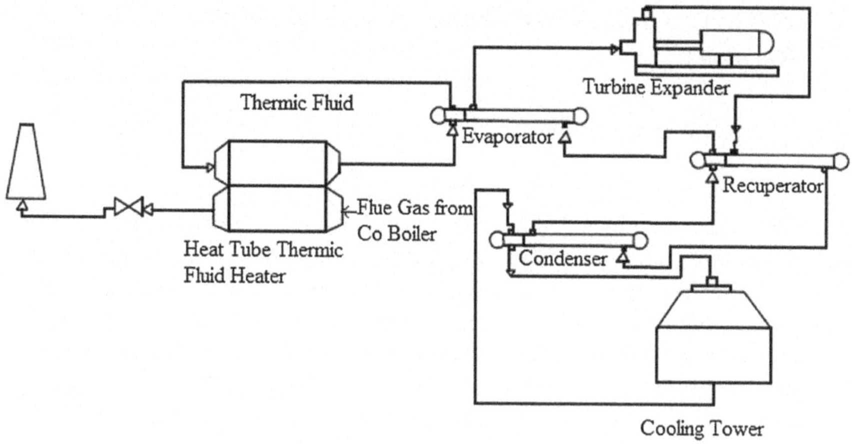

Fig. 5 Schmatic of a typical 1250 kWe organic Rankine cycle

Table 3 Cost benefit analysis

S. No.	Particular	Unit	Value
1	Rotary kiln capacity	TPD	200
2	Project cost	INR	150,000,000
3	Electricity saving	kWh/year	7832590.8
4	Monetary benefit	INR	42,628,581
5	Debt equity ratio	Ratio	10.77
6	Simple payback period	Year	3.52
7	NPV	INR	1,381,849
8	IRR	%age	11.202
9	ROI	%age	36.32
10	DSCR	Ratio	0.72
11	Process downtime	Weeks	2–3

Step 3: The dry saturated vapour expands through a turbine, generating power. This decreases the temperature and pressure of the vapour, and some condensation occurs.

Step 4: The wet vapour then enters a condenser where it is condensed at a constant pressure and temperature to become a saturated liquid. The pressure and temperature of the condenser are fixed by the temperature of the cooling coils as the fluid is undergoing a phase change (Table 3).

4.4 270 kWe ORC Power Project Based on Kiln Shell Heat Recovery

4.4.1 Technical Description

The organic Rankine cycle (ORC) is named for its use of an organic, high molecular mass fluid with a liquid–vapour phase change, or boiling point, occurring at a lower temperature than the water–steam phase change. The fluid allows Rankine cycle heat recovery from lower temperature sources such as biomass combustion, industrial waste heat, geothermal heat, solar ponds. The low-temperature heat is converted into useful work that can itself be converted into electricity. A prototype was first developed and exhibited in 1961 by solar engineers Harry Zvi Tabor and Lucien Bronicki. For a 200 TPD sponge iron kiln, the estimated flue gas flow rate would be 28,000 NM^3/h.

Step 1: The working fluid (i.e. water) is pumped from low to high pressure, as the fluid is a liquid at this stage and the pump requires little input energy.

Step 2: The high-pressure liquid enters into AQC and SP boiler where it is heated at constant pressure by using waste hot flue gas from clinker cooler and preheater, respectively, to become a dry saturated vapour.

Step 3: The dry saturated vapour expands through a turbine, generating power. This decreases the temperature and pressure of the vapour, and some condensation occurs.

Step 4: The wet vapour then enters a condenser where it is condensed at a constant pressure and temperature to become a saturated liquid. The pressure and temperature of the condenser are fixed by the temperature of the cooling coils as the fluid is undergoing a phase change (Fig. 6; Table 4).

5 Overcoming Barriers to WHR Policy

5.1 Technical Barriers

The principal hurdle for WHR systems is the heat recovery itself. Though the waste heat recovery equipment is commercially established and relatively standardized, each heat recovery situation presents unique challenges.

Some of the technical issues that affect WHR economics include:

- The waste heat sources at a plant are dispersed and difficult to reach or consolidate.
- Seasonal operations and low-volume operations reduce the economic benefits of WHP.

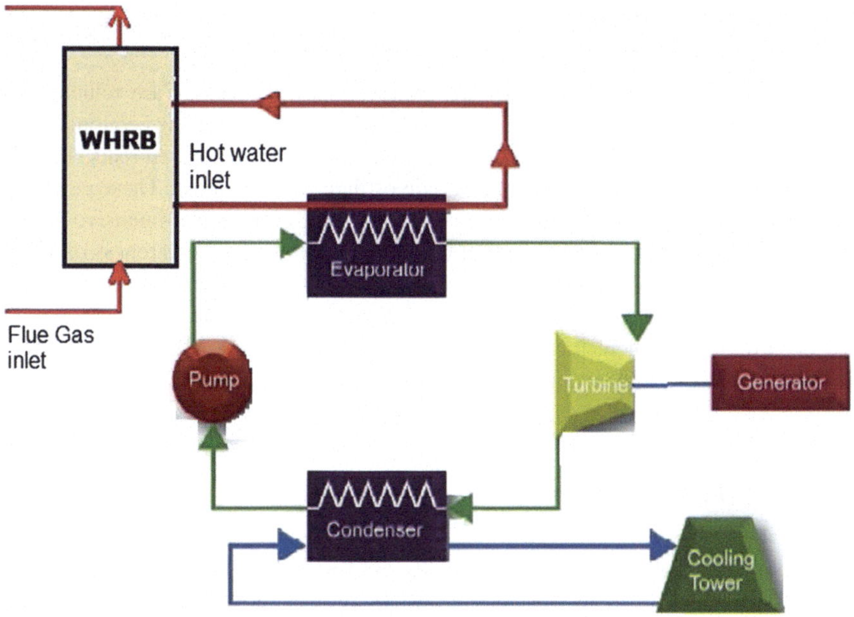

Fig. 6 Scematic of a 270 kWe organic Rankine cycle

Table 4 Cost benefit analysis (IISWBM 2015)

S. No.	Particular	Unit	Value
1	Rotary kiln capacity	TPD	300
2	Project cost	INR	28,386,000
3	Electricity saving	kWh/year	1,422,805
4	Monetary benefit	INR	7,686,650
5	Debt equity ratio	Ratio	14.86
6	Simple payback period	Year	3.62
7	NPV	INR	−3,325,894
8	IRR	%age	8.515
9	ROI	%age	39.05
10	DSCR	Ratio	0.62

- Lack of Technology Experts.
- Waste heat sources often contain chemical and/or mechanical contaminants that impact the complexity, cost and efficiency of the heat recovery process.
- Many times space limitations and equipment configurations make WHP systems difficult or impossible to site economically.

5.2 Business Barriers

As per business situation and company policy, many businesses are reluctant to make investments that do not increase production and ensure their economic survival. They are especially reluctant to take on projects with perceived risks, such as waste heat recovery projects that are outside of their core business. These concerns often lead to unrealistically high project hurdle rates for capital-intensive WHR projects. There is also a general lack of un-organized end-users' awareness of WHP technologies and how to implement them. Few technology demonstrations or case studies currently exist, and most site and project specific. There is resistance to accept new, unproven technology that could potentially jeopardize existing production processes, despite significant potential benefits.

5.3 Regulatory Barriers

Economic issues related to equipment costs and forecasted energy savings may be the greatest determinant of a successful WHP project; however, regulations and policies can have a substantial impact on project economics which may also lead to uncertainty/unavailability of land, water, coal and other natural resources, and even lack of WHR policy. For example, if the power cannot be used on site, projects will require a power purchase agreement with the utility. This is the case with WHP systems on natural gas pipeline compressor stations. Policy should promote open access, wheeling and banking, PPA, IA, settlements and net metering.

5.4 Financial Barriers

Two types of funding options could be incorporated in the policy, as placed below.

Energy Efficiency Fund: In order to finance various initiatives for development of WHR projects, an energy efficiency fund shall be created by the State Nodal Agency.

Budgetary Support: With regard to financing of the WHR-based energy projects, it shall be contemplated by the Government to make a provision of allocating annual budget for the development of WHR policy for the state or the country. A strong visionary approach is to be kept to make the projects cost-effective.

5.5 Coordination

State Nodal Agency, with the help and confidence, as applicable or relevant organization/Government Departments, would assist the Government in implementation of the provisions WHR policy.

Wherever the need arises, the relevant policy documents, directives and regulations are to be referred, such as:

(a) Investment and industrial policy
(b) Government policy on cogeneration and generation of electricity from renewable sources of energy,
(c) Electricity regulatory commission, notification
(d) Government land allotment policy.

SNA or State Nodal Agency will act as a single window for obtaining assistance from all line departments.

5.6 Institutional Framework

Policy has to be institutionalized, and its longevity and acceptance depend on its institutional framework. Relevant institutions of national repute, such as Indian Institutes of Technology, Indian Institute of Social Welfare and Business Management (IISWBM), Indian Institutes of Management, University of Calcutta, Jadavpur University, Indian Institute of Engineering Science and Technology (IIEST or erstwhile Bengal Engineering and Science University), all having their sustained presence and contribution towards development of the State, are expected to render useful services towards sustained implementation of WHR policy.

5.7 Social and Environmental Issues

The developer shall make suitable financial provisions for mitigation of adverse impacts according to the approved environment impact assessment plan and environment management plan.

Potential Avenues for Future Research

The scope of research includes continuous update of the WHR policy in coherence with notational issues and development in other states of the country.

5.8 Amendments and Interpretation of the Policy

Concerned government department will have the power to amend/issue clarifications, if any, on the matter related to the interpretation under this policy in consultation with appropriate authorities from time-to-time.

6 Suggested Road Map Towards Implementation of WHR Policy

A road map is suggested for industries to maximize utilization of waste heat recovery technologies.

The Government would encourage each industry to gain advantage of employing waste heat recovery technologies and techniques. The following guidelines would be useful to such industry units aspiring to get benefited through WHR technologies and techniques:

(a) To take support from top management.
(b) Corporate responsibility for managing energy as a resource is delegated to:
    - Board member
    - Special task force
    - Existing committee, e.g. energy
    - Nominated individual
(c) To understand the scenario in general and WHR potential in particular:
    - Quantify energy usage
    - Identify management strengths and weaknesses
    - Analyse stakeholder's needs
    - Anticipate barrier to implementation
(d) To plan and organize:
    - Develop a policy statement
    - Set objectives and targets
    - Develop detailed action plans
    - Organise roles and responsibilities
(e) Individual responsibility for controlling energy consumption, arising from specific organizational activities involving:
    - Business director
    - Departmental head
    - Budget holder
    - Team leader
    - Nearest end user

(f) Individual responsibility for paying for energy consumed, arising from specific organizational activities involving:

- Business director
- Departmental head
- Budget holder
- Finance director
- Premises manager
- Energy manager

(g) Responsibility for delivering advice and guidance on how consumption and expenditure should be managed is assigned to:

- Special task force
- Energy manager
- Environmental manager
- Premises manager
- External contractor

(h) To implement the outcome of previous exercises:

- Initiate priority actions and investments
- Carry out training and raise awareness
- Integrate energy in general and WHR in particular into business process
- Communicate performance

(i) The formal channels of communication to be used for dealing with energy management issues are:

- Normal line management
- Direct to special task force
- Direct to committee
- Direct to nominated individual

(j) The arrangements for reporting energy performance, including activities undertaken to reduce consumption, are:

- Annual reports to board
- Quarterly reports to committee
- Monthly reports to line manager
- Monthly reports to budget holders
- Published annual energy report

(k) To control and monitor the energy system in general and WHR system in particular:

- Carry out energy management audit.
- The period(s) over which energy performance will be reviewed are:
  - Quarterly internal departmental review
  - Annual in-house corporate review
  - External third-party review every 3–5 years

(l) Review progress:

- The arrangements for auditing performance are:
  - Audit conducted by those responsible for saving energy
  - Audit conducted by other in-house staff
  - Audit conducted by external third party

(m) Seek continuous improvement.

- P–D–C–A cycle often is found out to be suitable to seek continuous improvement from the policy implementation process in general, and for WHRP in particular
- Plan for implementation process
- Do the needful to implement
- Check the feedback and other responses on how implementation could be improved
- Act on modifying the process and if necessary, revise the plan for the next cycle.

7 Conclusion

WHR is mainly a by-product of process heating. The ever-increasing cost of conventional fuel used in process heating also leads to WHR. WHR potential only in the European Union (EU) for industrial sector was estimated to be about 20 TWh of electric energy. In developing economies, the potential will be much more and more relevant for its effect on economy, environment and energy security would directly affect the Human Development Index and thus the global map of progress. Technically, acceptable and commercially viable WHR policy can be considered as an important process optimization tool. Effective WHR policy can further strengthen and standardize conventional process integration with WHR

Hence, WHR policy has immense relevance to the energy efficiency, conservation, economic growth, environmental and social well-being of any modern society. Emphasize has been given in the present chapter to elaborate from the basics to explain step-by-step the need and process for WHR policy with a road map as a ready reference to the stakeholder. Such practice also would help the countries to mutually benefit collaterally, such as public awareness, the process of maturity of other relevant policies. Particularly, for a developing country, like India and China, the effect would be far-reaching and affect global scenario.

References

A Review on Waste Heat Recovery in Industries Department of Mechanical Engineering, G.H. Raisoni College of Engineering Nagpur (MH), India as available at http://www.ijrat.org/downloads/Vol-5/april-2017/paper%20ID-54201728.pdf. Accessed on 15 Nov 2017

Alexis L et al (2017) Institute for Global Environmental Strategies, IGES CDM project database. Accessed on 25 Sept 2017

China Energy Recovery (2017) http://www.chinaenergyrecovery.com/. Accessed on 4 Sept 2017

Combined Heat and Power Partnership (2012) Prepared by U.S. Environmental Protection Agency, May 2012

Energy Efficiency in Industrial Processes (EEIP) (2017) European Union initiative. www.ee-ip.org . Accessed on 20 June 2017

Gupta SK, Kaul SK. Waste heat recovery power plants in cement industry. Holtec Consulting Private Ltd., Gurgaon, India

http://suranaind.com/index.php?option=com_content&view=article&id=104&Itemid=161. Accessed on 12 Mar 2017

http://www.districtenergy.org/blog/2013/05/23/minnesota%E2%80%99s-new-waste-heat-recovery-law/; https://www.revisor.mn.gov/bills/text.php?number=HF729&session_year=2013&session_number=0&version=latest. Accessed on 6 May 2017

https://www.psmarketresearch.com/market-analysis/waste-heat-recovery-market. Accessed on Apr 2017

https://www.psmarketresearch.com/market-analysis/waste-heat-recovery-market

https://cfpub.epa.gov/si/ntislink.cfm?dirEntryID=48935

International Energy Agency (2012) Energy policies of IEA Countries. The Republic of Korea, Review [online]. https://www.iea.org/publications/freepublications/publication/Korea2012_free.pdf. Accessed on 16 Aug 2017

The report "Waste Heat Recovery System Market by Application (Preheating and Steam & Electricity Generation), End-Use Industry (Petroleum Refining, Metal Production, Cement, Chemical, Paper & Pulp, and Textile) - Global Trends & Forecasts to 2021" as found at http://www.marketsandmarkets.com/PressReleases/waste-heat-recovery-system.asp. Accessed on 14 Nov 2017

The State-of-the-Art Clean Technologies (SOACT) (2010) for Steelmaking Handbook (2nd edn)

Waste Heat recovery; Technology and opportunities in US Industry, US Department of Energy, March 2018, as available at https://www1.eere.energy.gov/manufacturing/intensiveprocesses/pdfs/waste_heat_recovery.pdf. Accessed on 15 Nov 2017

Waste Heat Recovery Power Plants in Cement Industry-S K Gupta / S K KaulHoltec Consulting Private Ltd., Gurgaon, India, available at http://www.holtecnet.com/holtecdocs/TechnicalPapers/p_2011_6.pdf. Accessed on 14 Nov 2017

WHR Policy for Sponge Iron and Rolling Mills in West Bengal (2015) The policy prepared by IISWBM, Kolkata, India

www.iea.org/Textbase/npsum/Korea2012SUM.pdf. Accessed on 7 June 2017

Part III
Enactment of Sustainable Energy

Role of Biomass for Sustainable Energy Solution in India

Kuntal Jana, Pinakeswar Mahanta and Sudipta De

Abstract Before the mining of fossil fuels, biomass was the main source of energy for heating and cooking including hot fuel gas production. With rapid industrialization and use of fossil fuels with high calorific values, use of biomass decreased rapidly. However, in the present context of both climate change and energy security, importance of biomass is regaining as a sustainable energy solution. Life cycle of CO_2 emission during the secondary energy production from biomass is lesser than that of the fossil fuels. It is even negative, say for bioenergy with carbon capture. Biomass with good calorific value is abundantly available in countries with rapidly increasing energy demands like India, China, Brazil. Utilization of biomass may increase energy access in rural areas and long-term energy security in India. However, proper selection of biomass, their logistics, and conversion pathways will play an important role. Sustainability of the biomass-based energy system should be assessed for its future feasibility. In this chapter, role of biomass for sustainable energy is assessed specifically for India. Present challenges of energy generation are reviewed. Availability of different types of biomass and logistics are explored. Then, various possible conversion technologies for the conversion of biomass for fuels, electricity are discussed. Possible energy system design for biomass inputs including direct firing, co-firing, gasification and polygeneration are discussed in this chapter. Sustainability assessment of these energy systems is discussed in this chapter. Finally, challenges and prospects related to biomass-based distributed energy solution are presented.

Keywords Biomass · Biofuels · Bioenergy · India · Sustainability

---

K. Jana · P. Mahanta (✉)
Centre for Energy, Indian Institute of Technology Guwahati, Guwahati 781039, India
e-mail: pinak@iitg.ernet.in

P. Mahanta
Department of Mechanical Engineering, Indian Institute of Technology Guwahati, Guwahati 781039, India

S. De
Department of Mechanical Engineering, Jadavpur University, Kolkata 700032, India

A. Gautam et al. (eds.), *Sustainable Energy and Transportation*,
Energy, Environment, and Sustainability,
https://doi.org/10.1007/978-981-10-7509-4_12

1 Introduction

Energy demand of India is increasing to maintain the present economic growth. Maintaining the energy demand and clean and affordable energy access to all, specifically in rural areas, are the major challenges to policy makers. Apart from these challenges, global warming due to climate change is another critical issue associated with energy sector. As energy sector of India is overwhelmingly dependent on coal and oil, energy supply is highly carbon intensive. Hence, reduction of greenhouse gas (GHG) emission is a challenging task for India as committed in Paris climate accord (COP21) (UNFCCC 2015). As India mostly depends on fossil resources in spite of limited reserve of fossil resources (mostly oil and gas), energy security of India is weak in a longer time frame. In case of transportation fuel, India is highly dependent on imported crude oil. In 2011–12, import of crude oil was 171,729 thousand tonnes. Hence, significant amount of financial burden and insecurity due to geopolitical issues become barriers to India's sustainable energy future.

Biomass may play an important role in this regard. Plenty of under-utilized or even unutilized biomass is available in India. Utilization of biomass as energy resource may be a sustainable energy solution for India. In rural and remote areas where grid connectivity is not economically feasible or geographically feasible, energy systems with locally available biomass becomes an effective solution. Hence, it increases energy access leading to better energy security of the locality. As biomass is considered as carbon neutral fuel, energy supply by using biomass is less carbon intensive than fossil fuel-based systems. Apart from electricity, biomass also provides partial solution for transportation fuel need. Alternative fuels such as biodiesel, bioethanol, mixed alcohol, biogas can be produced from different biomass to replace conventional diesel and gasoline. This increases energy security of the country and simultaneously reduces the carbon footprint of transport sector.

From technology development to supply chain logistics, there are many challenges associated with successful deployment of bioenergy. Promotion of sustainable bioenergy through technology development and policy making leads to long-term energy sustainability of a country. It provides access to secured clean energy for a sustainable future.

2 Challenges and Policies

Economy of India is growing fast. As a result, demand of secondary energy is increasing rapidly to meet this economic growth. During the year 2015/16, the shortage of electricity was 2.1% and the peak deficit was 3.2% (CEA 2016). Hence, capacity addition is required to meet the energy demand. Inclusive growth and access to energy for all are another challenge to the policy maker. More than 230 million people have no access to electricity in the India. However, 841 million of population of India rely on traditional use of biomass for cooking (IEA 2015). As cooking by combustion of biomass is detrimental to health, clean cooking is needed through various options like LPG, electricity or improved burning of solid fuels.

Energy security is another significant challenge for India. India strongly depends on imported crude oil. Domestic production of oil and gas decreased in the previous year (i.e., 2015/16). On the other hand, import of crude oil increased significantly to 199 million tonnes in 2015/16 (MoPNG 2016). Hence, India needs to search for an alternative option to replace the petroleum to reduce the dependency on imported oil. Fuel from biomass may be a promising option to replace imported oil.

Greenhouse gas emission associated with energy sector is another growing concern regarding survival of life on earth due to global warming (IPCC 2016). As coal contributes to the maximum power generation in India, Indian power sector is highly carbon intensive (MoSPI 2015). On the other hand, due to combustion of fossil oil, transport sector is also highly carbon intensive. Alternate to the coal is possible by harnessing suitable renewable energy. Biofuel can be an option to reduce the carbon intensity of transport sector.

To promote renewable energy and bioenergy, the Government of India has adopted several policies for renewable energy, rural electrification and biomass-based projects (TERI 2016). Most significant of those are as follows:

(i) National Biofuel Policy: in this policy, focus for biofuel production is based on feedstocks without affecting the food security, i.e. mainly from the residues of forest and agriculture, molasses, etc. In this policy, farmers, landless labourers are encouraged to undertake plantations that provide the feedstock for biofuel. Financial incentives are also required to promote biofuels. Research, demonstration and development will be supported for all aspects of biofuel through this policy.

(ii) National Biogas and Manure Management Programme: this scheme is for setting-up of family-type biogas plants, mainly for rural and semi-urban households.

(iii) National Biomass Cookstove Programme: the government launched the National Biomass Cookstove Programme for implementation in the Twelfth Five-year Plan period. In this policy, cookstoves for clean cooking are developed and promoted.

(iv) The Government of India has launched various programmes for the electrification of rural areas in the country, including the Pradhan Mantri Gramodaya Yojana in 2001/02, Accelerated Rural Electrification Programme in 2003/04 and Accelerated Electrification of One Lakh Villages and One Crore Households in 2004/05.

(v) Rural Electrification Policy, 2006: this policy promotes the usage of renewable technologies for electrification in remote areas where centralized grid connection is not technically or financially feasible.

Table 1 Central financial assistance and fiscal incentives for biomass power project and bagasse co-generation projects (MNRE 2016a, b)

Project type	Capital subsidy	Capital subsidy
	Special category states (north-east region, Sikkim, J&K, HP and Uttaranchal)	Other states
Biomass power projects	Rs. 2.5 million × (capacity in MW) (maximum support of Rs. 15 million per project)	Rs. 2.0 million × (capacity in MW) (maximum support of Rs. 15 million per project)
Bagasse co-generation by private sugar mills	Rs. 1.8 million × (capacity in MW) (maximum support of Rs. 15 million per project)	Rs. 15 million × (capacity in MW) (maximum support of Rs. 15 million per project)
Bagasse co-generation projects by cooperative/public sector sugar mills at the rate of per MW of surplus power (maximum support Rs. 60 million per project)		
40 bar and above	Rs. 4.0 million	Rs. 4.0 million
60 bar and above	Rs. 5.0 million	Rs. 5.0 million
80 bar and above	Rs. 6.0 million	Rs. 6.0 million

(vi) Remote Village Electrification Programme: this programme is for basic lighting facilities in un-electrified villages and hamlets that are not included in the Rajiv Gandhi Grameen Vidyutikaran Yojana (RGGVY). Presently, it is discontinued and a new scheme: 'Deendayal Upadhyaya Gram Jyoti Yojana (DDUGJY)' has started.

To promote biomass-based power and co-generation, the Government of India has policy for financial assistance and incentives. The amount of subsidy for these projects in different states is given in the Table 1. Apart from these subsidies, income tax holidays, concession on excise duty, exemption in sales tax are available from the Government of India.

3 Biomass as Energy Resource

Biomass is one of the greatest sources of energy in the forms of food, fuel, fibre etc. It is widely used for cooking in households, as low-cost housing materials, etc., in developing countries. Presently, the goal is to produce secondary energy, i.e. electricity and fuels from biomass due to lower carbon emission compared to fossil fuels. It is also a good source of fuel for distributed energy solution in rural areas (Hiloidhari et al. 2014).

Table 2 Energy potential from residues (EJ) (Bauen et al. 2004)

Region	Crop	Forest	Dung	Total
World	24	36	10	70
North America	4	9	0.7	14
Europe	3	5	1	9
Asia-Pacific	0.8	0.8	0.4	2

Theoretically, the potential of biomass is huge, though quantifying the potential is difficult due to many unstable parameters. Complex nature of biomass production, economics of production, varying productivity rate, wide range of conversion technologies, ecological, social and many other parameters makes it difficult to estimate the future potential of biomass (Rosillo-Calle et al. 2007). Also, depending on the socio-economic condition of local people, availability may vary. In theory, 800 EJ of energy can be harvested from agricultural land without affecting the food supply of the world (Faaij et al. 2002). Bauen et al. (2004) estimated the potential of residues for energy production as given in Table 2. It is noted that significant amount of energy can be obtained from biomass residues. Hence, biomass has the potential to be a significant energy resource all over the world.

Ministry of New and Renewable Energy (MNRE) of Government of India assessed the potential of biomass for energy production in India. About 32% of country's primary energy demand is met by biomass. More than 70% population of India depends on biomass for energy needs mainly for cooking and heating. According to MNRE, estimated production of biomass in India is about 500 million tonnes per year. However, out of this, 120–150 million tones of biomass is surplus. This biomass may be utilized for secondary energy production in the country. By utilizing surplus biomass, 18,000 MW power can be produced and 7000 MW of co-generation by using sugarcane bagasse is possible (MNRE 2016a, b). State-wise potential of surplus biomass and bagasse-based co-generation is shown in Fig. 1. From the figure, it is noted that the maximum power and co-generation can be obtained from the state of Punjab, Maharashtra and Uttar Pradesh. This is due to maximum availability of agricultural residue and bagasse from sugar mills in these states.

To utilize the existing potential of bioenergy, MNRE has targeted to commission around 10,000 MW of power generation by 2022. The region-wise target is shown in Fig. 2. The maximum power is to be generated from northern region and then from western region. Apart from power, biogas may also be a good source of energy for cooking or combustion engine. The target of the National Biogas and Manure Management Programme is to set up biogas plant for rural and semi-urban households. Locally available digestible biomass like cow dung, kitchen waste is the input to this biogas plant for anaerobic digestion. In this programme, biogas plant is set up to provide fuel for clean cooking and lighting and digested slurry may be used as fertilizers. A total of 4.75 million biogas plants have already been installed till 31.03.2014 in India. The state-wise potential and cumulative achievement as on March 2015 of number of biogas plants are shown in Fig. 3.

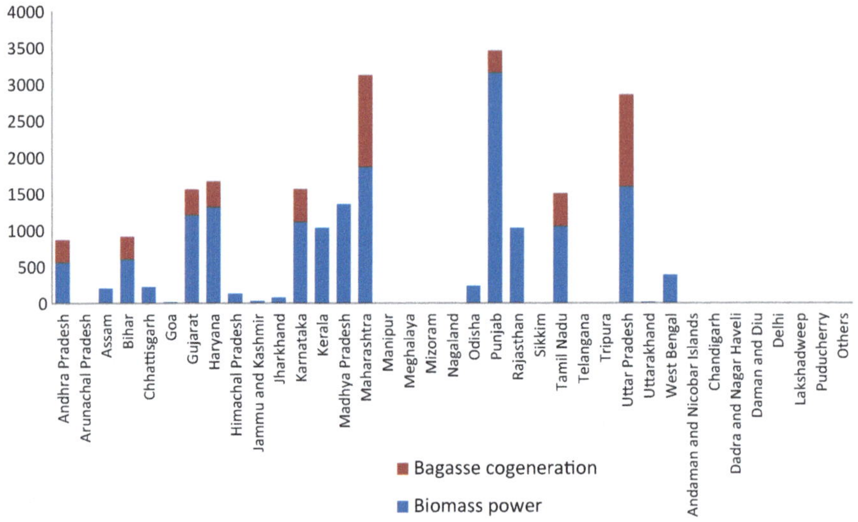

Fig. 1 Sate-wise potential of biomass-based power and bagasse-based co-generation (MW) (MNRE 2016a)

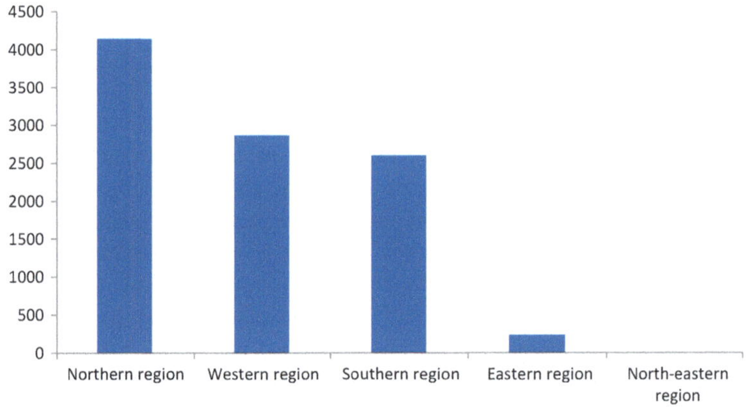

Fig. 2 Tentative region-wise biomass power (MW) target to be achieved by 2022 (MNRE 2016b)

4 Classifications of Biomass

From the previous section, it is observed that biomass has significant potential as primary source of energy. Potential of biomass varies according to its physical properties and chemical composition (Lestander 2012). Hence, utilization pathways for various biomasses are different according to source and classification.

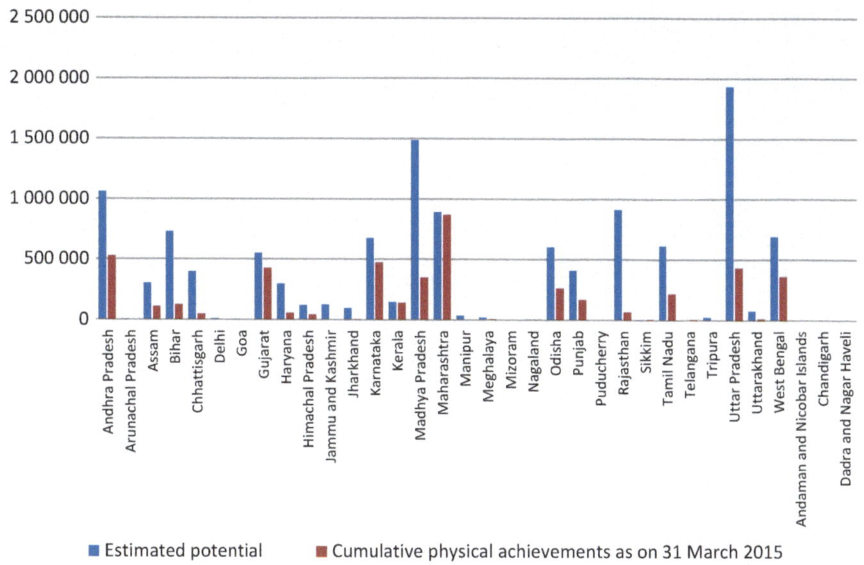

Fig. 3 Number of plants under National Biogas and Manure Management Programme (MNRE 2016a)

Biomass can be classified in different ways. Depending upon the types, it can be divided in three ways as follows:

- *Farm products*: corn, sugar cane, sugar beet, wheat, etc., produce ethanol; rape seed, soybean, palm sunflower seed, jatropha, etc., produce biodiesel.
- *Ligno-cellulosic materials*: straw or cereal plants, husk, wood, scrap, slash, etc., produces ethanol, bioliquid and syngas or producer gas. This gas can be utilized for various products in the downstream processes as discussed subsequently.
- *Microalgae*: *Scenedesmus obliquus*, *Euglena* sp., Cyanobacteria are the example of microalgae. Microalgae are the best as the potential source for oil extraction than the traditional biofuel crops. This oil can be used as a replacement of diesel.

Depending upon the source of availability in India, biomass may be classified into five types as shown in Fig. 4. Depending on the source, physical and thermo-chemical properties may vary. Examples of various types of biomass are given in this figure. Performance of biomass-based energy system depends on the thermo-chemical properties of biomass. In Table 3, thermo-chemical properties of representative biomass are given. Sugarcane bagasse, cotton stalks represent the agro-industrial waste; rice straw, wheat straw, rice husk represent the agricultural waste; Eucalyptus represent the energy crops. From the figure, it is noted that thermo-chemical properties of biomass vary with the source of biomass.

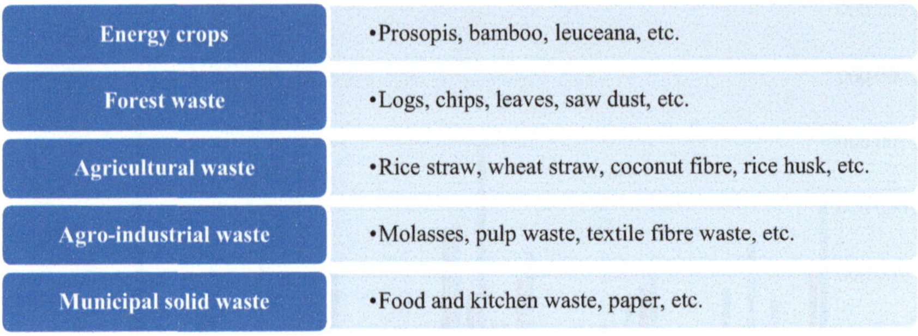

Fig. 4 Types of biomass resources available in India

Table 3 Thermo-chemical characteristics of selected biomass (Domalski et al. 1986)

	Sugarcane bagasse	Rice straw (weathered)	Wheat straw	Eucalyptus	Coconut fibre	Rice husk	Cotton stalks
Higher calorific value (MJ/kg)	17.33	14.56	17.51	19.42	20.05	16.14	15.83
Lower calorific value (MJ/kg)	16.24	13.76	16.49	18.23	19.02	15.27	14.79
FC	14.95	13.33	19.80	17.82	29.7	16.67	17.3
VCM	73.78	62.31	71.30	81.42	66.58	65.47	65.4
Ash	11.27	24.36	8.90	0.76	3.72	17.86	17.3
C	44.80	34.6	43.20	49.0	50.29	40.96	39.47
H	5.35	3.93	5.00	5.87	5.05	4.3	5.07
O	39.55	35.38	39.40	43.97	39.63	35.86	39.14
N	0.38	0.93	0.61	0.30	0.45	0.4	1.2
S	0.01	0.16	0.11	0.01	0.16	0.02	0.02
Cl	0.12	0	0.28	0.13	0.28	0.12	0
Ash	9.79	25	11.40	0.72	4.14	18.34	15.1

5 Availability and Logistics of Biomass

Biomass is one of the most abundantly available resources in India. To ensure long-run operation, availability and logistics of biomass should be uninterrupted. Biomass from crop production is one of the greatest sources for long-term planning of biomass-based project as given in Table 2. It does not affect the food security and ecological balance. However, the amount of residue production depends on cropped area and residue yield as shown in Eq. (1). Worldwide area for cultivation of different crops is shown in Fig. 5a. It is noted that maximum area is cultivated for

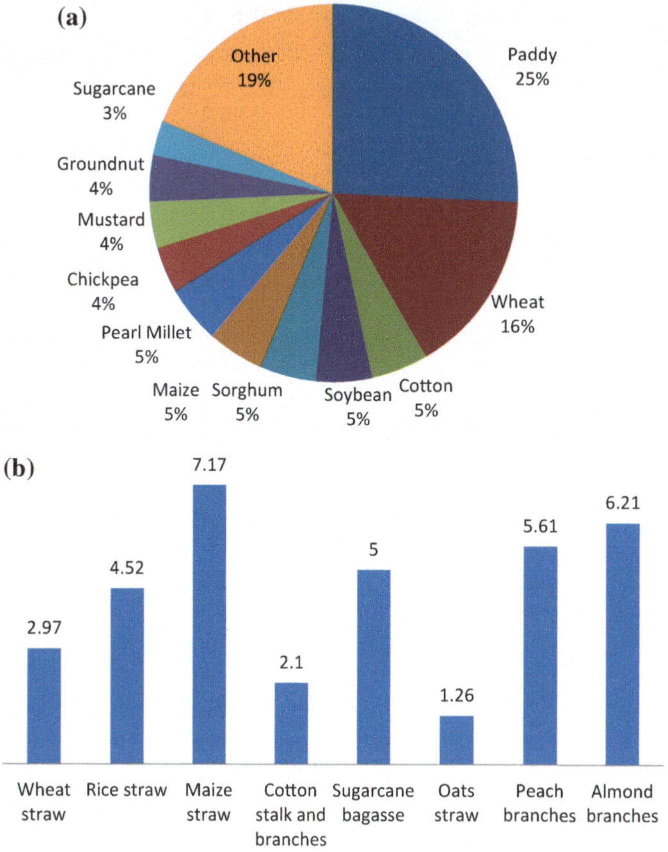

Fig. 5 **a** Percentage of worldwide area for cultivation of different crops (Cardoen et al. 2015). **b** Residue yield (t/ha) of representative biomass

paddy and wheat. Hence, energy system design with input from residue of paddy and wheat production will be sustainable solution from the viewpoint of availability. However, residue yield of wheat straw and rice straw along with other representative biomass is shown in Fig. 5b. Hence, residue production from different biomass may be estimated.

$$\text{Residue production} = \text{Residue yield} \times \text{Gross cropped area} \quad (1)$$

Similarly, in India, agriculture contributes biomass production mostly. State-wise area for crop production, generation of biomass and surplus of biomass in India is given in Table 4 (MNRE 2016a, b). It is noted that more than 500 million ton biomass is produced per year, and out of that, more than 130 million ton of biomass is surplus. According to the Ministry of New and Renewable Energy (MNRE) of India, estimated potential of biomass-based power and bagasse

Table 4 State-wise availability of biomass in India (MNRE 2016a, b)

State	Area (kha)	Crop production (kt/y)	Biomass generation (kt/y)	Biomass surplus (kt/y)
Andhra Pradesh	2540.2	3232.0	8301.7	1172.8
Assam	2633.1	6075.7	6896.3	1398.4
Bihar	5833.1	13817.8	20441.8	4286.2
Chattisgarh	3815.5	6142.8	10123.7	1907.8
Goa	156.3	554.7	827.2	129.9
Gujarat	6512.9	20627.0	24164.4	7505.5
Haryana	4890.2	13520.0	26160.9	9796.1
Himachal Pradesh	710.3	1329.2	2668.2	988.3
Jammu and Kashmir	368.7	648.7	1198.7	237.7
Jharkhand	1299.8	1509.0	2191.2	567.7
Karnataka	7277.3	38638.5	23766.8	6400.6
Kerala	2041.7	9749.7	9420.5	5702.6
Madhya Pradesh	9937.0	14166.9	26499.6	8033.3
Maharashtra	15278.3	51343.3	36804.4	11803.9
Manipur	72.6	159.4	318.8	31.9
Meghalaya	0.8	14.0	42.0	8.4
Nagaland	27.1	87.6	149.2	27.2
Orissa	2436.6	3633.3	5350.4	1163.4
Punjab	6693.5	27813.7	46339.8	21267.0
Rajasthan	12537.5	93654.8	204887.6	35531.1
Tamil Nadu	2454.0	24544.6	15976.6	6658.7
Uttar Pradesh	12628.2	46800.8	50416.7	11725.9
Uttaranchal	66.4	135.8	159.9	51.6
West Bengal	5575.6	21062.8	23316.0	2959.7
Total	105786.8	399262.1	546422.6	139355.8

co-generation is shown in the Fig. 1. From these estimations, it is noted that potential of biomass as energy resource is reasonable.

To maintain the continuous supply of biomass to the energy conversion plant, logistics of biomass should be properly designed (Gold and Seuring 2011). In Fig. 6, components of a generic logistics for biomass are shown. From the figure, it is observed that the biomass collected from the field is processed and loaded for transportation. Baler may be used for this purpose. Subsequently, biomass is transported to the plant. Then, the same is fed to the energy conversion plant by using suitable arrangements, say, conveyor belt. Hence, the total supply chain should be maintained properly for uninterrupted operation.

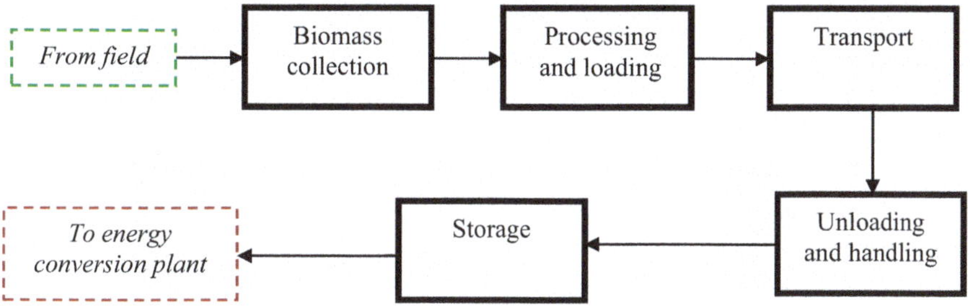

Fig. 6 A generic logistics of biomass

6 Secondary Outputs from Biomass

There are many products and utilities that can be obtained from biomass as indicated by Fig. 7 (Ragauskas et al. 2006). It is known that biomass is primarily used for heating. Due to its good calorific value, environmental benefits and availability, biomass can be used for power generation by suitable energy conversion technology. Biomass is a good source of carbon (C) and hydrogen (H). Hence, apart from heating and power, synthetic fuels may be produced from biomass by various technologies as discussed later. Biomass is also good source of fibre, construction material and chemicals.

Importance of biofuel is increasing in the present context of limited petroleum resource, carbon emission and energy security. Transport sector of India mostly depends on oil. For example, high-speed diesel oil consumption in the transport

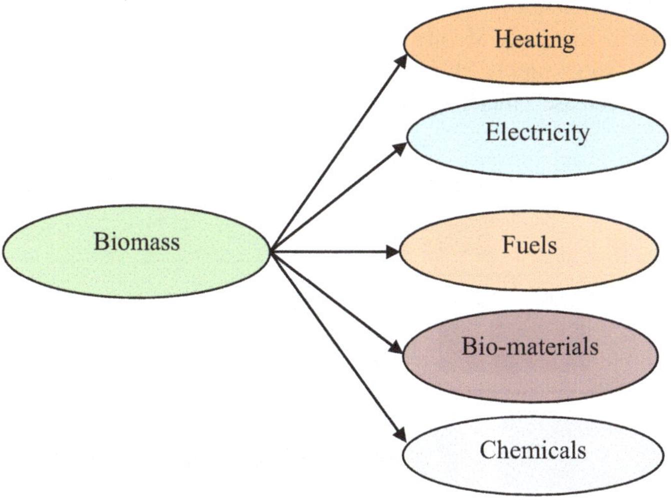

Fig. 7 Biomass to secondary products

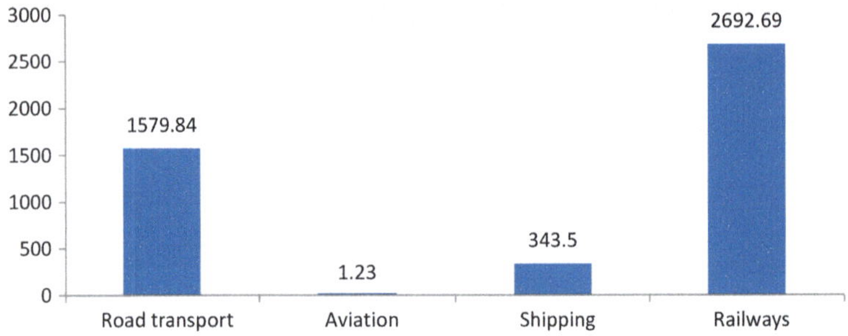

Fig. 8 Consumption (in thousand tonnes) of high-speed diesel oil in transport sector of India in 2014/15 (MoPNG 2016)

sector of India is shown in Fig. 8 (MoPNG 2016). Maximum amount of diesel is consumed in railway and road transport. However, this oil is mainly imported; hence, it reduces the energy security and increases the environmental and economical burden. Hence, Government of India made the National Biofuel Policy to promote biofuel production and use as discussed earlier.

Biofuel can be classified as primary and secondary biofuels as shown in Fig. 9 (Nigam and Singh 2011). Primary biofuel is directly used for mainly cooking say, firewood, crop residues, etc. Secondary biofuels are produced through several processing, and it is mainly used as fuel for engine. Secondary biofuels may be classified as first generation, second generation and third generation. In first-generation biofuel, seeds, grains, sugars are the input biomass. Bioethanol or butanol is produced by fermentation of starch from grains or sugars. Biodiesel is produced by transesterification of rapeseed, sunflower, jatropha, etc. Hence, first-generation biofuels affect the food security. In second-generation biofuels, lingo-cellulosic biomass is used to produce alcohol, Fischer–Tropsch fuels, dimethyl ether or biomethane. Mainly agro-residues, forest residues are used as substrate. Hence, food supply chain is not affected by this generation of biofuels. However, to increase efficiency, reduce cost and increase production rate, third

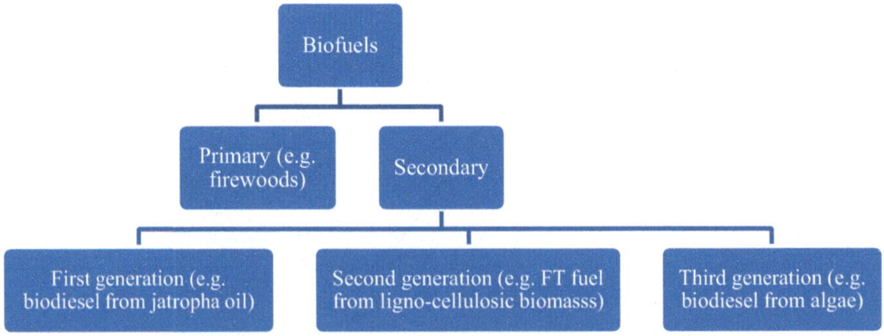

Fig. 9 Classification of biofuels

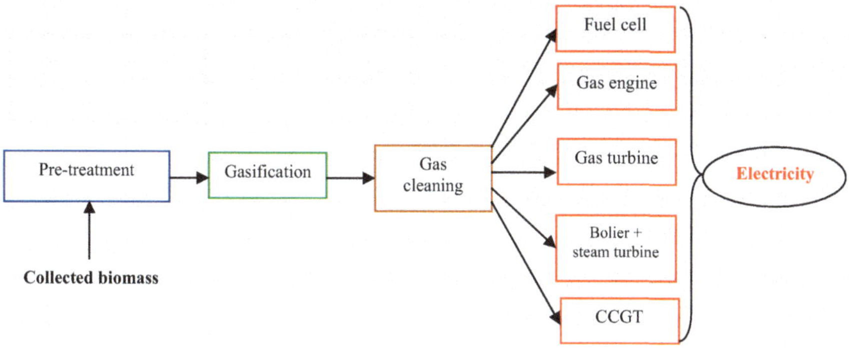

Fig. 10 Route of biomass to electricity based on gasification

generation of biofuel is targeted. In this generation, algae, sea weeds, microbes, etc., are used to produce biodiesel, bioethanol and hydrogen.

As stated earlier, biomass can be utilized for electricity production through gasification and utilization of produced gas in different energy conversion devices as shown in Fig. 10 (Asadullah 2014). Solid oxide fuel cell can produce electricity directly from the syngas with specified composition. However, syngas needs to be cleaned within specified limit. It can be used in gas turbine or gas engine as fuel for combustion. Power can be produced by external combustion of produced gas in boiler and utilization of generated steam in steam turbine or it can be used in combined cycle gas turbine (CCGT) to improve the efficiency. For these cases, tar and other impurities of the gas should be within the specified limit. However, selection of prime movers depends on scale, technologies and economics of power generation.

7 Conversion Pathways

To produce the secondary outputs as stated earlier, there are several technologies available. Biomass is not available in usable form in nature. Hence, it needs to be processed and converted into suitable form. Two basic conversion technologies are available, i.e. thermo-chemical route and biochemical route.

7.1 Thermo-chemical Route

Gasification is one of the most common thermo-chemical processes for conversion of biomass. A basic schematic of gasification is shown in Fig. 11. In this process, biomass is converted to gaseous products (e.g. syngas, producer gas) with

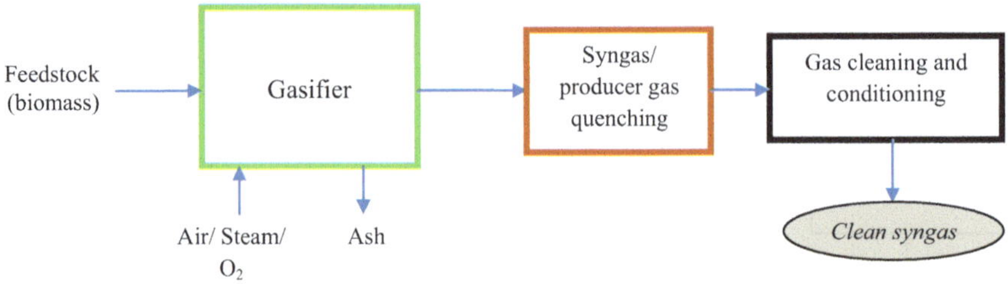

Fig. 11 Basic schematic of gasification process

reasonable calorific value. Air, steam or O_2 are used as oxidizing agent for conversion process. The reactions occur in the gasifier are given below.

$2C + O_2 \leftrightarrow 2CO$	Partial combustion of char
$C + H_2O \leftrightarrow H_2 + CO$	Water gas reaction
$2H_2 + O_2 \leftrightarrow 2H_2O$	Hydrogen combustion
$CO + H_2O \leftrightarrow CO_2 + H_2$	CO shift reaction
$CO_2 + C \leftrightarrow 2CO$	Boudouard reaction
$CH_4 + H_2O \leftrightarrow CO + 3H_2$	Steam-methane reforming
$C + 2H_2 \leftrightarrow CH_4$	Methanation reaction
$H_2 + S \leftrightarrow H_2S$	H_2S formation
$2N_2 + 3H_2 \leftrightarrow 2NH_3$	NH_3 formation

Hence, the produced gas contains H_2, CO, CO_2, CH_4, tar, sulphur compounds, etc. After gasifying process, syngas is cooled for downstream process. Then, it is cleaned and conditioned according to downstream utilization of syngas. A schematic diagram of the gasification process is illustrated in Fig. 11.

Depending upon the type of gasifier, composition of feedstock and thermodynamic states during the conversion process, the product gases may vary. These gasification reactors are selected according to the types of biomass input and desired product and also from economic point of view. Gasifiers are mainly classified into three groups, i.e. moving bed (or fixed bed) gasifier, fluidized bed gasifier, entrained flow gasifier. Classification of different types of gasifier is shown in Fig. 12 (Basu 2010).

In moving bed/fixed bed gasifiers, feedstock moves to the downward direction and gasified by gasifying agent (air, steam, oxygen). However, the direction of air, oxygen flow (i.e. gasifying agent) depends on design of the gasifier. In moving bed process, the oxygen consumption is lower comparative to the other processes. In downdraft gasifier, produced gas and solids both move downwards. In this gasifier, gasifying agent is fed into the lower section with biomass. The pyrolysis and combustion products flow downwards. Hot gas and remaining char move downwards. Tar content in the product gas is lower compared to other gasifiers. However, calorific value of syngas is lesser than that of other types of gasifiers. In fluidized bed, good mixing between oxidant and biomass augments heat and mass

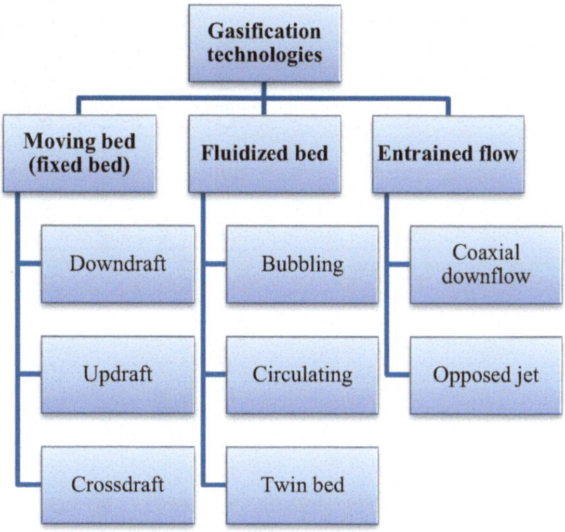

Fig. 12 Types of gasification technologies

transfer process. The gasifying medium is fed to the bottom of the gasifier while biomass is fed from the top. This reactor operates at less than 'ash agglomeration temperature' to avoid the melting and agglomeration of ash. Variety of fuels can be gasified in this process due to good mixing. In this gasifier, tar production can be minimized by using suitable bed materials. Entrained flow gasifier operates at higher pressure and temperature. It is suitable for coal gasification and IGCC plants. Summary of three common type gasifiers is given in Table 5.

7.1.1 Produced Gas Cleaning and Conditioning

After gasification process, this gas is to be cleaned and conditioned for downstream application. Removal or reduction of tar within specified limit is a challenging task, and various technologies are in developing stages like catalytic cracking, thermal cracking, etc. Acid gas removal unit may be used for removal of CO_2 and sulphur compounds depending on the downstream utilization procedure. Different processes are used for acid gas removal based on following principles:

- Absorption by chemical washes using amines, potassium carbonate; by physical washes using rectisol, selexol, purisol; by physico-chemical washes using sulfinol, amisol.
- Adsorption by molecular sieves, pressure swing adsorption, zinc oxide/copper oxide.
- Diffusion through permeable or semi-permeable membrane.
- Chemical conversion by catalyst may be done using CO shift, high-temperature shift, low-temperature shift, COS hydrolysis, methanation

Table 5 Summary of common type gasifiers (Basu 2010)

Type of gasifier	Feed type	Feed size (mm)	Product gas temperature (°C)	Feedstock flexibility	Pressure	Steam requirement	Cold gas efficiency (%)	Capacity	Challenges
Fixed bed	Pellet, briquette	<51	450–650	Less	Less	High	80	Small	Tar reduction
Fluidized bed	Small particles	<6	800–1000	High	Low–high	Moderate	89	Medium	Carbon conversion
Entrained flow	Fine particles	<0.15	>1260	For coal only	high	Low	80	Large	Raw syngas cooling

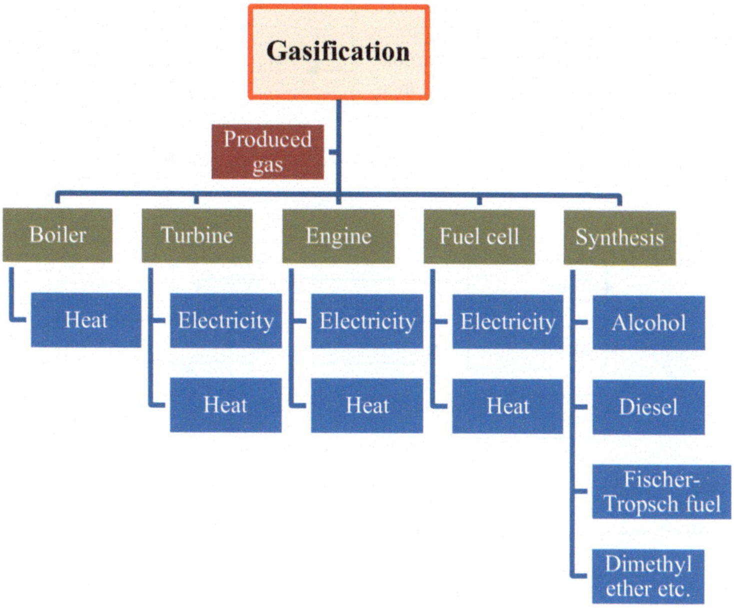

Fig. 13 Potential products of gasification process

The produced gases of the gasifier can be used in multiple options after suitable cleaning and conditioning as shown in Fig. 13. It may be used directly in boiler for heating. To produce power and heating, it may be used in gas turbine, gas engine or fuel cell. Syngas may produce different liquid fuels through various synthesis processes like Fischer–Tropsch synthesis, mixed alcohol synthesis or ethanol synthesis (Jana and De 2017).

Apart from gasification, there are other thermo-chemical processes for conversion of biomass as shown in Fig. 14 (Dufour 2016). In these processes, biomass is converted to bio-oil which is used as gasoline or diesel in combustion engine.

7.2 *Biochemical Route*

One of the most common liquid fuels that can be produced through biochemical conversion of biomass is ethanol. Routes of ethanol production in biochemical process are shown in Fig. 15a. Ethanol can be produced from sugar-based biomass, say sugarcane, directly by fermentation process. In this process, production yield of ethanol is more and more amount of GHG emission that can be reduced compare to other processes (Nigam and Singh 2011). Starch is another source of biomass for ethanol production. Maize, wheat may be used for this purpose. Before the fermentation process, hydrolysis of starch is required. Inputs to these processes are

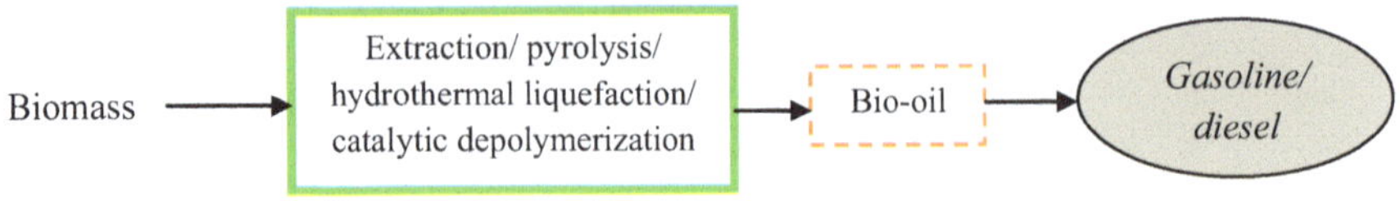

Fig. 14 Thermo-chemical conversion of biomass (except gasification process)

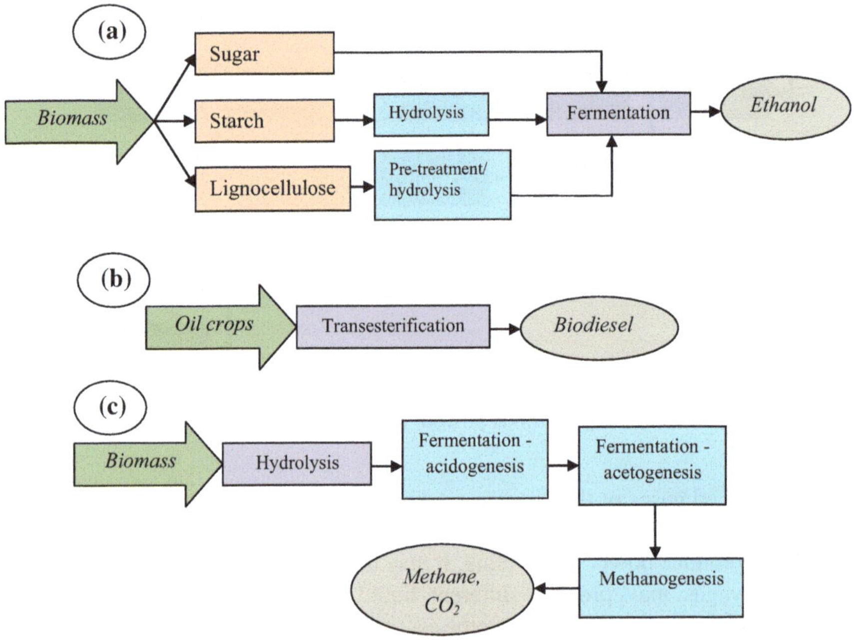

Fig. 15 Biochemical conversion of biomass to biofuels **a** fermentation, **b** transesterification, **c** digestion

food stuff. Hence, it may affect the food security of the country. However, ethanol may be produced from ligno-cellulosic biomass, though it is more complex method. In this process, first biomass is pretreated and then it is saccharified to release the fermentable sugars. After that, fermentation and distillation occur to obtain the desired ethanol.

As vegetable oil cannot be used directly in engine due to its higher viscosity and low volatility, it needs to be processed as shown in Fig. 15b. Transesterification is a solution for this purpose. Transesterification occurs with alcohol in the presence of catalyst to obtain diesel and glycerol. Soybean oil, palm oil, rapeseed and animal fats may be used as source of oil.

India has a special plan on National Biogas and Manure Management Programme as stated earlier. Objective of the programme is to produce biogas locally. By using bacterial degradation of biomass, biogas can be produced. This degradation is called as anaerobic digestion, which occurs in the absence of oxygen

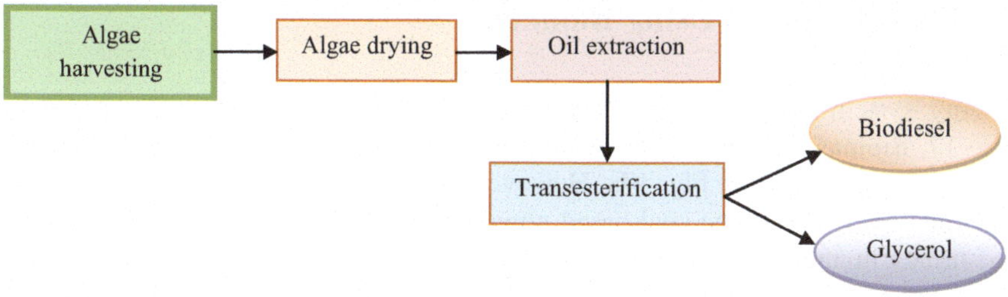

Fig. 16 Algae-based biofuel production

as shown in Fig. 15c. Three kinds of biomass may be used for digestion, i.e. (i) biomass from farm, e.g. liquid manure, feed waste (ii) household or municipality waste, e.g. organic waste, market waste, food waste (iii) industrial by-products, e.g. glycerine, food processing waste. Biogas generally contains 50–75% methane, 25–50% CO_2 with other elements like H_2O, O_2, traces of sulphur, H_2S. (Gomez 2013). However, performance of biomass digester depends on carbon to nitrogen ratio, temperature, pH value, chemical and biological oxygen demand, etc. About 98% of methane can be achieved with proper gas treatment. Hence, it can be used as a replacement of natural gas.

Algae-based fuel production is another option to replace the crude oil and carbon emission reduction. Algae can grow by utilizing CO_2 from atmosphere or CO_2 from high concentrated source like flue gas of power plant. Hence, it has the capability of capturing and fixing CO_2. Microalgae can produce proteins, lipids and carbohydrates over a short period of time in reasonable amount. Many algae are rich in oil compounds which can be used for diesel production. Similar to other biomass, algae can be converted into biofuels through thermo-chemical and biochemical routes. In Fig. 16, biochemical conversion of algae to biodiesel and glycerol is shown. In this process, firstly, algae are harvested. Then, it is dried. After that, oil is extracted and transesterification of oil is done as discussed earlier to produce biodiesel and glycerol. In Table 6, summary of different conversion technologies and products for various Indian biomasses are given.

Table 6 Summary of biomass to biofuel conversion

Indian biomass type	Example	Conversion technology	Product
Farm products	Corn, sugarcane	Fermentation	Ethanol
	Rapeseed, Jatropha	Transesterification	Biodiesel
Ligno-cellulosic material	Straw, wood	Hydrolysis and fermentation	Ethanol, mixed alcohol
		Thermo-chemical	Syngas, FT fuel, bioliquid
Algae	*Scenedesmus obliquus*, *Euglena* sp.	Transesterification	Biodiesel, glycerol
Digestible	Cow dung, kitchen waste	Anaerobic digestion	Biogas

8 Energy System Using Biomass

Energy system using biomass may be designed in various ways as shown in Fig. 17. Biomass can be fed directly to the boiler of steam-based power plant for direct firing as shown in Fig. 17a. However, for this case, design of boiler is to be modified compared to the coal-based power plant. Biomass can be fired with coal (co-firing) in the boiler as shown in Fig. 17b. In this process, specific carbon emission will be reduced compare to fossil fuel-based power plant. Integrated gasification with combined cycle (IGCC) is another option to use biomass efficiently as shown in Fig. 17c. In this process, product gas of the gasifier is firstly used in gas turbine cycle and then in steam turbine cycle. Hence, more thermodynamic efficiency can be achieved through combined cycle. Power can be produced from biofuels by using combustion engines, e.g. gas engine, CI engine, SI engine as shown in Fig. 17d. However, substantial modification in the present technology is required to adopt the biofuels.

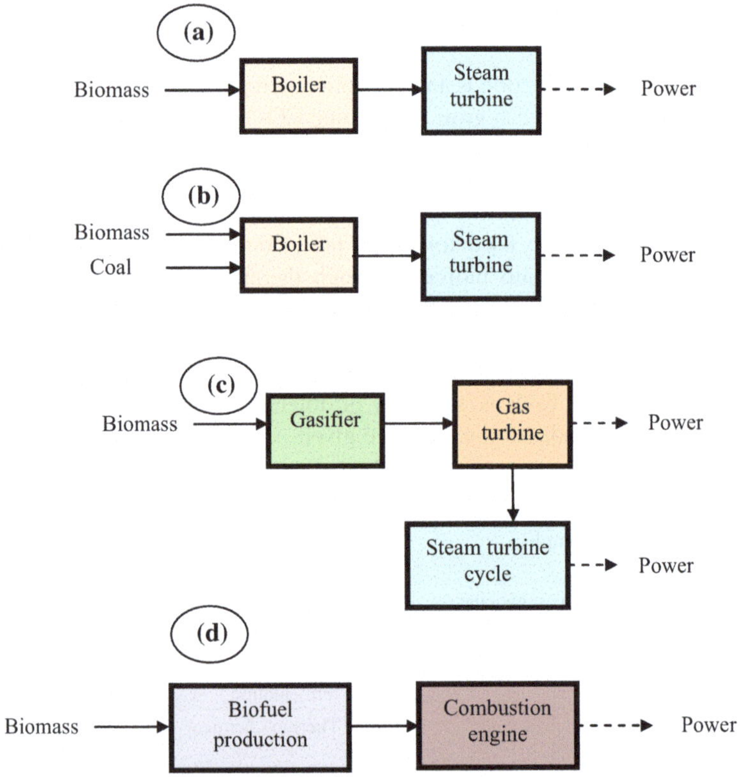

Fig. 17 Energy system using biomass **a** direct firing, **b** co-firing, **c** IGCC, **d** combustion engine

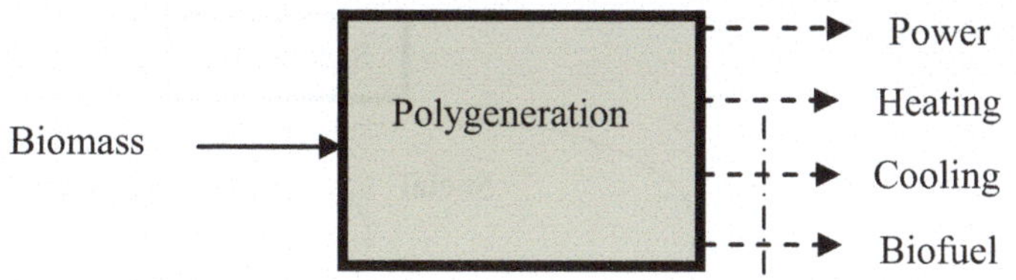

Fig. 18 Biomass-based polygeneration

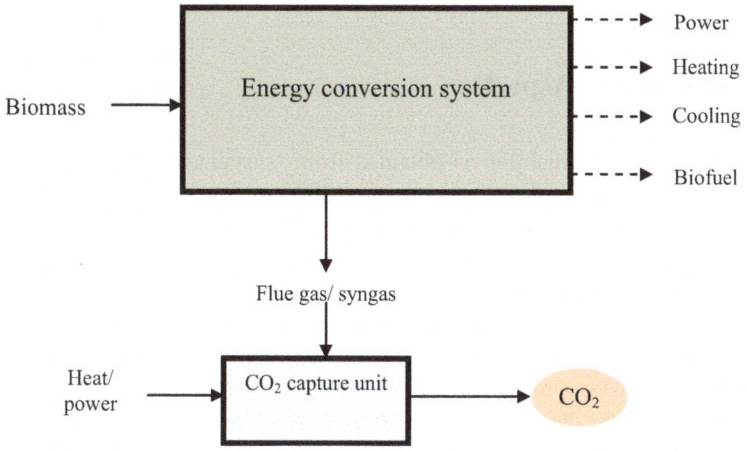

Fig. 19 Biomass-based energy system with carbon capture

Polygeneration is another novel energy system to utilize biomass in an efficient way. Polygeneration is the process of producing multiple utilities from a single plant by using suitable process integration as shown in Fig. 18. Outputs of polygeneration may be power, heating, cooling, biofuel, chemicals, etc. (Jana et al. 2017).

Biomass-based energy system may be equipped with carbon capture unit as shown in Fig. 19. By integrating carbon capture process, net carbon negative energy system is possible (Jana and De 2016a). However, to capture CO_2, additional amount of heat and power is required. Hence, it reduces the overall efficiency of the plant (Jana and De 2016b).

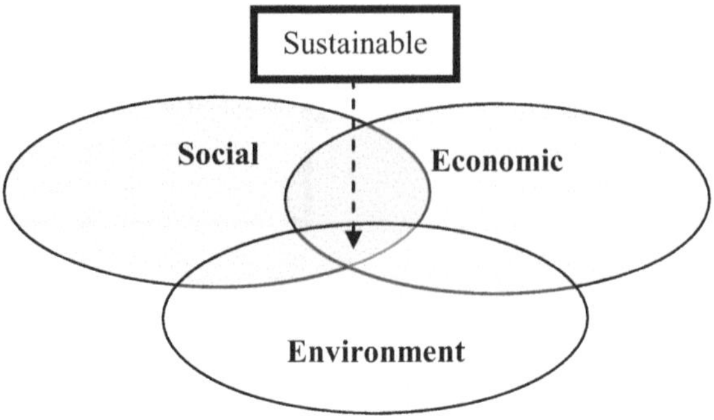

Fig. 20 Dimensions of sustainability aspects

9 Sustainability Aspects

The definition of sustainability is obtained from 'sustainable development'. It is defined by the Brundtland Commission in 1987 as 'development that meets the needs of the present generation without compromising the ability of future generations to meet their own needs' (United Nations 1987). Due to the multi-dimensional impacts of bioenergy, the sustainability assessment is important to find the long-term sustainable solution through bioenergy (Scarlat and Dallemand 2011). Presently, most of the sustainability indicators measure the performance by social, economic and environmental indicators as shown in Fig. 20. The indicators using these dimensions are shown in Table 7. These indicators are developed by Global Bioenergy Partnership (GBEP 2011) to assess the sustainability and to compare with other options from the viewpoint of sustainability. Environmental indicators include the impacts on atmosphere, hydrosphere and lithosphere. However, social indicators include various changes in land use, income, life standard.

Table 7 List of Global Bioenergy Partnership (GBEP) indicators (GBEP 2011; Hayashi et al. 2014)

S. No.	Dimension	Indicators
1	Environmental	Lifecycle GHG emission
2		Soil quality
3		Harvest levels of wood resources
4		Emission of non-GHG air pollutants
5		Water use and efficiency
6		Water quality
7		Biological diversity and landscape
8		Land use and land use change related to bioenergy feedstock production

(continued)

Table 7 (continued)

S. No.	Dimension	Indicators
9	Social	Allocation and tenure of land for new production
10		Price and supply of a national food basket
11		Change in income
12		Jobs in the bioenergy sector
13		Change in unpaid time spent by women and children collecting biomass
14		Bioenergy used to expand access to modern energy services
15		Change in mortality and burden of disease attributable to indoor smoke
16		Incidence of occupational injury, illness and fatalities
17	Economic	Productivity
18		Net energy balance
19		Gross value added
20		Change in consumption of fossil fuel and traditional biomass
21		Training and re-qualification of the workforce
22		Energy diversity
23		Infrastructure and logistics for distribution of bioenergy
24		Capacity and flexibility of use of bioenergy

10 Conclusions

Biomass-based energy is considered as one of the sustainable solutions, specifically for a country like India. More than 139 million tones of biomass is available as surplus in India. With this amount of biomass, 18 GW power can be generated and 7 GW of co-generation can be obtained by sugarcane bagasse according to the estimation of Government of India. Due to various barriers, Indian Government has taken different policies like National Biofuel Policy, National Biogas and Manure Management Programme, National Biomass Cookstove Programme to promote the use of biomass as energy resources. To convert different Indian biomass, several thermo-chemical and biochemical conversion technologies are available for producing electricity, liquid and gaseous fuel and other value-added products.

Though new policies are emerging and research and development are growing on bioenergy with supply chain improving, still, there are many challenges to overcome in the field of bioenergy. Challenges related to technology are mostly related to the improvement of efficiency, reduction in cost, reduction in operational hazards like tar removal, pretreatment of biomass, production of value-added co-products. Maintaining the supply chain and good logistics are two very important challenges for the success of biomass-based energy system without affecting food security. Improved policies are also required in the area of change in land use. Also, suitable policies for biofuels, financial incentives, commercial deployment of plants, etc., are also required.

Acknowledgements Dr. Kuntal Jana gratefully acknowledges to the Science and Engineering Research Board (SERB, DST), India for awarding National Post-Doctoral Fellowship.

References

Asadullah M (2014) Barriers of commercial power generation using biomass gasification gas: a review. Renew Sustain Energy Rev 29:201–215

Basu P (2010) Biomass gasification and pyrolysis—practical design and theory. Elsevier, UK

Bauen A, Woods J, Hailes R (2004) Bioelectricity vision: achieving 15% of electricity from biomass in OECD countries by 2020. WWF, Brussels, Belgium. Available at: www.panda.org/downloads/europe/biomassreportfinal.pdf. Last accessed on 05 July 2017

Cardoen D, Joshi P, Diels L, Sarma PM, Pant D (2015) Agriculture biomass in India: part 1—estimation and characterization. Resour Conserv Recycl 102:39–48

CEA (Central Electricity Authority) (2016) Executive summary power sector. Available at www.cea.nic.in/reports/monthly/executivesummary/2016/exe_summary-03.pdf

Domalski EA, Jobe Jr TL, Milne TA (1986) Thermodynamic data for biomass conversion and wastes incineration. A product of Solar Technical Information Program, Prepared by the National Bureau of Standards, United States

Dufour A (2016) Thermochemical conversion of biomass for the production of energy and chemicals. ISTE and Wiley, United States

Faaij APC, Schlamadinger B, Solantausta Y, Wagener M (2002) Large scale international bio-energy trade. In: Proceedings of 12th European conference and technology exhibition on biomass for energy, industry and climate change protection, Amsterdam

GBEP (2011) The global bioenergy partnership sustainability indicators for bioenergy. Global Bioenergy Partnership, Rome

Gold S, Seuring S (2011) Supply chain and logistics issues of bio-energy production. J Clean Prod 19:32–42

Gomez CDC (2013) Biogas as an energy option: an overview. In: Wellinger A, Murphy J, Baxter D (eds) The biogas handbook. Woodhead Publishing Limited, UK

Hayashi T, van Ierland EC, Zhu X (2014) A holistic sustainability assessment tool for bioenergy using the Global Bioenergy Partnership (GBEP) sustainability indicators. Biomass Bioenerg 66:70–80

Hiloidhari M, Das D, Baruah DC (2014) Bioenergy potential from crop residue biomass in India. Renew Sustain Energy Rev 32:504–512

IEA (2015) World energy outlook. International Energy Agency, Paris

IPCC (2016) Fifth assessment report (AR5). Intergovernmental Panel on Climate Change

Jana K, De S (2016a) Environmental impact of an agro-waste based polygeneration without and with CO_2 storage: life cycle assessment approach. Biores Technol 216:931–940

Jana K, De S (2016b) Utilizing waste heat of the flue gas for post combustion CO_2 capture—a comparative study for different process layouts. Energy Sources Part A 38(7):960–966

Jana K, De S (2017a) Thermo-chemical ethanol production from agricultural waste through polygeneration: performance assessment through a case study, pp 137–156. In: Agarwal AK, Agarwal RA, Gupta T, Gurjar BR (eds) Biofuels. Springer, Singapore. doi:https://doi.org/10.1007/978-981-10-3791-7_8

Jana K, Ray A, Majoumerd MM, Assadi M, De S (2017) Polygeneration as a future sustainable energy solution—a comprehensive review. Appl Energy 202:88–111

Lestander TA (2012) Chemical composition of biomass. In: Dahlquist E (ed) Biomass as energy source: resources, systems and applications. CRC Press, Boca Raton

MNRE (Ministry of New and Renewable Energy) (2016a) Annual report 2015/16. MNRE, Government of India, New Delhi

MNRE (Ministry of New and Renewable Energy) (2016b) Details available at http://mnre.gov.in/file-manager/UserFiles/Tentative-State-wise-break-up-of-Renewable-Power-by-2022.pdf. Last accessed on 8 Jan 2016

MoPNG (Ministry of Petroleum and Natural Gas) (2016) Annual report 2015–16. MoPNG, New Delhi

MoSPI (Ministry of Statistics and Programme Implementation) (2015) Energy statistics 2015. MoSPI, Government of India, New Delhi

Nigam PS, Singh A (2011) Production of liquid biofuels from renewable resources. Prog Energy Combust Sci 37:52–68

Ragauskas AJ, Williams CK, Davison BH, Britovsek G, Cairney J, Eckert CA, Frederick Jr. WJ, Hallett JP, Leak DJ, Liotta CL, Mielenz JR, Murphy R, Templer R, Tschaplinski, T (2006) The path forward for biofuels and biomaterials. Science 311(5760):484–489

Rosillo-Calle F, de Groot P, Hemstock SL, Woods J (2007) The biomass assessment handbook. Earthscan, UK

Scarlat N, Dallemand J-F (2011) Recent developments of biofuels/bioenergy sustainability certification: a global overview. Energy Policy 39:1630–1646

TERI (2016) TERI energy and environment data diary and yearbook 2015/16 (Updated edition)

UNFCCC (2015) Synthesis report on the aggregate effect of the intended nationally determined contributions. United Nations Framework Convention on Climate change

United Nations (1987) Our common future (the Brundtland Report). World Commission on Environment and Development. Oxford University Press, Oxford

Biomass-Based Distributed Energy Systems: Opportunities and Challenges

Sudip Ghosh

Abstract Biomass is not only considered as an important energy career for transition from fossil fuel to renewable energy, this is by far the most sustainable form of energy system. Given its inherent carbon neutrality, biomass-based energy conversion systems have the potential of becoming carbon-negative systems. This feature of biomass, together with its widespread availability, propels the development and deployment of biomass-based distributed energy systems. Among the different pathways of biomass energy conversion, gasification is considered as the most versatile route which can yield multiple end products, ranging from liquid fuels, methane, hydrogen, electricity, process heat and even refrigeration. Several technology options are currently available commercially: some are under different stages of development and some are futuristic, yet seemingly viable. This chapter presents an overview of the different technology options. Their status and potentials as distributed systems are reported and analysed. Socio-economic perspectives favouring or hindering their penetration, with particular reference to Indian and Asian scenario, have also been discussed herein.

Keywords Energy · Biomass · Gasification · Distributed generation
Cogeneration · CHP

1 Introduction

With increase in global population, rapid infrastructural growth and economic expansion, the demand for energy is also increasing. On the one hand, the sources of fossil fuels are depleting fast, and on the other hand, their use is causing serious environmental damages. Therefore, the energy supply and demand structure needs to be revised accordingly to achieve global sustainability. Energy researchers are

S. Ghosh (✉)
Department of Mechanical Engineering, Indian Institute of Engineering Science and Technology, Shibpur, Howrah 711103, West Bengal, India
e-mail: ghoshsudip@mech.iiests.ac.in; sudipghosh.becollege@gmail.com

A. Gautam et al. (eds.), *Sustainable Energy and Transportation*,
Energy, Environment, and Sustainability,
https://doi.org/10.1007/978-981-10-7509-4_13

paying increasing attention towards exploring and developing renewable energy technologies.

In another two decades, global marketed energy consumption is likely to rise by 60–70% and this growth will largely be driven by the demands of the non-OECD countries, who are likely to report an annual average growth of 3.3% compared to that of 1.1% for the OECD countries (World Energy Outlook 2015). India and China will have the largest changes in energy demand, followed by Africa, Southeast Asia, Middle East and Latin America, while the EU, the US and Japan are likely to have a negative growth rate each, as shown in Fig. 1. Global CO_2 emission is also predicted to grow at a moderate rate of 0.5–0.6% during this time.

However, global fuel mix for electricity generation is projected to undergo a considerable change in the coming decades, with the renewables overtaking coal as the major generation fuel. Among the renewable sources, hydro and wind (onshore wind) will continue to play major roles in power generation, particularly in grid-based utility market. However, for small- and medium-scale generation, mostly for off-grid and distributed generation, biomass is likely to have the largest contribution, as indicated in Fig. 2, staying ahead of solar PV, solar thermal energy (STE), off-shore wind and geothermal. It is therefore imperative that biomass-based generation technologies will continue to be in focus and development of newer, more efficient yet economically more rewarding technologies would remain a priority for a long time into the future.

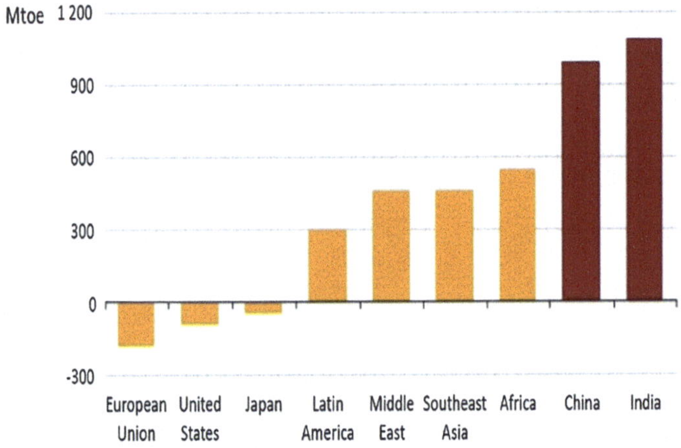

Fig. 1 Change in energy demand; 2014–2040 (World Energy Outlook 2015)

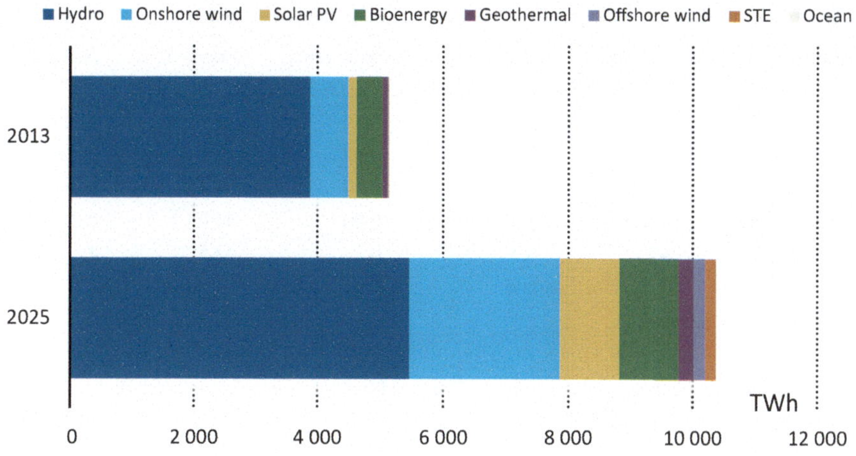

Fig. 2 Renewable energy generation (Tracking Clean Energy Progress, IEA 2016)

2 Energy from Biomass

In a broader sense, biomass is any organic (and decomposable) matter, derived from plants or animals. Biomass is available on a renewable basis and is traditionally used in the forms of wood, charcoal, agricultural resides and animal dung for cooking and heating in the residential sector. Natural biomass resources are normally classified as follows (Buragohain et al. 2010; Roy 2013):

Woody biomass: woods from forest and energy plantation, wood chips, sawdust, leaves, etc.
Agro-residual biomass: rice, coffee, coconut husks and shells, straw, jute stick, etc.
Waste biomass: cooking waste, animal waste, industrial and municipal waste, etc.
Aquamarine biomass: aquatic weeds and plants, algae, seagrass beds, etc.

Unlike fossil fuels, biomass is rich in oxygen. The oxygen content, in many instances, is even more than the carbon content. Typical composition of different biomass sources is shown in Table 1.

Biomass is considered a carbon neutral considering that the carbon dioxide (CO_2) that is emitted during energy conversion gets reabsorbed by the plantation or energy crops (DTI 2007), although a slightly positive life-cycle CO_2 contribution could be noted because of the emissions released during crop harvesting, production, supply and transport phases (PB Power 2008). The apparent carbon neutrality or the savings on account of greenhouse gas (GHG) are one of the major driving forces behind the development of wider penetration of biomass energy systems. The other contributing factors are its availability across almost all geographical regions on the land, feasibility of economic energy plantation and, importantly, the absence of the problem of intermittency (Gill et al. 2005).

Table 1 Composition of major biomass resources (Buragohain et al. 2010)

Biomass	Carbon	Hydrogen	Nitrogen	Oxygen	Ash
(A) Ultimate analysis *Composition in weight percent (dry basis)*					
Bamboo dust	39.88	5.5	0.89	47.92	5.81
Rice husk	37.03	5.25	0.09	40.94	16.69
Sawdust	52.28	5.2	0.47	40.85	1.2
Coconut fibre	45.68	5.89	0.99	44.63	2.81
Bagasse	47	6.5	0	42.5	4

Biomass	Fixed carbon	Volatile matter	Ash
(B) Proximate analysis			
Bamboo dust	9.3	74.2	16.5
Rice husk	13.2	65.3	19.2
Sawdust	25.0	72.4	2.6
Coconut fibre	29.7	66.6	3.7
Bagasse	11.9	86.3	1.8

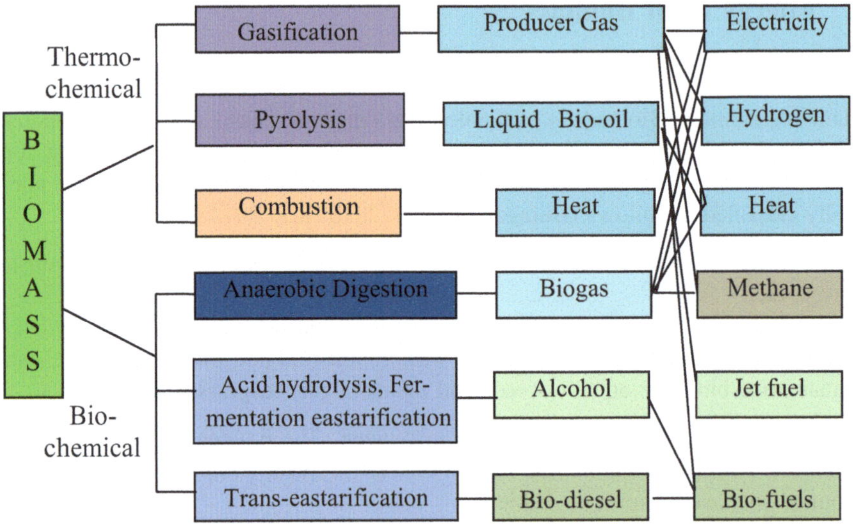

Fig. 3 Conversion routes of biomass

Biomass can be converted into different useful products, such as electricity, heat, transport fuel and chemical feedstock via different process routes (McKendry 2002). Figure 3 shows the different pathways of conversion of biomass (Roy 2013). Among these pathways, gasification is considered as the most versatile route which can yield multiple end products. When applied to power or cogeneration, the usual conversion route is gasification, followed by combustion in an IC engine or a gas turbine. Apart from gasification, the other traditional route to power and

Table 2 Efficiency comparative of biomass power generation

Pathway	Configuration	Capacity (MW)	Efficiency (% LHV)	Status
Combustion boiler	Steam engine	<0.5	<10	Old, outdated traditional
	Steam turbine	1–100	10–30	
Gasification (fixed bed)	Gas engine	0.01–1	15–30	Commercial demonstration
	Gas turbine	0.5–5	20–32	
Gasification (AFB/CFB)	Gas turbine	5–20	25–32	Demonstration
BIGCC	GT-ST	10–120	30–45	Demonstration

Sources SMUD ReGen Report (2004), Paisley and Welch (2003), EU BASIS Report (2015)

cogeneration has been the direct biomass combustion in a boiler to produce steam to run a Rankine cycle plant or supply process steam or both.

Table 2 gives an indicative comparative of the efficiency ranges of different technology options currently available.

3 Competing Technologies

Producer gas-fed IC engines have attained some degree of technological maturity over the last few decades. The engines can be operated on 100% producer gas (PG) mode as well as on dual fuel (DF) mode, using diesel as supplementary fuel. Although biomass-based gas engines are very robust and commercially available, efficiency of such system is very low, in the range of 18–25% for simple small-scale plants (EU BASIS Report 2015; Mondal et al. 2015), and cost of electricity generation is also higher compared to conventional coal-based plants. Also, substantial gas cleaning is required for engine-based applications of producer gas. Schematic diagram of a typical biomass gasifier–gas engine set-up is shown in Fig. 4.

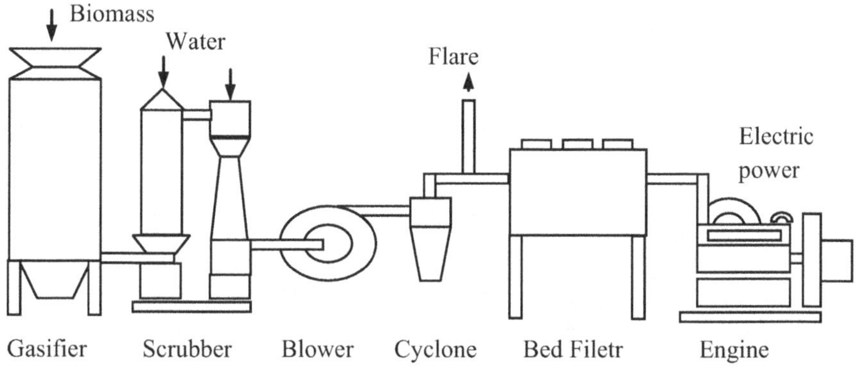

Fig. 4 Typical gasifier–gas engine assembly

Energy content of the producer gas can efficiently be recovered in a combined gas–steam combined cycle and thus forming a biomass-integrated gasification combined cycle (BIGCC). Efficiency level of IGCC plants is considerably higher compared to gasifier–gas engine plants. They can offer power generation efficiency in the range 35–40% (Franco and Giannini 2005), and advanced technology options could yield even better efficiency level at appropriate scale of power output (Paisley and Welch 2003). However, extensive gas cleaning is required to remove tar, particulate matter, SO_2 and ash from the producer gas (Oakey et al. 2004). Also, design modification of the GT combustor is essential for use of low calorific value gas in IGCC. The last decade of the twentieth century saw the first demonstration of BIGCC when a 6 MWe plant started operation at Varnamo, Sweden, in 1996 (Ståhl and Neergaard 1998). It employed circulating fluidized bed (CFB) gasification with hot gas clean-up. The plant recorded a net electrical efficiency of 32%. This plant also earned the credit of demonstrating combined heat and power (CHP) generation in BIGCC, having a thermal capacity of 9 MW. However, the plant was shut down after few years of operation faced with several operational problems. Figure 5

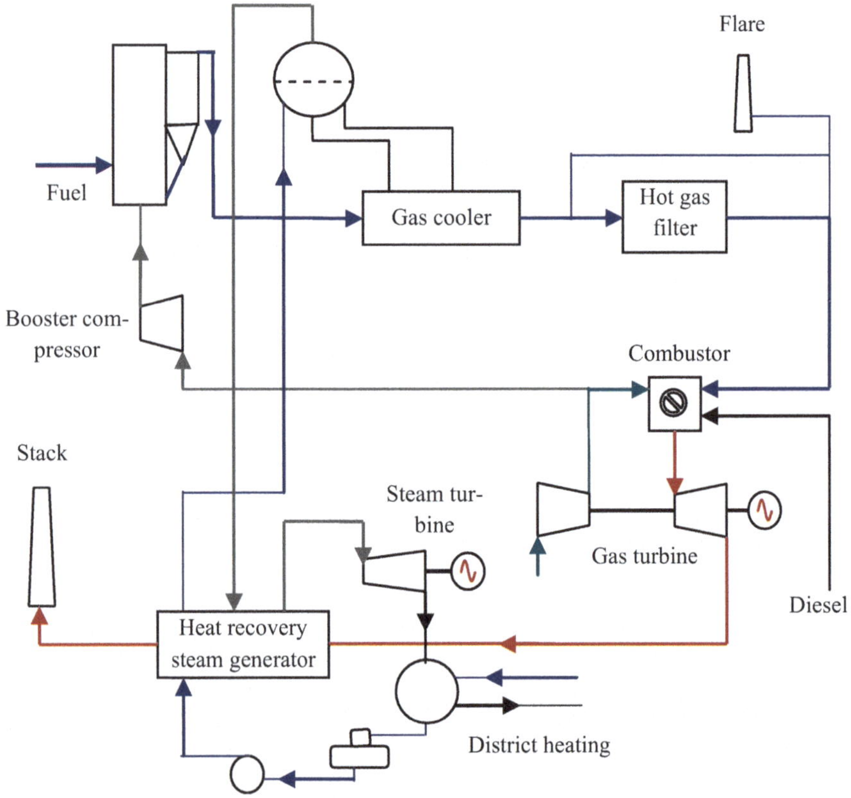

Fig. 5 Schematic diagram of BIGCC plant at Varnamo, Sweden

shows the schematic diagram of the Varnamo BIGCC plant. However, biogasification-based IGCC plants have hardly been implemented in community-scale applications because of the capital cost involvement and the high-end technology.

Fixed and fluidized bed gasifiers are commonly employed in gasification. Entrained flow gasifiers were primarily developed for coal gasification. They have been tested and demonstrated for biomass application also, but these types of gasifiers are not suitable for biomass due to the particle sizing restrictions. Fluidized bed proved to be better modes of combustion and gasification for a variety of biomass feeds. The scale of plant also influences the choice of gasification technology, since all the types are not suitable and practicable for bigger scale. The fixed bed systems are better adaptable for low capacity units, typically at kW level of power output and, at the most, up to 1 MW output. Fluidized bed gasification has been employed, or planned to be employed, in wide range of systems with capacity ranging from less than 10 MW to more than 100 MW (IRENA 2012). The first BIGCC demonstration plant at Varnamo also employed circulating fluidized bed (CFB) gasification.

The design of the gasifier, the biomass feed and its moisture content as well as the air equivalence ratio influence largely the producer gas composition and its calorific value. Air gasification results in high nitrogen percentage (40–50% or even beyond) in the producer gas leading to lower calorific value of gas and bigger sizes of the downstream components (Mondal and Ghosh 2016; Nagel et al. 2011). Steam gasification yields hydrogen-rich gas, but sustaining steam gasification requires additional heat input for the steam generator and the gasification reactor. Solid biomass is pretreated before it is fed to the gasifier. Pretreatment includes drying, cutting and sizing and briquetting. Most gasifiers usually tolerate a moisture content of 10–15%, and a higher value leads to lower calorific value gas mixture causing difficulty during ignition and combustion. Husky and dusty biomass feeds are briquetted before they are fed to the gasifier.

Producer gas derived from biomass gasification process contains different types of contaminants which are harmful for the downstream components and need to be cleaned before its end use. Alkali and sulphur compounds, tars and particulate matters are the main impurities present in the producer gas. The level of gas cleaning depends on the application. High level of cleaning is required if producer gas is intended to be used in the reciprocating engines, gas turbines and fuel cells. Conventionally, a low- or medium-temperature gas cleaning train is employed where scrubbers (both wet and venturi), cyclone separators and filters (fabric, bed and bag) are used to clean raw producer gas (Wickwire 2007). Table 3 shows the types of impurities present in the producer gas and the corresponding gas treatments or gas cleaning arrangements. In hot gas clean-up systems, ceramic or metallic filters and different kinds of absorber materials and catalytic converters or partial oxidation are employed.

While BIGCC technology holds great promise of high-efficiency power and cogeneration, the high-end technology, together with its high initial cost, does not yet favour its application in distributed generation market. All operating BIGCC

Table 3 Producer gas impurities and their common treatments

Impurities	Types and description	Treatment
Tar	A mix of condensable hydrocarbons; single-ring aromatics, PAH, oxygenated hydrocarbons	Wet scrubbing, sand bed filters
Particulate matter	Small-sized solids; mainly unconverted biomass char and ash	Cyclone separators, ESP, fabric filters and wet scrubbers
Alkali compounds	Mainly Sodium and Potassium salt vapours present in the biomass	Gas cooling and filtering, wet scrubbing
Ammonia	Formed due to the reaction of nitrogen (in fuel and air) with hydrogen (in fuel and water vapour)	Wet scrubbing, catalytic conversion and hydrocarbon reforming
Sulphur	In the form of sulphides (H_2S, COS)	Cold scrubbing with absorbers, active carbon/metal oxide beds

plants are of multi-MW level, and several operating plants have capacities higher than 10 MW. BIGCC is unlikely to be a commercial entity at sub-MW level.

In a conventional BIGCC, directly fired gas turbines (DFGTs) are used and the GT rotor is directly exposed to the combustion products. In an indirectly heated or externally fired gas turbine (EFGT) cycle, hot combustion product does not come in direct contact with the turbine blades. EFGT cycle can handle low-cost and inferior-grade solid fuels without requiring elaborate gas cleaning processes and without any design modification of the GT combustor and the fuel compressor (Mondal et al. 2015).

The EFGT cycle can be configured in two ways, open or closed, as shown in Fig. 6. Open cycle usually employs air as working fluid, while in a closed cycle, an inert gas (such as CO_2, He and N_2) could be used and recycled as the working fluid. Often, a combustor–heat exchanger (CHX) duplex unit is used to heat up the working fluid, as shown in Fig. 7, and in open-cycle configuration, warn exhaust air from the expander can be used as combustion air in the combustor of the CHX unit. Exhaust heat of the CHX unit can be further utilized through a bottoming heat recovery unit for cogeneration. Closed-cycle EFGT offers another advantage that of delinking working fluid pressure from the combustor pressure. Conventional inert gases used as working fluids are CO_2, He and N_2. Worldwide, many installations have come up, both with open cycle and with closed cycle and using different fuels, including coal, natural gas and blast furnace gas. A good review of such plants could be found in Al-attab and Zainal (2015). Table 4 shows some of the installations that are based on biomass-based EFGT (Al-attab and Zainal 2015).

EFGT technology opens up a wider scope of development for higher-efficiency systems at relatively low-energy penalty (cost penalty as well) compared to the conventional systems requiring extensive gas cleaning. With added heat recovery options of low or moderate complexity, higher power or cogeneration efficiency is readily obtainable. The systems are implementable at low kW level to MW level.

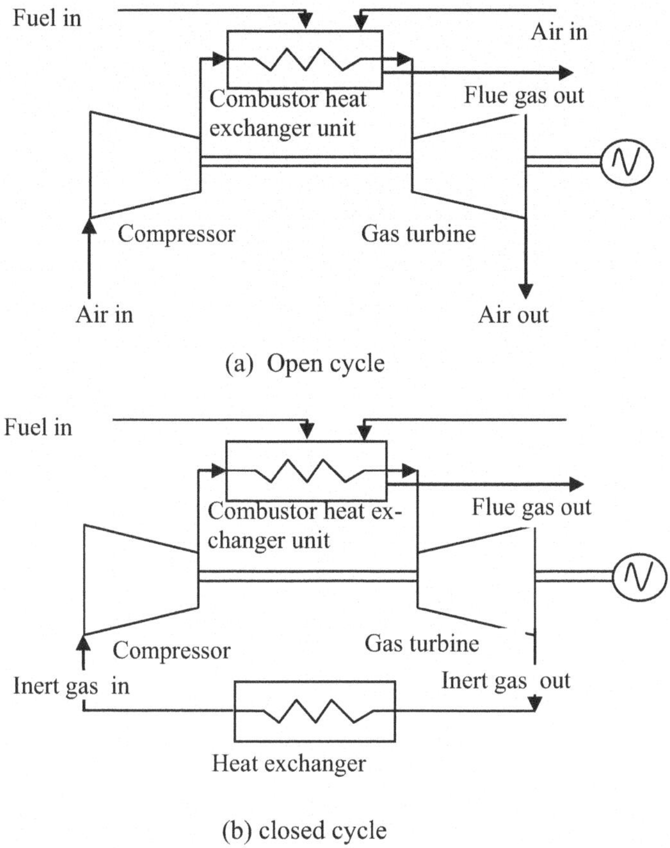

Fig. 6 EFGT cycle configurations

As the technology matures, employing upgraded heat exchanger and micro-GT suitable for TIT of 1000 °C and beyond and with the integration of suitable heat recovery power or cogeneration cycles, EFGT-based systems would yield high-efficiency distributed power generation and cogeneration systems. Initial development in closed-cycle EFGT systems took place with coal and natural gas and even with nuclear reactor heat.

By the middle of last century, several closed-cycle EFGTs were successfully demonstrated across USA and the Europe. The first such unit was the 2.3 MWe Escher Wyss plant facility in Zurich, installed in late 1930s, although its working fluid was air (Al-attab and Zainal 2015). The plant used coal and also oil as fuel. A closed-cycle EFGT with helium was developed in 1960 for a cryogenic facility for gas liquefaction without any net output power. Several EFGT-CHP plants of larger capacities were also developed in the later part of the last century (Bammert 1987; Al-attab and Zainal 2015). Notable among them are the Coburg plant with

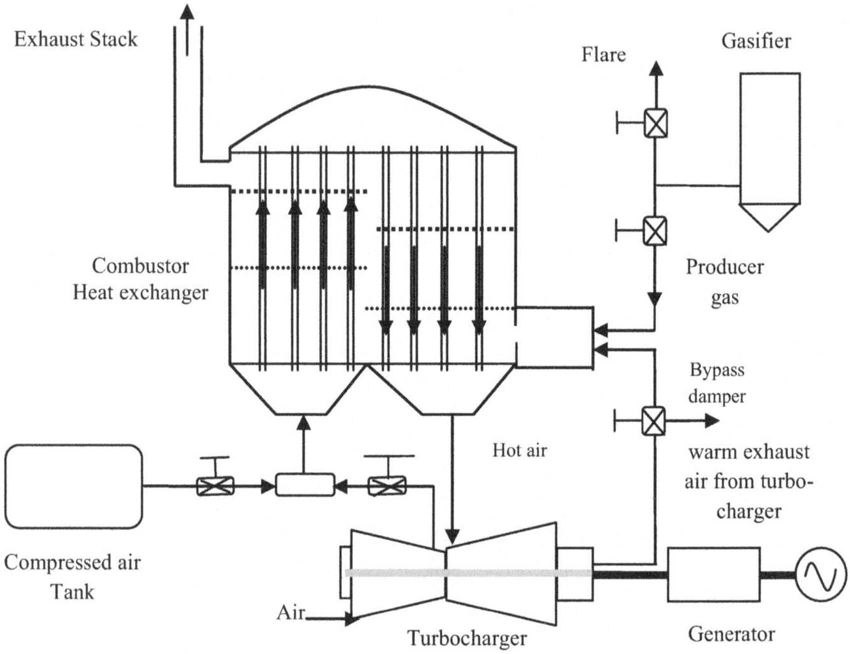

Fig. 7 EFGT system with turbocharger and CHX unit

Table 4 Some biomass based EFGT installations

Cycle and GT types	Power MWe	TIT (°C)	Pressure ratio	Combustor Heat exchanger	Location
Open cycle micro-GT	0.5	800	–	Tubes submerged in FB combustor	Belgium
Open cycle micro-GT	0.05	800	4.5:1	Shell and tube with recuperator	UK
Open cycle micro-GT	0.08	700	4:1	Shell and tube with recuperator	Italy
Open cycle Turbocharger	None	690	2.2:1	Shell and tube	Malaysia
Open cycle radial GT	1.8	900	–	Not specified	Philippines
Open cycle radial GT	2.0	–	–	Not specified	Germany

6.6 MWe electrical and 16 MWth, the Oberhausen plant with 13.7 MWe and 28 MWth outputs and the Gelsenkirchen plant with 17.25 MWe and 29 MWth. However, all these first-generation EFGT plants had a low TIT, 700 °C or lower as non-availability of suitable heat exchanger put severe restriction on this front.

Table 5 Cost of electricity for biomass power generation

Pathway	Configuration	Capacity (MW)	COE (cUSD/kWh	Technology status
Combustion boiler	Rankine cycle	30–100	7–21	Traditional commercial
Combustion boiler	Organic rankine cycle	0.5–2	11–25	Demonstration early commercial
Gasification	Gas engine	0.01–10	6.5–14	Demonstration commercial
BIGCC	GT-ST combined cycle	30–100	10.5–13.5	Demonstration commercial
Biogas CHP	Gas engine	0.25–2.5	16–22	Commercial demonstration

Source Chowdhury (2014), IRENA (2012)

4 Developing Technologies

Different types of heat exchangers have been developed for EFGT systems. Most of the producer gas-fired systems consider heat exchanger constructed of stainless steel. The experimental facility at USM, Malaysia, also considered SS heat exchanger (Al-attab and Zainal 2010). Challenges exist in design and development of a heat exchanger which can deliver TIT at the level of 1000 °C and beyond. One approach is to use ceramic heat exchange surfaces that can withstand high temperature of about 1100 °C, in order to yield a TIT that can exceed 1000 °C. Ceramic HE will have a long operational life span, but the development cost is very high. With further development and larger production volume, this heat exchangers might be economical in the future giving acceptable payback period.

Apart from the heat exchanger, the gas combustor and the GT are the crucial components of an EFGT system, and together, they influence the cost as well as the performance of such plants. Micro-GT is the usual type of gas turbine that is coupled to the generator to produce electricity.

Vehicular turbochargers are also used in an EFGT cycle for small-scale generation (Fig. 7). A vehicular turbocharger is used specially in diesel engine to improve the engine's performance. However, it may be used as expander in an EFGT cycle. One of the major issues of the usage of turbocharger is the disadvantages of oil lubrication system compared to air or magnetic bearings used in conventional turbomachines. Al-attab and Zainal (2015) have reported current status of some notable turbocharger-based GT systems. Smaller and compact turbocharger-based EFGT-CHP plant has also been developed recently, under a project named Exheat (Al-attab and Zainal 2015), the schematic view of which is shown in Fig. 8. The unit is rated at 6 kWe boasts of an electrical efficiency of about 24%.

Several improvement options are also available for EFGT systems. Mondal and Ghosh (2013) proposed integration of reciprocating compressor (with air storage), replacing the conventional axial compressor in the air cycle. Direct biomass-fired

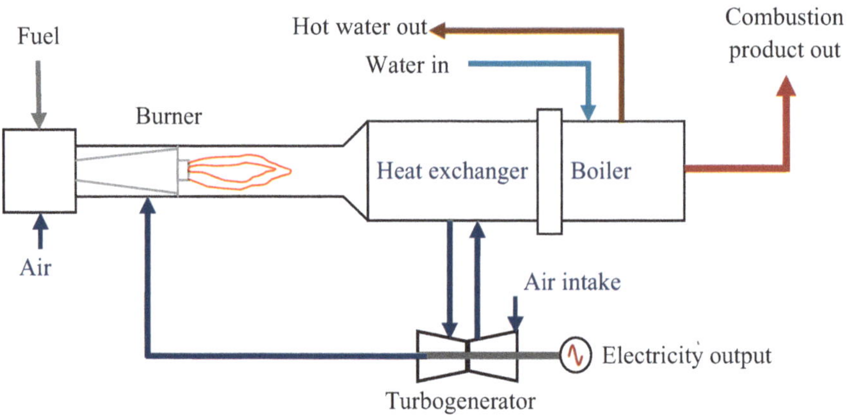

Fig. 8 Compact turbocharger-based EFGT-CHP plant

heat exchanger has also been proposed by them. Simulation studies indicate potential efficiency enhancements due to the changes in the configuration. Integrating a bottoming steam cycle or an organic Rankine cycle (ORC) has also been investigated (Mondal and Ghosh 2017a). The bottoming integration would certainly render added complexity in the EFGT system, but the performance enhancements are likely to justify the same. Study suggests that direct biomass-fired combustor heat exchanger, applied to an air–steam combined cycle (Fig. 9), could potentially extend the overall power efficiency of an EFGT system to nearly 50% mark (Mondal and Ghosh 2017b).

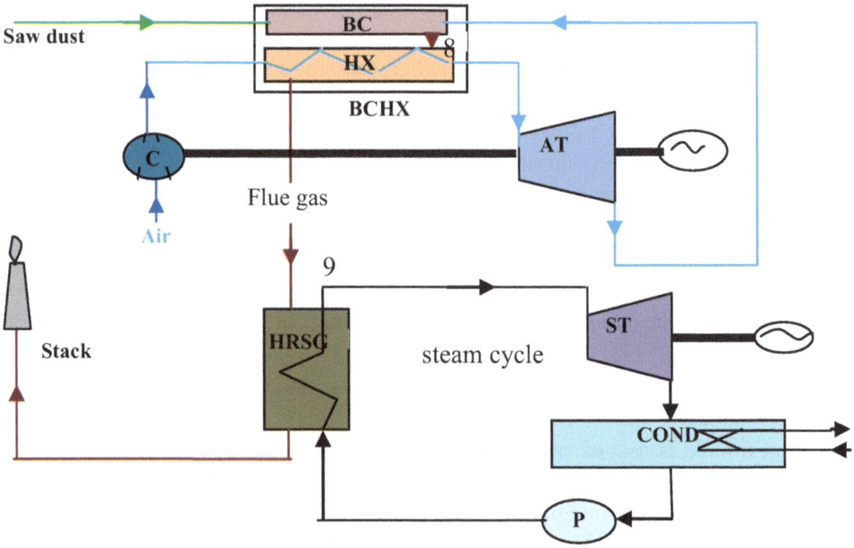

Fig. 9 Direct biomass-fired EFGT combined cycle plant

Another technology that holds promise of useful and successful integration with biomass gasification is that of the high-temperature fuel cells. Solid oxide fuel cell (SOFC) and molten carbonate fuel cell (MCFC) are the two types that can tolerate biomass-derived producer gas, provided impurities are sufficiently removed from the gas. Over the last few decades, considerable development has taken place in this front. While material development and improved production technologies propelled faster development of the fuel cells, high-temperature gas cleaning and prospect of internal reforming and catalytic conversion made producer gas feeding more sustainable for the cell stacks. SOFC, being of solid oxide construction and having higher operating temperature, offers added advantages of better heat integration and heat recovery in the system. Natural gas-based SOFC-GT hybrid system has attained some degree of maturity and is planned for commercialization (Zhang et al. 2006; Cheddie 2010). In recent years, several works attempted long-term tests of SOFC stack with biogas and biomass-derived producer gas (Mehr et al. 2015; Nagel et al. 2011). Being inherently efficient, the SOFC stack renders biomass gasification-based power or cogeneration considerably efficient even at kW scale, matching the level attained by advanced BIGCC systems and surpassing them if heat recovery integration is considered (Bang-Møller et al. 2011; Roy and Ghosh 2017). Figure 10 shows a schematic of an integrated biomass gasification combined cycle employing a SOFC and an organic Rankine cycle (Roy and Ghosh 2017).

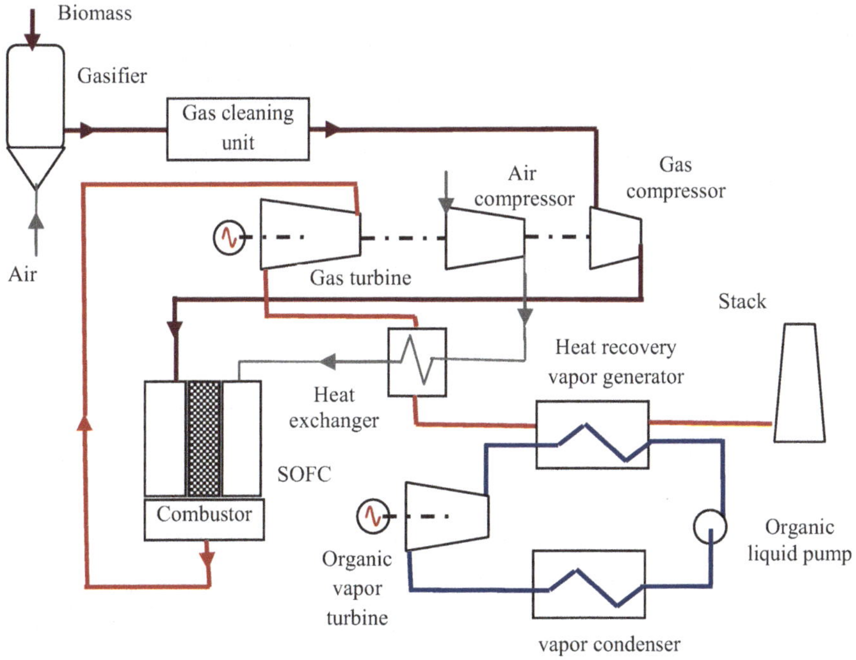

Fig. 10 Integrated biomass gasification–SOFC combined cycle

5 Economics of Distributed Generation

The different technologies are at different stages of development and maturity. The ones that are quite matured offer lowest cost of deployment, e.g. direct biomass combustion steam cycle for power or CHP, biogas-driven gas engines or gasifier-based gas engines. The developing technologies that are under demonstration and awaiting commercialization call for relatively higher cost of deployment.

Equipment cost, O&M cost and feedstock cost together with the system efficiency decide the estimates for cost of electricity, although other factors like lead time for implementation, cost of borrowing and government policies and subsidies significantly contribute to the calculation of levelized unit cost of electricity (LCOE) for any utility project, as depicted in Fig. 11.

Gasifier–gas engine is seen to offer competitive COE among the presently available technology options, ranging from 6.5 to 14 cUSD/kWh, for small-scale plants suitable for distributed generation. Recent studies suggest the developing EGFT technologies would be able to match the economics of gasifier–gas engine while yielding better plant performances. Table 6 shows predicted LCOE for several improved EFGT schemes (Mondal and Ghosh 2016, 2017a, b) (Table 5).

Systems employing high-temperature fuel cells would have cost estimates highly sensitive to the unit costs of SOFC and MCFC, both being highly cost-intensive in the present situation. Reliable LCOE analysis is thus seldom found in the literature. Recent studies adopted different ranges of capital cost estimates for SOFC, varying from 1250 $/kW to as high as 15,000 $/kW. Future projections put the values at 500–1000 $/kW, with US DOE, and setting a target of 400 $/kW by 2020 (Pruitt et al. 2013; Wei 2015; Gandiglio et al. 2016). Specific cost of MCFC is still in the

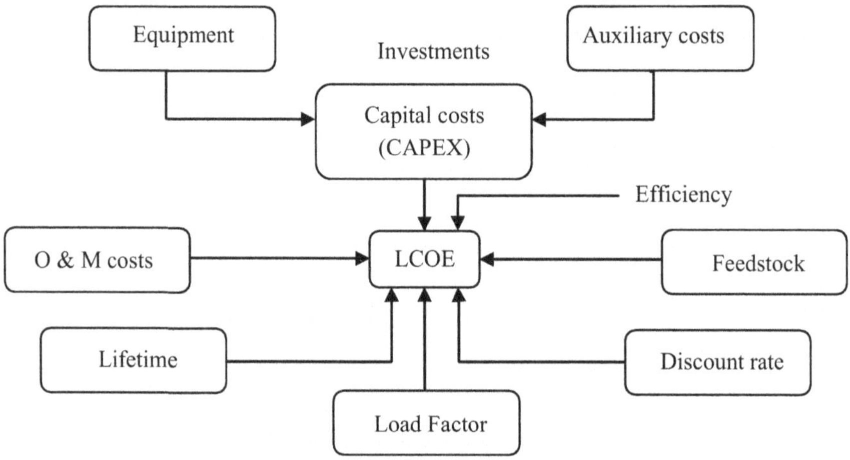

Fig. 11 Influencing factors for cost of electricity (IRENA 2012)

Table 6 Performance and cost comparative for EFGT technologies

Parameter	BG-EFCC	BG-FCCC	BG-EFCC-ORC	BG-EFCC-RC	BFCC
Efficiency (%)	39.24	37.28	42.99	41.75	46.01
ESBC (kg/kWh)	0.56	0.57	0.46	0.52	0.48
LCOE (cUSD/kWh)	11.5	11.2	11.1	11.1	11.2

BF Biomass fired, *BG* biomass gasifier, *ESBC* electrical specific biomass consumption, *EFCC* externally fired combined cycle, *EFCC* externally fired combined cycle cogeneration, *LCOE* levelized unit cost of electricity, *ORC* organic rankine cycle, *RC* reciprocating compressor

range of 3000–3200 US$/kW (Mamaghani et al. 2015; Campanari et al. 2014), though a more recent US DOE report (Ahmed et al. 2015) considered a specific cost of 2000 $/kW for low production volume and predicts a further cost reduction as the production volume increases for MCFC stack (Soltani et al. 2013; Ahmed et al. 2015).

6 Socio-environmental Issues

Biomass, being considered a carbon-neutral resource, renders carbon neutrality to the biomass-fed systems, as far as energy conversion is concerned. A new approach towards attaining climate change goals to create and develop carbon-negative systems, as maintaining carbon emission within allowable limits alone, is considered not enough to achieve the objective of restricting global temperature rise. Biomass-based systems are placed favourably when carbon negativity is considered. Capturing emission from biomass combustion and gasification system would play a major role in this direction. Availability of biomass resources on a sustainable basis is another factor that promotes penetration of biomass technologies in distributed generation markets. Prospect of energy plantation provides further support in ensuring uninterrupted supply of biomass feeds, while improving the economics and living standard of rural population.

There, however, are several challenges facing the biomass technology development and penetration. Biomass resources often need elaborate drying and preprocessing, apart from time, energy and cost associated with biomass harvesting and transport. Most often, energy plantation is found to support only the transport biofuel industries, leaving the distributed generation players in critical resource supply situations. High initial cost of equipment for some emerging technologies (such as EFGT, fuel cell) hinders their commercialization efforts and government supports in the form of discounts in lending rate, and taxes become essential.

The developing and emerging economies of the world have large biomass resources. Despite that, their contribution to global biomass power is relatively small, the bulk of the biomass capacity being located in Europe and North America (IRENA 2012). Government policy is therefore an imperative factor for development and deployment of biomass-based systems.

The Ministry of New and Renewable Energy (MNRE) in India is currently promoting biomass and biogas projects for rural penetration. The ministry reported over 1800 gasifier power systems is being deployed throughout the country with a total capacity of over 150 MWe. Total biomass-based generation capacity currently stands at over 4800 MW (Source: MNRE Website, Govt of India). There have been issues related to gas quality, and significant progress has been made addressing the issues ensuring safe and reliable system deployment.

7 Conclusion

Biomass-based distributed energy systems have very high potential in a society where carbon emission is likely to play increasingly significant role in technology development and environmental cost assessment. Among the different pathways of biomass energy conversion, gasification is seen as the most versatile route which can yield electricity, heat, refrigeration and other end products such as liquid fuels, synthetic natural gas and hydrogen. Several technology options have attained maturity over the years and are currently available commercially. This chapter presented an overview of the different biomass energy conversion technologies, focussing largely on the ones suitable for distributed generation. Their status and potentials as distributed systems are discussed and analysed. Economic as well as socio-environmental perspectives favouring or hindering their development and penetration have also been discussed, with particular reference to developing economics and the Indian scenario.

References

Ahmed S, Papadias D, Ahluwalia R, Hua T, Roh HS (2015) Performance and cost analysis for a 300 kW tri-generation Molten carbonate fuel cell system. In 2015 US DOE Hydrogen and Fuel Cells Program and Vehicle Technologies Office, Annual Merit Review and Peer Evaluation Meeting

Al-attab KA, Zainal ZA (2010) Performance of high-temperature heat exchangers in biomass fuel powered externally fired gas turbine systems. Renew Energy 35(5):913–920

Al-attab KA, Zainal ZA (2015) Externally fired gas turbine technology: a review. Appl Energy 138:474–487

Bammert K (1987) Operating experiences and measurements on turbo sets of CCGT-cogeneration plants in Germany. Journal of engineering for gas turbines and power, 109(1), 22–31

Bang-Møller Christian, Rokni Masoud, Elmegaard Brian (2011) Exergy analysis and optimization of a biomass gasification, solid oxide fuel cell and micro gas turbine hybrid system. Energy 36(8): 4740–4752

Buragohain B, Mahanta P, Moholkar VS (2010) Biomass gasification for decentralized power generation: the Indian perspective. Renew Sustain Energy Rev 14:73–92

Campanari S, Chiesa P, Manzolini G, Bedogni S (2014) Economic analysis of CO_2 capture from natural gas combined cycles using Molten Carbonate Fuel Cells. Applied Energy 130:562–573

Chowdhury NR (2014) Advances in trends in woody biomass gasification. Energy engineering and management, Tecnico Lisboa.

Denver F (2010) Cheddie, integration of a solid oxide fuel cell into a 10 MW gas. Turbine Power Plant Energies 3:754–769

Economic Analysis of Biomass Energy (2007) DTI UK working paper 1 (biomass strategy). Department of Trade and Industry (DTI), London, UK

Franco A, Giannini N (2005) Perspectives for the use of biomass as fuel in combined cycle power plants. Int J Therm Sci 44:163–177

Gandiglio M, Drago D, Santarelli M (2016) Techno-economic analysis of a solid oxide fuel cell installation in a biogas plant fed by agricultural residues and comparison with alternative biogas exploitation paths. Energy Procedia 101:1002–1009

Gill B, MacLeod N, Clayton D, Cowburn R, Roberts J, Hartley N (2005) Biomass task force. Report to government. London, UK: Defra

IRENA (2012) Renewable energy technologies: cost analysis series. International Renewable Energy Agency (IRENA) working paper, vol 1: Biomass for power generation

Mamaghani AH, Najafi B, Shirazi A, Rinaldi F (2015) Exergetic, economic, and environmental evaluations and multi-objective optimization of a combined molten carbonate fuel cell-gas turbine system. Applied Thermal Engineering 77:1–11

McKendry P (2002) Energy production from biomass (part 2): conversion technologies. Biores Technol 83:47–54

Mehr AS, Mahmoudi SMS, Yari M, Chitsaz A (2015) Thermodynamic and exergoeconomic analysis of biogas fed solid oxide fuel cell power plants emphasizing on anode and cathode recycling: a comparative study. Energy Convers Manag 105:596–606

Mondal P, Ghosh S (2013) Thermal performance of an indirectly heated bio-gasification based combined cycle plant employing reciprocating compressor. Int J Emer Tech and Adv Eng 3 (3):186–192

Mondal P, Ghosh S (2016) Bio-gasification based externally fired combined cogeneration plant: thermo-economic and exergo-economic analyses. Int J Exergy 20(4):496–516

Mondal P, Ghosh S (2017a) Exergo-economic analysis of a 1 MW BIGCC plant with indirectly heated GT cycle and supercritical ORC. Clean Technol Environ Policy 19(5):1475–1486

Mondal P, Ghosh S (2017b) Techno-economic performance evaluation of a direct biomass-fired combined cycle plant employing air turbine. Clean Technol Environ Policy 19(2):427–436

Mondal P, Mondal K, Ghosh S (2015) Bio-gasification based distributed power generation system employing indirectly heated GT and supercritical ORC: energetic and exergetic performance assessment. Int J Renew Energy Res 5(3)

Nagel FP, Ghosh S, Pitta C, Schildhauer TJ (2011) Biomass integrated gasification fuel cell systems–concept development and experimental results. Biomass Bioenerg 35(1):354–362

Oakey J, Simms N, Kilgallon P (2004) Gas turbines: gas cleaning requirements for biomass-fired systems. Mater Res 7(1):17–25

Paisley MA, Welch MJ (2003) Biomass gasification combined cycle opportunities using the future energy Silvagas gasifier coupled to Alstom's industrial gas turbines. In: Proceedings of ASME Turbo Expo 2003, Georgia, 16–19 June 2003

Pruitt KA, Braun RJ, Newmanc AM (2013) Establishing conditions for the economic viability of fuel cell-based, combined heat and power distributed generation systems. Appl Energy 111:904–920

Report on conversion efficiency of biomass. EU BASIS Report, July 2015. Available online at http://www.basisbioenergy.eu/

Roy PC (2013) Role of biomass energy for sustainable development of rural India: case studies. Int J Emerg Technol Adv Eng 3:577–582

Roy D, Ghosh S (2017) Energy and exergy analyses of an integrated biomass gasification combined cycle employing solid oxide fuel cell and organic Rankine cycle. Clean Techn Environ Policy 19:1693–1709

Soltani S, Mahmoudi SMS, Yari M, Morosuk T, Rosen MA, Zare V (2013) A comparative exergoeconomic analysis of two biomass and co-firing combined power plants. Energy Convers Manag 76:83–91

Ståhl K, Neergaard M (1998) IGCC power plant for biomass utilisation, Värnamo, Sweden. Biomass Bioenergy 15(3):205–211

Technology Assessment for Biomass Power Generation (2004) SMUD ReGen Report, UC Davic

Tees, Renewable Energy Plant Non Technical Summary (2008) PB Power London, UK

Tracking Clean Energy Progress (2016) International Energy Agency. Website: http://www.iea.org/etp/tracking2016/

Wei M (2015) A total cost of ownership model for solid oxide fuel cell systems in power and combined heat and power applications. Fuel Cell Seminar, 17 Nov 2015, Los Angeles, CA

Wickwire S (2007) Biomass Combined Heat and Power catalog of technologies. Washington, DC

World Energy Outlook (2015) International Energy Agency. Website: http://www.worldenergyoutlook.org/weo2015/

Zhang X, Li J, Li G, Feng Z (2006) Dynamic modeling of a hybrid system of the solid oxide fuel cell and recuperative gas turbine. J Power Source 163:523–531

Individual Consumption and Its Effects on Carbon Emissions: A Comparison of Developed and Developing Countries

Pragya Gupta and Madhumati Dutta

Abstract The objective of this study is to determine the nature of household consumption of goods and services and its implications in terms of differences in carbon emissions between developed and developing countries. In this study, data on per capita consumption expenditure (PCCE) for 71 countries (37 developed and 34 developing) and 12 consumption categories were used. The average PCCE by each consumption category over both developed and developing countries was estimated for the years 2004–2011 and compared with that of 1995. The twelve consumption categories were ranked (for developed and developing countries) according to CO_2 emission intensities (kg/USD) as well as the percentage share of each category in total PCCE (in USD) in 1995 and for the period 2004–11. The two sets of ranks were matched to compare the implications of household consumption on carbon emissions. The results indicate that there are similarities as well as differences in terms of the burden of responsibility of causing CO_2 emissions between consumption categories. While 'housing' and 'transport' occupy first and third places, respectively, for both groups, the second place is occupied by 'food and non-alcoholic beverages' in developed and 'clothing and footwear' in developing countries. Also, consumption categories 'communication' in developed and 'education' in developing countries are lowest emitting categories. The trends also revealed that *changes* in the consumption pattern were more significant in developing countries, both in terms of the rate of growth of PCCE and changes in the percentage share of different consumption categories in total PCCE.

Keywords Per capita consumption expenditure · Developed countries
Developing countries · CO_2 emissions

P. Gupta (✉)
Indian Institute of Engineering Science and Technology, Shibpur, India
e-mail: gupta.pragya18@gmail.com

M. Dutta
Department of Humanities and Social Sciences, Indian Institute of Engineering Science and Technology, Shibpur, India

A. Gautam et al. (eds.), *Sustainable Energy and Transportation*,
Energy, Environment, and Sustainability,
https://doi.org/10.1007/978-981-10-7509-4_14

1 Introduction

A country's consumption expenditure increases with the rise in population as well as in the average individual's consumption expenditure. It is well known that rising population is responsible for higher CO_2 emissions in both developed and developing countries (Stern 2007; IPCC 2007; UNEP 2010). However, the importance of the role of per capita consumption demand in causing climate change has been increasing (UNDESA 2007; Lorek and Spangenberg 2001; OECD 2008). The underlying demand for products has implications on carbon emissions. Historically, 80% of GHG/CO_2 emissions are contributed by 20% of the population with the highest consumption levels (Satterthwaite 2009). The environmental impact from household activities has grown in the past three decades and is expected to intensify in the next 20 years, particularly in the areas of energy, transport and waste (OECD 2002). The share of residential sector in total CO_2 emissions is 17% (Nejat et al. 2015). Although technological innovations have reduced the energy and material intensity of many consumer goods, the benefits have been offset by the increased volume of goods/services used and discarded as well as the composition of consumer demand (OECD 2002). As the process of economic development gains pace, the structure of consumption also changes. This has implications on carbon emissions. These changes will differ between developed and developing countries.

2 Literature Review and Objectives

Efforts have been made in the past to calculate the direct and indirect CO_2 emissions from households of individual countries (Herendeen 1978; Pachauri and Spreng 2002; Reinders et al. 2003; Nijdam et al. 2005; Wiedmann et al. 2006; Kerkhof et al. 2009; Gough et al. 2012; Büchs and Schnepf 2013; Steen-Olsen et al. 2016). These studies use variations of input–output methodology to calculate direct and indirect CO_2/GHG emissions due to household consumption. Some of the studies are briefly discussed.

Liu et al. (2011) concluded that indirect carbon emissions of Chinese rural and urban households are much higher than the direct emissions. In the study, out of six consumption categories, indirect emissions of urban households were highest from 'miscellaneous commodities and services', while in rural areas, 'food and tobacco' was the highest emitting consumption category. Pachauri and Spreng (2002) calculated direct and indirect energy requirements of Indian households. They concluded that energy intensity of 'food, beverages and tobacco' was highest and increased over the period of time, followed by 'transport and communication', 'clothing and footwear', 'education and recreation', 'housing and household effects', 'other services' and 'medical care and hygiene'. Cohen et al. (2005) obtained similar results for Brazil. However, housing ranked second in terms of energy intensity in his study. Reinders et al. (2003) calculated direct and indirect energy requirements of households of 11 European Union countries using their household consumption expenditure data for 12 consumption categories. The results of the study indicated that 'electricity, gas and other fuels for housing' emits highest CO_2, followed by 'food and beverages', 'fuel for transport', 'housing' (this category is equivalent to the 'household goods and services' category in the current study), 'recreation and culture', 'furnishing and transport', 'hotels, cafes and restaurants', 'clothing and footwear', 'miscellaneous goods and services' and 'education' and 'communication'. Shui and Dowlatabadi (2005) ranked consumer activities based on CO_2 emissions for the USA. The ranking is in the following order: housing operations, personal travel, home energy, transportation operation, food and beverages, apparel, others, personal insurance, entertainment and health care. The study conducted by Nijdam et al. (2005) also concluded that energy requirements and environmental load on housing were highest in the Netherlands, followed by food, leisure, personal care, furnishing, labour and clothing. Ivanova et al. (2016) calculated the environmental pressure from household consumption for 43 countries and found that mobility, shelter and food had the highest 'environmental footprint'. Similar results were obtained by Gough et al. (2012), Kerkhof et al. (2009), Wiedmann et al. (2006) and Herendeen (1978).

The survey revealed that CO_2/GHG emissions from 'housing' in developed countries were higher, followed by 'food and non-alcoholic beverages' and 'alcoholic beverages and tobacco'. In developing countries, CO_2/GHG emissions from 'food and non-alcoholic beverages' and 'alcoholic beverages and tobacco' were higher, followed by 'housing'.

Most of the above studies have been carried out on developed countries. In comparison, fewer studies are conducted on developing countries. For example, Das and Paul (2014) and China—Wang and Yang (2014), Xu et al. (2016). Secondly, even fewer studies have compared emissions from household consumption between developed and developing countries (one such effort has been made by Lenzen et al. 2006). Also, no specific conclusions could be drawn about CO_2 emissions from consumption categories other than 'housing', 'food and non-alcoholic beverages' and 'alcoholic beverages and tobacco' because they were not comparable with the consumption categories taken into consideration by other studies.

This paper attempts to fill these research gaps. The objective of the present study is to compare trends in the structure and growth of per capita household consumption expenditure between developed and developing countries and differences in its implications on CO_2 emissions across various product groups.

3 Methodology

Annual data on per capita household consumption expenditure for 12 consumption categories for 71 countries for the years 1990, 1995, 2004–2011 measured at current prices in USD (US Dollars), published by Euromonitor International (2012) were used. However, data for the year 1990 were not taken into consideration as data for some of the countries for this year were missing. The list of the countries is given in Appendix (Table 10). These 71 countries were divided into 'developed' and 'developing' countries based on the Human Development Index ranking of 2013 (UNDP 2013). Accordingly, countries with 'Very High Human Development Index', i.e. countries with ranking between 1 and 47, were considered as 'developed', and rest of the countries, i.e. from ranking 48 onwards, were considered as 'developing'. This division resulted in 38 developed and 33 developing countries. Data on per capita consumption expenditure were available for 12 consumption categories, namely alcoholic beverages and tobacco, food and non-alcoholic beverages, clothing and footwear, housing, household goods and services, health goods and medical services, communication, leisure and recreation, hotels and catering, transport, education, and miscellaneous goods and services.

After the grouping of 'developed' and 'developing' countries, per capita household expenditure of the two groups was calculated for each of the 12 consumption categories. For this purpose, average of per capita household expenditure of 'all countries' in each of the categories was calculated. This was followed by the calculation of average for all the years taken into consideration, i.e. 1995 and 2004–2011. The results of these calculations are shown in Tables 1, 2, 3, 4, 5 and 6. Henceforth, all the expenditure values will be average of 'all countries' (in each country category) unless otherwise mentioned and will be referred to as per capita consumption expenditure (PCCE). As mentioned above, PCCE values were given in current prices. These values were converted into 1997 prices using Consumer Price Indices for the USA. Now, average per capita expenditure patterns across various consumption categories over time are discussed in detail. Expenditure trends are discussed for two time periods 1995 and 2004–11.

Table 1 Emission rankings of consumption categories

Consumption category	Emission ranking	
	Developed countries	Developing countries
Alcoholic beverages and tobacco	8	9
Food and non-alcoholic beverages	3	10
Clothing and footwear	4	2
Housing	1	1
Household goods and services	6	8
Health goods and medical services	12	3
Communication	11	11
Leisure and recreation	7	6
Hotels and catering	5	7
Transport	2	5
Education	9	12
Miscellaneous goods and services	10	4

Source Calculated from Table 11

Table 2 Average per capita consumption expenditure (in USD and 1997 prices)

Year	Developed countries	Developing countries
1995	10,941.58	1554.24
2004	12,277.66	1600.99
2005	12550.29	1770.34
2006	12879.97	1907.97
2007	14133.95	2213.79
2008	14727.51	2483.18
2009	13419.23	2323.61
2010	13665.48	2540.21
2011	14493.48	2678.46

Source Calculated from Euromonitor International (2012)

Table 3 Annual growth rate of per capita consumption expenditure (%)

Year	Developed countries	Developing countries
1995–2004	12.21	3.01
2004–05	2.22	10.58
2005–06	2.63	7.77
2006–07	9.74	16.03
2007–08	4.2	12.17
2008–09	−8.88	−6.43
2009–10	1.84	9.32
2010–11	6.06	5.44

Source Calculated from Euromonitor International (2012)

Studies discussed in Sect. 2 do not calculate carbon emissions in terms of emissions per unit of expenditure nor do they calculate emissions for developed and developing countries as a group. Secondly, consumption categories in these studies are not comparable with the consumption categories taken into consideration in the present study. Since the data on emissions per unit expenditure are not available for developed and developing countries as a group, a study carried out by Weber (2008) is used to rank consumption categories in developed countries and another study carried out by Grunewald et al. (2012) is used to rank consumption categories in developing countries (Table 1).

The above-mentioned studies calculated CO_2 emission intensities (emissions per unit of a currency) of an elaborate set of consumption categories for USA (Weber 2008) and India (Grunewald et al. 2012) that were most comparable with the present study.

Emission intensities calculated by Weber (2008) and Grunewald et al. (2012) were for the year 2004.[1] The difference was that the emissions data for the USA

[1]Emissions by India and the USA have been compared in detail in the next phase of this research and are not within the scope of this paper.

Table 4 (a) Percentage share of per capita consumption expenditure in developed countries, (b) percentage share of per capita consumption expenditure in developing countries

(a)									
Commodity group	1995	2004	2005	2006	2007	2008	2009	2010	2011
Alcoholic beverages and tobacco	3.67	3.61	3.54	3.46	3.46	3.48	3.64	3.57	3.55
Food and non-alcoholic beverages	14.39	12.95	12.7	12.59	12.57	12.9	13.07	12.83	12.78
Clothing and footwear	6.62	5.36	5.34	5.35	5.35	5.16	5.09	4.98	4.92
Housing	21.77	22.07	22.16	22.22	21.95	22.42	22.85	22.87	22.94
Household goods and services	6.48	5.97	5.93	5.89	5.9	5.78	5.7	5.64	5.57
Health goods and medical services	4.65	5.25	5.26	5.13	5.09	5.22	5.59	5.7	5.77
Communication	1.91	3.11	3.11	3.1	3.05	3.01	2.94	2.93	2.95
Leisure and recreation	9	9.37	9.31	9.26	9.32	9.16	9.18	9.17	9.11
Hotels and catering	7.54	7.67	7.68	7.73	7.79	7.76	7.68	7.63	7.61
Transport	12.08	12.32	12.49	12.53	12.52	12.38	11.7	11.96	12.05
Education	1.38	1.56	1.59	1.61	1.61	1.6	1.65	1.68	1.7
Miscellaneous goods and services	10.51	10.75	10.88	11.11	11.39	11.11	10.92	11.04	11.05
Total	100	100	100	100	100	100	100	100	100
(b)									
Alcoholic beverages and tobacco	3.02	2.94	2.98	3.03	3.08	3.07	3.08	3.15	3.14
Food and non-alcoholic beverages	31.29	27.71	27.11	26.9	26.81	27.19	27.25	26.95	26.76
Clothing and footwear	8.2	6.65	6.4	6.31	6.16	6.05	5.88	5.79	5.69
Housing	16	17.94	18.09	17.93	17.9	17.8	17.69	17.44	17.41
Household goods and services	7.43	6.88	6.83	6.72	6.59	6.41	6.27	6.2	6.14
Health goods and medical services	3.5	4.14	4.2	4.27	4.33	4.37	4.57	4.66	4.74
Communication	1.94	3.68	3.83	3.95	4.08	4.13	4.27	4.38	4.4
Leisure and recreation	4.59	4.15	4.2	4.19	4.18	4.09	4.02	4.03	4.05
Hotels and catering	4.56	4.83	5.04	5.18	5.34	5.5	5.9	6	5.99
Transport	10.93	11.36	11.7	11.83	11.78	11.76	11.26	11.43	11.5
Education	2.34	3.07	3.04	3.07	3.03	3	3.09	3.14	3.21

(continued)

Table 4 (continued)

(a)									
Commodity group	1995	2004	2005	2006	2007	2008	2009	2010	2011
Miscellaneous goods and services	6.2	6.65	6.57	6.62	6.72	6.62	6.73	6.85	6.97
Total	100	100	100	100	100	100	100	100	100

Source Calculated from Euromonitor International (2012)

Table 5 Rate of growth of per capita consumption expenditure (%)

Commodity group	1995–2004		2004–11	
	Developed countries	Developing countries	Developed countries	Developing countries
Alcoholic beverages and tobacco	10.24	0.08	16.25	78.78
Food and non-alcoholic beverages	1.04	−8.77	16.43	61.55
Clothing and footwear	−9.09	−16.48	8.19	43.25
Housing	13.77	15.48	22.63	62.35
Household goods and services	3.39	−4.53	10.13	49.18
Health goods and medical services	26.76	21.92	29.56	91.34
Communication	83.04	95.63	11.79	99.99
Leisure and recreation	16.78	−6.88	14.72	63.39
Hotels and catering	14.24	9.08	17.05	107.76
Transport	14.48	7.06	15.41	69.3
Education	26.73	35.04	29.07	75.19
Miscellaneous goods and services	14.84	10.49	21.23	75.32

Source Calculated from Euromonitor International (2012)

were in terms of kg CO_2/USD, while the same for India were in terms of kt CO_2/100,000 Indian Rupees (kt = kilo tonne). Hence, the emissions data for India were converted into kg CO_2/USD. To convert kt CO_2/100,000 Indian Rupees into kg CO_2/USD, Indian emission values were multiplied by a factor of 144.7.[2] Sum of total emissions (emissions per USD multiplied by the monthly PCCE) of various commodities in each consumption category divided by the sum of monthly PCCE was calculated to arrive at the emissions per USD of a particular consumption category (Table 11).

[2]1 kt = 10^6 kg and PPP conversion factor for the year 2004 was \$1 = Rs.14.47.

Table 6 (a) Rate of growth of per capita consumption expenditure in developed countries (%), (b) rate of growth of per capita consumption expenditure in developing countries (%)

Commodity group	1995–2004	2004–05	2005–06	2006–07	2007–08	2008–09	2009–10	2010–11
(a)								
Alcoholic beverages and tobacco	10.24	0.35	0.36	9.77	4.77	−5.07	−0.04	5.77
Food and non-alcoholic beverages	1.04	0.22	1.77	9.54	6.96	−7.96	0.08	5.77
Clothing and footwear	−9.09	1.8	2.94	9.63	0.49	−10.39	−0.21	4.79
Housing	13.77	2.67	2.92	8.37	6.46	−7.39	1.98	6.52
Household goods and services	3.39	1.58	1.94	9.85	2.13	−10.43	0.78	5.01
Health goods and medical services	26.76	2.4	0.06	8.87	6.96	−2.77	3.85	7.55
Communication	83.04	2.27	2.25	8.03	2.74	−11.26	1.49	6.95
Leisure and recreation	16.78	1.63	2.07	10.41	2.46	−8.95	1.81	5.46
Hotels and catering	14.24	2.31	3.39	10.48	3.8	−9.96	1.18	5.91
Transport	14.48	3.62	2.93	9.69	3.04	−14.16	4.24	6.99
Education	26.73	4.21	4.15	9.68	3.61	−6.31	4.1	7.29
Miscellaneous goods and services	14.84	3.48	4.74	12.48	1.7	−10.65	2.93	6.32
(b)								
Alcoholic beverages and tobacco	0.08	12.17	9.61	17.97	11.64	−6.02	11.62	5.25
Food and non-alcoholic beverages	−8.77	8.17	6.92	15.68	13.74	−6.21	8.1	4.71
Clothing and footwear	−16.48	6.5	6.19	13.25	10.18	−9.1	7.67	3.73
Housing	15.48	11.49	6.83	15.81	11.57	−7.02	7.79	5.27
Household goods and services	−4.53	9.65	6.11	13.78	9.13	−8.53	8.13	4.4
Health goods and medical services	21.92	12.12	9.43	17.74	13.09	−2	11.32	7.35
Communication	95.63	15.19	11.12	19.73	13.7	−3.27	12.01	5.94
Leisure and recreation	−6.88	12.12	7.43	15.69	9.91	−8.08	9.53	5.95
Hotels and catering	9.08	15.45	10.9	19.5	15.49	0.34	11.34	5.25
Transport	7.06	13.86	8.98	15.54	12	−10.42	10.94	6.08
Education	35.04	9.82	8.54	14.74	11.08	−3.77	11.07	7.89
Miscellaneous goods and services	10.49	9.23	8.63	17.72	10.59	−5	11.32	7.32

Source Calculated from Euromonitor International (2012)

4 Per Capita Consumption Expenditure Trends During 1995 to 2004–11

Per capita household consumption expenditure has steadily increased in both developed and developing countries (Table 2). PCCE in developed countries increased from USD 10,941.58 in 1995 to USD 14,493.48 in 2011, whereas in developing countries, it increased from USD 1554.24 in 1995 to USD 2678.46 in 2011. Average (average of the years 2004–11) PCCE for the period 2004–11 was USD 13,518.45 in developed countries and USD 2189.82 in developing countries. The resulting growth rate of PCCE from 1995 to 2004–11 was 23.6% for developed countries and 40.9% for developing countries. Thus, rise in PCCE was much higher in developing countries as compared to developed ones. Tables 2 and 3 show that PCCE had increased in all the years in both developed and developing countries except in the year 2009. This was the year of economic recession that had mostly affected developed countries. Although developing countries were not affected by this economic recession as severely as developed ones, it did influence some sectors of the developing economies (mainly those sectors which were dependent largely on imports from the developed countries). Percentage fall in PCCE in developed countries was higher than that in the developing ones. This downturn in the global economy might explain the fall in PCCE. However, reasons for this downfall are not within the scope of this study. Between 1995 and 2011, yearly fluctuations in PCCE were more in developed countries for all the consumption categories. In comparison, in developing countries, PCCE on 'food and non-alcoholic beverages', 'clothing and footwear', 'household goods and services' and 'leisure and recreation' reduced between 1995 and 2004, and there was a steep rise in PCCE between 2006 and 2008, especially for 'housing', 'transport' and 'food and non-alcoholic beverages'. The rise was steepest for 'food and non-alcoholic beverages'.

It was interesting to observe that though PCCE had increased over the years (Fig. 1, Tables 12 and 13), the percentage share of the expenditure had not increased for all the consumption categories. Instead, for some, it reduced. Table 4a, b shows the annual percentage share of PCCE in developed and developing countries. These tables revealed that the percentage share of various consumption categories has changed over the years, indicating that the relative importance of the relevant consumption group in the total household budget has changed. Although PCCE of developed countries is much higher than that of the developing ones, it is evident (Fig. 2 and Table 4a, b) that developed countries spent more on 'alcoholic beverages and tobacco', 'housing', 'health goods and medical services', 'leisure and recreation', 'hotels and catering', 'transport' and 'miscellaneous goods and services'.

In contrast, developing countries spent proportionately more on 'food and non-alcoholic beverages', 'clothing and footwear', 'household goods and services', 'communication' and 'education'. Further, the data revealed that both developed

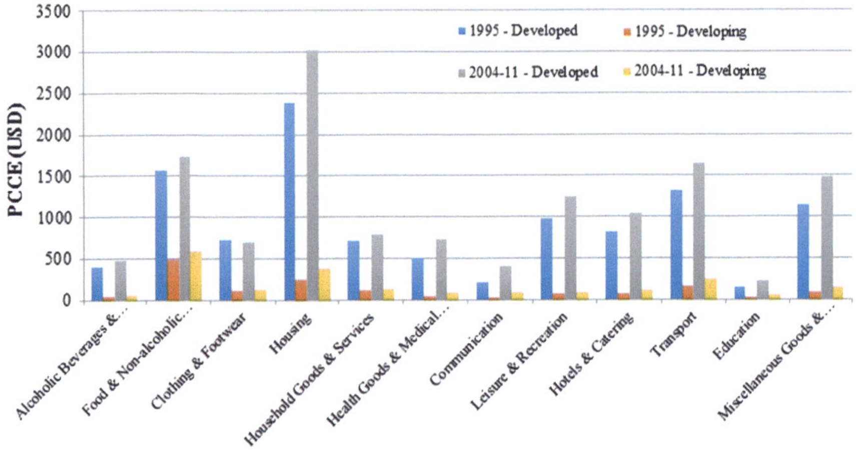

Fig. 1 Per capita consumption expenditure in 1995 and 2004–11 (in USD and 1997 prices). *Source* Derived from Euromonitor International (2012)

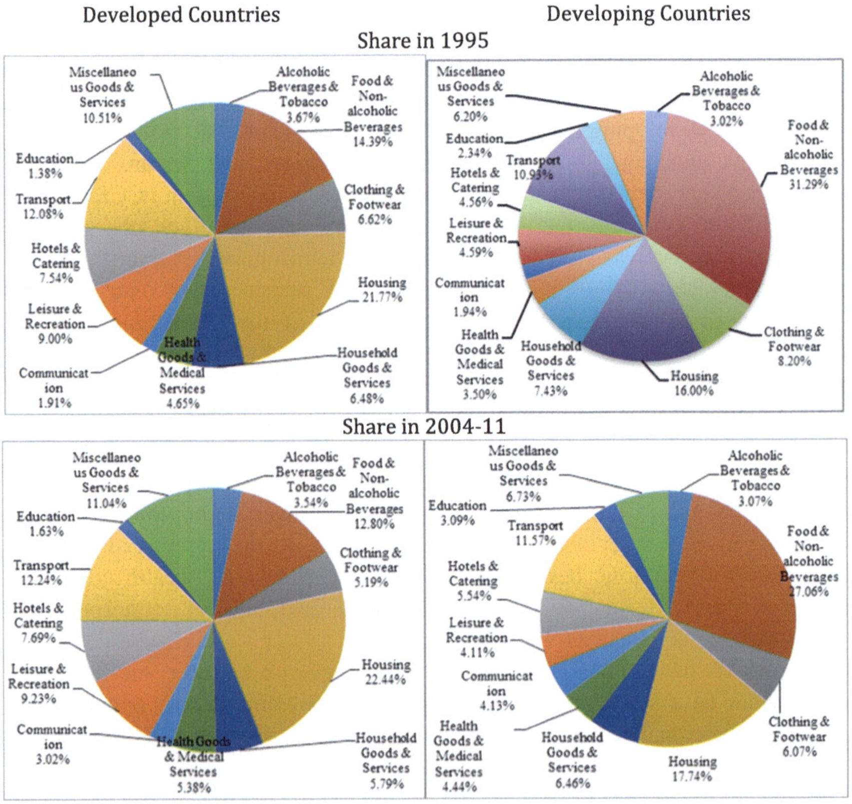

Fig. 2 Percentage share of per capita consumption expenditure of various consumption categories

and developing countries spent maximum on 'food and non-alcoholic beverages', 'housing' and 'transport'. Developed countries spent highest on 'housing', followed by 'food and non-alcoholic beverages' and 'transport', while developing countries spent highest on 'food and non-alcoholic beverages', followed by 'housing' and 'transport'. However, the share of 'food and non-alcoholic beverages' in total expenditure reduced over the years. Though the percentage share of expenditure on 'food and non-alcoholic beverages' reduced, it remained the second most important item of consumption. Developed countries spent the least on 'education'. In comparison, developing countries spent the least on 'communication' in 1995 and 'alcoholic beverages and tobacco' in 2004–11.

The rate of growth of PCCE has also varied across the 12 consumption categories in both groups of countries. These growth rates are shown in Tables 5 and 6a, b. The annual rate of growth of PCCE over the years has been higher in developing countries than that in the developed ones as shown in Table 3. However, during 1995–2004 and 2010–11, the rate of growth was lower in developing countries. The rate of growth of PCCE was higher in developed countries for majority of consumption categories during 1995–2004 (Table 7). The rate of growth of PCCE was highest for 'communication' and lowest for 'clothing and footwear' in both developed and developing countries (Table 5). In fact, the growth rate was negative for 'clothing and footwear'.

During 2004–11, the rate of growth of PCCE was highest for 'health goods and medical services' and 'education', followed by 'housing' in developed countries. In developing countries, the rate of growth of PCCE was highest for 'hotels and catering', followed by 'communication' and 'alcoholic beverages and tobacco'. 'Clothing and footwear' remained the weakest growing consumption category in both developed and developing countries. Changes observed in the PCCE of different consumption categories are summarized in Table 7.

Based on the percentage share of various consumption categories in developed and developing countries, consumption categories were ranked for the year 1995 and for the period 2004–11. Wilcoxon rank-sum test was used to determine the statistical significance of difference in expenditure ranking of the consumption categories between developed and developing countries and across the two time periods considered in the study. It is a nonparametric test used for comparing ordinal ranks. The test was carried out at 5% significance level. The expenditure ranks were significant across country groups (p-value for developed and developing countries for 1995 = 0.00048; p-value for developed and developing countries for 2004–11 = 0.00048) and across time (p-value for developed countries for the years 1995 and 2004–11 = 0.000977; p-value for developing countries for the years 1995 and 2004–11 = 0.000488).

These rankings are shown in Table 8. From the rankings, it is revealed that developing countries have experienced more changes in the relative importance of the consumption categories than the developed countries. In developed countries,

Table 7 Changes in per capita consumption expenditure patterns in developed versus developing countries

Developed countries	Developing countries
Consumption categories for which percentage share of PCCE increased from 1995 to 2004–11	**Consumption categories for which percentage share of PCCE increased from 1995 to 2004–11**
Housing, health goods and medical services, communication, leisure and recreation, hotels and catering, transport, education, miscellaneous goods and services	Housing, health goods and medical services, communication, hotels and catering, transport, education, miscellaneous goods and services
Consumption categories for which percentage share of PCCE decreased from 1995 to 2004–11	**Consumption categories for which percentage share of PCCE decreased from 1995 to 2004–11**
Food and non-alcoholic beverages, clothing and footwear, household goods and services	Food and non-alcoholic beverages, clothing and footwear, household goods and services, leisure and recreation
Consumption categories with higher percentage expenditure share	**Consumption categories with higher percentage expenditure share**
Alcoholic beverages and tobacco, housing, health goods and medical services, leisure and recreation, hotels and catering, transport, miscellaneous goods and services	Food and non-alcoholic beverages, clothing and footwear, household goods and services, communication, education
Consumption categories with higher rate of growth of PCCE during 1995–2004	**Consumption categories with higher rate of growth of PCCE during 1995–2004**
Alcoholic beverages and tobacco, food and non-alcoholic beverages, clothing and footwear, household goods and services, health goods and medical services, leisure and recreation, hotels and catering, transport, miscellaneous goods and services	Housing, communication, education
Consumption categories with higher rate of growth of PCCE during 1995 to 2004–11	**Consumption categories with higher rate of growth of PCCE during 1995–2004**
–	All consumption categories

Source Figure 2 and Table 5

the relative importance of 'household goods and services' and 'health goods and medical services' increased, while that of 'clothing and footwear' decreased. The relative importance of the rest of the consumption categories remained the same. In developing countries, the relative importance of 'health goods and medical services', 'hotels and catering', 'communication' and 'miscellaneous goods and services' increased, while the relative importance of 'alcoholic beverages and tobacco', 'clothing and footwear' and 'leisure and recreation' decreased.

Table 8 Changes in the relative importance of consumption categories

Rank	Developed countries		Developing countries	
	1995	2004–11	1995	2004–11
1	Housing	Housing	Food and non-alcoholic beverages	Food and non-alcoholic beverages
2	Food and non-alcoholic beverages and housing	Food and non-alcoholic beverages	Housing	Housing
3	Transport	Transport	Transport	Transport
4	Miscellaneous goods and services	Miscellaneous goods and services	Clothing and footwear	Miscellaneous goods and services
5	Leisure and recreation	Leisure and recreation	Household goods and services	Household goods and services
6	Hotels and catering	Hotels and catering	Miscellaneous goods and services	Clothing and footwear
7	Clothing and footwear	Household goods and services	Leisure and recreation	Hotels and catering
8	Household goods and services	Health goods and medical services	Hotels and catering	Health goods and medical services
9	Health goods and medical services	Clothing and footwear	Health goods and medical services	Communication
10	Alcoholic beverages and tobacco	Alcoholic beverages and tobacco	Alcoholic beverages and tobacco	Leisure and recreation
11	Communication	Communication	Education	Education
12	Education	Education	Communication	Alcoholic beverages and tobacco

5 Matching Expenditure and Emission Ranks

In order to study the effect of changing consumption pattern on carbon emissions in developed and developing countries, emission rankings were matched with expenditure rankings. This is shown in Table 9a, b.

In developed countries, 'housing', 'food and non-alcoholic beverages' and 'transport' had high expenditure and emission ranks in 1995 as well as 2004–11. The rate of growth of PCCE of these three consumption categories was lower as

Table 9 (a) Matching expenditure and emission ranks in developed countries, (b) matching expenditure and emission ranks in developing countries

Consumption categories	Rank		
	Expenditure—1995	Expenditure (2004–11)	Emissions
Alcoholic beverages and tobacco	10	10	8
Food and non-alcoholic beverages	2	2	3
Clothing and footwear	7	9	4
Housing	1	1	1
Household goods and services	8	7	6
Health goods and medical services	9	8	12
Communication	11	11	11
Leisure and recreation	5	5	7
Hotels and catering	6	6	5
Transport	3	3	2
Education	12	12	9
Miscellaneous goods and services	4	4	10
(b)			
Alcoholic beverages and tobacco	10	12	9
Food and non-alcoholic beverages	1	1	10
Clothing and footwear	4	6	2
Housing	2	2	1
Household goods and services	5	5	8
Health goods and medical services	9	8	3
Communication	12	9	11
Leisure and recreation	7	10	6
Hotels and catering	8	7	7
Transport	3	3	5
Education	11	11	12
Miscellaneous goods and services	6	4	4

Source Calculated from Tables 1 and 8

compared to some of the other categories. In developing countries, 'housing', 'clothing and footwear' and 'transport' had high expenditure and emission ranks in 1995 and 2004–11. Though 'food and non-alcoholic beverages' was the highest expenditure consumption category, it was least emitting in the developing countries. In developing countries, changes in consumption pattern from 1995 to 2004–2011 were more significant. People spent proportionately more on 'health goods and medical services', 'hotels and catering', 'communication' and 'miscellaneous goods and services' during 2004–11 than in 1995. Amongst these four consumption categories, 'health goods and medical services' was the most carbon emitting and 'communication' was the least carbon emitting with second highest rate of growth of PCCE during 2004–11. On the other hand, people in developing countries spent proportionately less on 'alcoholic beverages and tobacco' and emissions per USD from this category were also low. Also, developing countries spent proportionately lesser amount on 'clothing and footwear' and 'leisure and recreation' during 2004–11 than in 1995. Amongst these two consumption categories, while the rate of growth of PCCE was lowest for 'clothing and footwear'. In developed countries, there were not many significant changes in the PCCE patterns (in terms of percentage share) from 1995 to 2004–11 (Table 8), except that people spent proportionately more on 'health goods and medical services' and 'household goods and services' and lesser on 'clothing and footwear'. However, 'health goods and medical services' was the lowest emitting consumption category. At the same time, it had the highest rate of growth of PCCE during 2004–11. Emissions per USD from 'household goods and services', which had a lower rate of growth of PCCE, were much higher. Developed countries spent less on 'clothing and footwear', which was more carbon-emitting consumption category and its rate of growth of PCCE was the lowest.

6 Conclusions

Changes in consumer preferences (as indicated by changes in percentage share of consumption categories from 1995 to 2004–11), 'household goods and services' became more CO_2-emitting consumption category in developed countries, while 'health goods and medical services' and 'miscellaneous goods and services' were more CO_2 emitting in developing countries.

A ranking of 12 consumption categories based on their per capita household consumption expenditure (in 1995 and during 2004–11) and CO_2 emission intensities (amount of CO_2 (in kg) emitted per USD spent) allows to assign the responsibility of causing higher carbon emissions of each country group on different consumption categories. Though total PCCE of developing countries was not as high as that of the developed countries, there were significant structural changes in the consumption patterns of developing countries, as their consumption basket shifted food to non-food, which includes high CO_2 emitting and higher growth rates consumption categories like 'housing', 'clothing and footwear' and 'health goods and medical services'.

On the other hand, PCCE in developed countries was much higher but there were not too many changes in their structure of consumption. Although their consumption basket also shifted from food to non-food (percentage share of non-food expenditure increased, 'food and alcoholic beverages' still remained high CO_2-emitting consumption category along with 'housing' and 'transport'.

Thus, for reducing CO_2 emissions from household consumption, different product groups need to be targeted in developed and developing countries.

Appendix

See Tables 10, 11, 12, 13.

Table 10 List of developed and developing countries

Developed countries	Developing countries
Australia	Algeria
Austria	Argentina
Belgium	Azerbaijan
Canada	Belarus
Chile	Bolivia
Croatia	Brazil
Czech Republic	Bulgaria
Denmark	China
Estonia	Colombia
Finland	Ecuador
France	Egypt
Germany	India
Greece	Indonesia
Hong Kong, China	Jordan
Hungary	Kazakhstan
Ireland	Kuwait
Israel	Malaysia
Italy	Mexico
Japan	Morocco
Latvia	Nigeria
Lithuania	Pakistan
Netherlands	Peru
New Zealand	Philippines

(continued)

Table 10 (continued)

Developed countries	Developing countries
Norway	Romania
Poland	Russia
Portugal	Saudi Arabia
Singapore	South Africa
Slovakia	Thailand
Slovenia	Tunisia
South Korea	Turkey
Spain	Turkmenistan
Sweden	Ukraine
Switzerland	Venezuela
Taiwan	Vietnam
United Arab Emirates	
UK	
USA	

Table 11 Emission intensities in developed and developing countries (in kg CO_2/USD)

Commodity group	Developed countries	Developing countries
Alcoholic beverages and tobacco	0.4196	0.3699
Food and non-alcoholic beverages	0.7354	0.3317
Clothing and footwear	0.5814	0.6332
Housing	2.3795	1.4183
Household goods and services	0.4955	0.5794
Health goods and medical services	0.34	0.6312
Communication	0.3449	0.151
Leisure and recreation	0.4557	0.5956
Hotels and catering	0.5102	0.5858
Transport	0.8123	0.6043
Education	0.4167	0.1105
Miscellaneous goods and services	0.4108	0.6157

Source Weber (2008) and Grunewald et al. (2012)

Table 12 Annual per capita consumption expenditure in developed countries for various consumption categories (in USD and 1997 prices)

Commodity group	1995	2004	2005	2006	2007	2008	2009	2010	2011
Alcoholic beverages and tobacco	401.53	442.64	444.19	445.79	489.32	512.66	486.68	486.48	514.54
Food and non-alcoholic beverages	1573.63	1589.93	1593.47	1621.74	1776.37	1899.93	1748.74	1750.15	1851.13
Clothing and footwear	723.78	657.96	669.81	689.52	755.89	759.59	680.71	679.29	711.83
Housing	2380.64	2708.54	2780.76	2861.94	3101.42	3301.7	3057.8	3118.22	3321.6
Household goods and services	709.04	733.08	744.68	759.16	833.92	851.68	762.86	768.84	807.37
Health goods and medical services	508.76	644.9	660.35	660.74	719.36	769.39	748.1	776.93	835.56
Communication	208.74	382.08	390.74	399.52	431.6	443.44	393.52	399.37	427.14
Leisure and recreation	984.72	1149.92	1168.7	1192.91	1317.13	1349.49	1228.72	1250.92	1319.22
Hotels and catering	824.21	941.55	963.32	995.96	1100.31	1142.14	1028.39	1040.54	1102.07
Transport	1321.11	1512.45	1567.23	1613.23	1769.6	1823.43	1565.2	1631.5	1745.54
Education	150.7	190.99	199.02	207.28	227.34	235.55	220.69	229.75	246.51
Miscellaneous goods and services	1149.15	1319.71	1365.7	1430.46	1608.96	1636.28	1462.06	1504.85	1599.95

Source Calculated from Euromonitor International (2012)

Table 13 Annual per capita consumption expenditure in developing countries for various consumption categories (in USD and 1997 prices)

Commodity group	1995	2004	2005	2006	2007	2008	2009	2010	2011
Alcoholic beverages and tobacco	47.01	47.05	52.78	57.85	68.25	76.19	71.61	79.92	84.12
Food and non-alcoholic beverages	486.33	443.68	479.92	513.15	593.6	675.13	633.2	684.51	716.76
Clothing and footwear	127.44	106.44	113.36	120.37	136.32	150.2	136.54	147	152.48
Housing	248.74	287.24	320.24	342.11	396.2	442.03	410.99	442.99	466.33
Household goods and services	115.43	110.2	120.84	128.22	145.89	159.22	145.63	157.48	164.4
Health goods and medical services	54.43	66.36	74.4	81.42	95.86	108.41	106.24	118.28	126.97
Communication	30.11	58.91	67.85	75.4	90.27	102.64	99.28	111.2	117.81
Leisure and recreation	71.27	66.37	74.42	79.95	92.5	101.66	93.45	102.35	108.44
Hotels and catering	70.83	77.26	89.2	98.92	118.21	136.52	136.98	152.51	160.51
Transport	169.92	181.92	207.14	225.75	260.83	292.13	261.69	290.33	307.98
Education	36.34	49.07	53.89	58.49	67.12	74.55	71.74	79.68	85.97
Miscellaneous goods and services	96.37	106.48	116.31	126.35	148.73	164.48	156.27	173.95	186.68

Source Calculated from Euromonitor International (2012)

References

Büchs M, Schnepf SV (2013) Who emits most? Associations between socio-economic factors and UK households' home energy, transport, indirect and total CO_2 emissions. Ecol Econ 90: 114–123

Cohen C, Lenzen M, Schaeffer R (2005) Energy requirements of households in Brazil. Energy Policy 33(4):555–562

Das A, Paul SK (2014) CO_2 emissions from household consumption in India between 1993 and 94 and 2006–07: a decomposition analysis. Energy Economics 41:90–105

Euromonitor International (2012) World consumer income and expenditure patterns 2013

Gough I, Abdallah S, Collins JR, Smith C (2012) The distribution of total greenhouse gas emissions by households in UK, and some implications for social policy

Grunewald N, Harteisen M, Lay J, Minx J, Renner S (2012) The carbon footprint of Indian households. http://www.iariw.org/papers/2012/GrunewaldPaper.pdf

Herendeen RA (1978) Total energy cost of household consumption in Norway, 1973. Energy 3 (5):615–630

Ivanova D, Stadler K, Steen-Olsen K, Wood R, Vita G, Tukker A, Hertwich EG (2016) Environmental impact assessment of household consumption. J Ind Ecol 20(3):526–536

IPCC (2007) Climate change 2007: synthesis report. Contribution of Working Groups I, II and III to the Fourth Assessment Report of the Intergovernmental Panel on Climate Change

Kerkhof AC, Nonhebel S, Moll HC (2009) Relating the environmental impact of consumption to household expenditures: an input–output analysis. Ecol Econ 68:1160–1170

Lenzen M, Wier M, Cohen C, Hayami H, Pachauri S, Schaeffer R (2006) A comparative multivariate analysis of household energy requirements in Australia, Brazil, Denmark, India and Japan. Energy 31(2–3):181–207

Liu L, Wu G, Wang J, Wei Y (2011) China's carbon emissions from urban and rural households during 1992–2007. J Clean Prod 19:1754–1762

Lorek S, Spangenberg JH (2001) Environmentally sustainable household consumption: from aggregate environmental pressures to indicators for priority fields of action. Wuppertal Institute for Climate, Environment and Energy

Nejat P, Jomehzadeh F, Taheri MM, Gohari M, Majid MZA (2015) A global review of energy consumption, CO_2 emissions and policy in the residential sector (with an overview of the top ten CO_2 emitting countries). Renew Sustain Energy Rev 43:843–862

Nijdam DS, Wilting HC, Goedkoop MJ, Madsen J (2005) Environmental load from Dutch private consumption: how much damage takes place abroad. J Ind Ecol 9:147–168

OECD (2002) Towards sustainable household consumption? Trends and policies in OECD countries

OECD (2008) Household behaviour and the environment: reviewing the evidence, pp 1–264

Pachauri S, Spreng D (2002) Direct and indirect energy requirements of households in India. Energy Policy 30:511–523

Reinders A, Vringer K, Blok K (2003) The direct and indirect energy requirement of households in the European Union. Energy Policy 31:139–153

Satterthwaite D (2009) The implications of population growth and urbanization for climate change. Environ Urbanization 21(2):545–567

Steen-Olsen K, Wood R, Hertwich EG (2016) The carbon footprint of Norwegian household consumption 1999–2012. J Ind Ecol 20:582–592

Stern N (2007) The economics of climate change: the stern review. Cambridge University Press

Shui B, Dowlatabadi H (2005) Consumer lifestyle approach to us energy use and the related CO_2 emissions. Energy Policy 33:197–208

UNDESA (2007) Sustainable consumption and production: promoting climate-friendly household consumption patterns. Division for Sustainable Development, Policy Integration and Analysis Branch

UNDP (2013) The rise of the south: human progress in a diverse world. Human Development Report

UNEP (2010) Assessing the environmental impacts of consumption and production: priority products and materials. In: Hertwich E, van der Voet E, Suh S, Tukker A, Huijbregts M, Kazmierczyk P, Lenzen M, McNeely J, Moriguchi Y (eds) A report of the Working Group on the Environmental Impacts of Products and Materials to the International Panel for Sustainable Resource Management

Wang Z, Yang L (2014) Indirect carbon emissions in household consumption: evidence from the urban and rural area in China. J Clean Prod 78:94–103

Weber CL (2008) Trade, consumption and climate change: an input-output study of the United States

Wiedmann T, Min J, Barrett J, Wackernagel M (2006) Allocating ecological footprints to final consumption categories with input–output analysis. Ecol Econ 56:28–48

Xu X, Han L, Lv X (2016) Household carbon inequality in urban China, its sources and determinants. Ecol Econ 128:77–86

Patterns of Energy Consumption in Manufacturing: Looking at the Eastern States of India

Gopa Ghosh and Madhumati Dutta

Abstract Energy is a vital input in achieving rapid economic growth in India and therefore in each of its states. Key policies such as 'Make in India' and 'Smart Cities Mission' are expected to only contribute to a rapid increase in energy consumption in most Indian States. Yet, India is also committed to reducing its greenhouse gas emissions. Hence, it needs to adopt policies that would achieve growth and, at the same time, reduce carbon emissions. Manufacturing is a significant component of the engine that spurs economic growth, and it is a major consumer of energy. This study aims to identify the core factors that have influenced energy consumption by the manufacturing industries of five eastern states (Bihar, Chhattisgarh, Jharkhand, Odisha, and West Bengal) of India in the period 2010–11 to 2014–15 and also intends to make an interstate comparison of energy consumption. We conduct an index decomposition analysis, more specifically the log mean Divisia index, to identify key factors behind the increase in energy consumption. The findings of the study suggest that besides energy intensity, 'level of activity' is a major contributing factor. The findings also indicate that there remains a lot of scope for improving energy-related policies with regard to the manufacturing industries of eastern India.

1 Introduction

According to the IPCC's estimates, the world can emit only about 2900–3300 Gt of CO_2 from all sources from the earliest period of the Industrial Revolution until 2100 to ensure that global temperature does not rise more than 2 °C. As it has already emitted about 1,900 Gt of CO_2, it can emit only between 1000 and 1400 Gt in the

G. Ghosh (✉) · M. Dutta
Indian Institute of Engineering Science and Technology, Shibpur, Howrah, India
e-mail: gopaghosh37@yahoo.com

M. Dutta
e-mail: madhumatidutta@yahoo.co.in

A. Gautam et al. (eds.), *Sustainable Energy and Transportation*, Energy, Environment, and Sustainability,
https://doi.org/10.1007/978-981-10-7509-4_15

period 2012–2100 (IPCC 2013). India is the third largest carbon emitting country, and although its share in global emissions is still 7%, significantly less than that of China (29%) or the USA (14%), its future potential as an emitter is enormous—over 1990–2012; the change in emissions was 236% (second only to China for which it was 262%), and in 2016, the change in emissions per capita has been 5.1%, whereas for China, it was negative (−0.7%) (Olivier et al. 2016; The Statistics Portal 2016).

Energy consumption, linked with the activity level of an economy, is one of the major causes of GHG (including CO_2) emissions. As the energy-intensive industrial sector is a crucial component of economic activity in India, policy for the abatement of carbon emissions has to include energy management in this sector. Industry was the second largest energy consuming sector in India in 2010, consuming 37% of total energy. During 2010–2014, the compound annual growth rate (CAGR) of India's CO_2 emissions from fuel combustion by the industrial sector was 7.40% per annum, which was quite high, and total CO_2 emissions from fuel combustion were 5.57% during the same period (IEA 2012, 2013, 2014, 2015a, and 2016). In post 2008–09, industrial performance remained lackluster due to high rates of interest, long-term infrastructure bottlenecks, and low domestic and external demand. The latest gross domestic product (GDP) estimates show that industry grew by just 1.0% in 2012–13 and slowed further in 2013–14, posting a modest increase of 0.4% (Government of India 2013–14).

'Make in India' is an initiative launched by the Government of India in 2014 to encourage national as well as multi-national companies to manufacture their products in India. To achieve this objective, energy demand in the industrial sector is projected to increase rapidly (4.4% annually) until 2040 (IEA 2015b). On the other hand, the 'Go Green' slogan requires businesses to be environmentally compliant (Kaur 2016).

Year 2015 was a historic year because it was the hottest recorded year since 1880 and 2015 closed with the adoption of the landmark Paris Agreement on Climate Change by 194 countries and the European Union (Olivier et al. 2016). India, too, submitted its new climate action plan on October 1, 2015 to the UN Framework Convention on Climate Change (UNFCCC), and according to the Intended Nationally Determined Contribution (INDC), India wants to reduce the emissions intensity of its GDP (from the 2005 level) by 33–35% by 2030 (UNFCCC 2015; Government of India 2015). Before INDC, India has also made a commitment to reduce the energy intensity of its GDP by 20–25% (from 2005 levels) within 2020 (Planning Commission 2011), and several policy measures were launched to achieve this goal.

In this context, therefore, it is important to identify the policies that each state of India should adopt to reduce the consumption of energy in the manufacturing sector and for that, it is necessary to obtain a clear view of the present status and historical trend of energy consumption at the state level. The Government of India has already taken an initiative to enhance energy efficiency as one of the challenges in the National Action Plan on Climate Change (2008). Different state and local governments have also addressed this goal, which requires the participation of industry located within these states or local governments. This study has used data over a five year period (2010–11 to 2014–15[1]) to obtain the current scenario of energy use, in order to determine the necessary policies that would facilitate the twin objectives of 'Make in India' and a low carbon future. Several crucial factors determine energy consumption. The change in energy consumption is described using a decomposition methodology to identify the crucial influencing factors. The present study focuses on the aggregate manufacturing sector of eastern India (i.e., Bihar, Chhattisgarh, Jharkhand, Odisha, and West Bengal) to know the status of energy consumption by these states, which are abundantly blessed with rich mineral resources, specifically coal—one of the major sources of energy in manufacturing. In 2015, approximately 79.4% (Jharkhand 26.4%, Odisha 24.7%, Chhattisgarh 17.9%, West Bengal 10.3%, and Bihar 0.01%) of India's reserved coal was in the eastern part of India (CSO 2016). This study also aims at comparing the manufacturing sector's energy consumption among the five above-mentioned states for the period 2010–11 to 2014–15.

The chapter is arranged in the following manner. The decomposition method used in this study and the required data for analysis is explained in Sect. 2. In Sect. 3, we present a brief overview of the trends in the industrial output growth and energy consumption of the aggregate manufacturing industries of the five eastern Indian states. The data analysis itself as well as a discussion of the results is presented in Sect. 4. In Sect. 5, we discuss the policies which have already been promoted to encourage the growth of the manufacturing industries and to reduce energy consumption in an efficient manner. We add some conclusions in Sect. 6.

2 The Methodology and Data

Policymakers require a clear view of the present status and historical trend of energy consumption to design appropriate policies toward efficiency in energy consumption in the manufacturing sector at the state level, and for this, we need to identify the underlying factors affecting energy demand. Various analytical methodologies are available in the literature for energy demand analysis, and three important approaches are (a) simple descriptive analysis, (b) econometric analysis, and (c) decomposition analysis (Bhattacharyya 2011). The decomposition methods have been used by Jenne and Cattell (1983), Marlay (1984), Park (1992), Liu et al. (1992), Choi et al. (1995), Ang and Zhang (2000), Ang (2004), and Shahiduzzaman and Alam (2012), and this study also uses decomposition analysis to identify the factors that have influenced changes in energy consumption of manufacturing sector of eastern India for the time period 2010–11 to 2014–15 and uses the results to design a policy toward energy efficient and low carbon future. The literature

[1]Available up to 2014–15.

indicates that there are two different types of methodologies (Hoekstra and Bergh 2003) available to decompose change of a variable into their determinant effects. They are structural decomposition analysis (SDA) and index decomposition analysis (IDA). The index decomposition analysis (IDA) method has been classified into two groups: methods linked to the Laspeyres index and methods linked to the Divisia index (Ang and Choi 1997; Ang 2004; Ang et al. 2009). In the literature on energy decomposition methods, there are three different methods available for the IDA method linked to the Divisia index—they are the arithmetic mean Divisia index (AMDI) method, the logarithmic mean Divisia index method I (LMDI I), and the logarithmic mean Divisia index method II (LMDI II). The above-mentioned three methodologies are distinguished according to their individual weight function. The AMDI method uses an arithmetic weight function, whereas the LMDI I and the LMDI II methods use the log mean weight function (Ang 2004). The method used before 2000 was mainly based on the Laspeyres Index and AMDI.[2] Researchers have found that these energy decomposition methods have an unexplained residual term in the decomposition results, and therefore, recent studies have used other methods to get proper results. According to the available literature, the LMDI I is perfect in decomposition and is the most preferred method (Ang 2004), and the LMDI I method has been adopted in this study. According to the LMDI I decomposition technique, the three factors which influence the consumption of energy are as follows: level of economic activity, economic structure, and energy intensity. If the economic structure of the manufacturing sector and conservation measures remains the same as in the base year, then energy consumption can change due to any change in economic activity. Again, any change in the economic structure of the manufacturing industry can affect energy consumption by the sector, when the level of economic activity and energy intensity remain the same as in the base year. Also, energy consumption can change due to revision of energy conservation measures when economic activity and the economic structure remain the same as in the base year. Liu et al. (1992), Choi et al. (1995), Ray and Reddy (2007), Sahu and Narayanan (2010), Ghosh (2012), and Ghosh et al. (2014) have used the LMDI I methodology to analyze energy demand patterns of the manufacturing sector of Singapore, Korea, and India. Mehodi and Aalami (2011) and Nasab et al. (2012) have used the LMDI technique for the transport and industrial sectors of developing countries and Iran, respectively. Besides these studies, Shahiduzzaman and Alam (2012) and Cian et al. (2013) have offered a sector-wise energy demand analysis of Australia and 40 major economies, respectively. Our study is expected to contribute to the literature on energy demand by providing useful information about the causal factors of energy consumption in the industrial sector of eastern India. Further, this study is relevant for two additional reasons. Firstly, to become an energy efficient country, India needs to carry out *state*-wise energy demand analysis. There is considerable literature available for India, but

[2]Methods used by Jenne and Cattell, Marlay, Reitler et al., Howarth et al., Park, Sun and Ang et al. are discussed in Ang (2004).

insignificant at the state level. Secondly, while there exists a literature on sector-wise analysis of energy demand, source-wise analysis is lacking.

2.1 Index Decomposition Analysis—The Log Mean Divisia Index I (LMDI I) Model

The following is the LMDI I energy efficiency accounting framework using the IDA approach (Ang et al. 2010).

Let us assume, an economy which is divided into 'n' no of energy consuming sectors, the total energy consumption at period 't' can be expressed as:

$$E_t = \sum_{i=1}^{n} E_{i,t} = \sum_{i=1}^{n} Y_t . \frac{Y_{i,t}}{Y_t} \frac{E_{i,t}}{Y_{i,t}} = \sum_{i=1}^{n} Y_t . S_{i,t} I_{i,t} \tag{1}$$

where

E_t	total energy consumption for all sectors at t
$E_{i,t}$	energy consumption of the ith sector at t
Y_t	total production level for all sectors at t
$Y_{i,t}$	production level of the ith sector at t
$S_{i,\,t} = Y_{i,t}/Y_t$	production share of the ith sector at t
$I_{i,t} = E_{i,t}/Y_{i,t}$	energy intensity of the ith sector at t.

Equation (1) expresses energy consumption in terms of the three basic factors in the IDA method. Taking into consideration these three factors, the change in energy demand over time, for example, from year 0 to year t, can be theoretically decomposed in the following additive manner:

$$\mathrm{E}_t - E_0 = \Delta E_{\mathrm{TOT}} = \Delta E_{\mathrm{OE}} + \Delta E_{\mathrm{SE}} + \Delta E_{\mathrm{IE}} \tag{2}$$

According to LMDI I approach (Ang et al. 2010; Ang 2012), we have

$$\Delta E_{\mathrm{OE}} = \Delta E_{i-\mathrm{OE}}^{0,t} = \sum_{i=1}^{n} w_{i,t} \ln \frac{Y_t}{Y_0} \tag{3}$$

$$\Delta E_{\mathrm{SE}} = \Delta E_{i-\mathrm{SE}}^{0,t} = \sum_{i=1}^{n} w_{i,t} \ln \frac{S_{i,t}}{S_{i,0}} \tag{4}$$

$$\Delta E_{\mathrm{IE}} = \Delta E_{i-\mathrm{IE}}^{0,t} = \sum_{i=1}^{n} w_{i,t} \ln \frac{I_{i,t}}{I_{i,0}} \tag{5}$$

where ΔE_{OE}, ΔE_{SE}, and ΔE_{IE} are the activity effect or output effect (OE), structural effect (SE), and efficiency effect or intensity effect (IE), representing the contribution of (a) the change of sectoral production level, (b) the change of sectoral composition, and (c) the change of sectoral energy intensity to the change of total energy consumption from the base year (0) to the current year (t). The above-mentioned effects are the key influential factors due to which energy consumption of manufacturing sector could be affected. Again, $w_{i,t}$ is the logarithmic weighting scheme, specified in the following, $w_{i,t} = \frac{E_{i,t} - E_{i,0}}{\ln E_{i,t} - \ln E_{i,0}}$, where $L(x, y) = (y - x)/\ln(y/x), x \neq y$.

In general terms, the activity tells us that when the structure and conservation measures of the ith sector remain the same as that of the base year, then change is only due to growth in output of that sector. It can be described by the contribution of a given sector to the overall gross domestic product (GDP). The activity factor is given by Y in Eqs. (1) and (3) captures the activity effect used in LMDI I method. The structural effect talks about change in the consumption of energy due to any change in the structure or composition of the ith sector, and there is no change in the activity effect and intensity effect. In other words, the structural effect is referring to shifts in the mix of products or activities. These shifts can be either intersectoral or intrasectoral. This effect could be captured by the change in product share of the ith sector in the total gross domestic product, and the product share would affect the use of energy consumption of that sector, and Eq. (1) provides us the product share, i.e., $S_i = Y_i/Y$. Equation (4) gives us the formula of the structural effect which is used in LMDI I. The intensity effect shows us the change in the consumption of energy solely due to change in any conservation measures, keeping the other two effects unchanged. In the Eq. (1), the intensity factor is given by I_i. This intensity factor is related to I_i in Eqs. (1) and (5) and captures the intensity effect which is used in the LMDI I method.

To execute the above analysis, we need data on total output of the ith manufacturing unit and fuels consumed in the factory sector by type of fuel in the five states (Bihar, Chhattishgarh, Jharkhand, Odisha, and West Bengal). The required data set has been obtained from various issues of the Annual Survey of Industries (ASI) for the years 2010–11 to 2014–15. Also, there are twenty-four 2-digit industry groups available in the National Industrial Classification[3] (NIC) of 2008. But all twenty-four 2-digit industry groups are not operating in the states being covered in this study. Therefore, data on all twenty-four 2-digit industry is not available for the five states, and that is why, to maintain similarity among the data for the 5 states, we have to exclude seven industry groups which are manufacture of wearing apparel (14), manufacture of leather and related products (15), manufacture of computer, electronic, and optical products (26), manufacture of motor vehicles, trailers, and semi-trailers (29), manufacture of other transport equipment (30), other manufacturing (32), and repair and installation of machinery and equipment (33). The ASI reports provide state-wise data on the consumption of major forms of

[3]NIC is the counterpart of the International Standard Industry Classification (ISIC) for India; the manufacturing sector of India has been classified into various 2-digit, 3-digit, and 4-digit industry groups.

energy, namely coal, electricity, petroleum products, and other fuels, by the manufacturing industry. Values in both rupees and quantity in original units are available for coal and electricity, while the data on petroleum products and other fuels is available in monetary terms. Hence, to maintain consistency in the data set, this study uses the monetary value (in lakhs) of consumption of all three sources of energy. Output data is also represented in monetary units (in Lakh). Hence, for both output and energy consumption, we need to calculate real values based on the corresponding wholesale price indices (WPI) taking 2004–05 as base year to factor out the effect of inflation.

3 Trends in Industrial Output Growth and Energy Consumption in the Five Eastern Indian States

During 2010–11 to 2014–15, Gujarat, Maharashtra, Tamil Nadu, Karnataka, and Uttar Pradesh were the top five states in terms of their contribution (totaling 58%) to total manufacturing output. At the lower end of spectrum are Odisha, Jharkhand, Chhattisgarh, Bihar, and Delhi. West Bengal occupies in the eighth position. In this same period, the CAGR[4] of manufacturing output was on an average of 3% per annum for the five states, whereas for India, it was 5% (11% for the industrially progressive states) per annum. Research on industrial growth has been preoccupied with the issue of growth rates which is necessary to assess the status of the manufacturing industry. Figure 1 shows the upward trend of output of the manufacturing sectors of the five states—the trends are more or less similar.

Among these five states, West Bengal's manufacturing output is significantly higher (at 5% of total production in India); while the levels are lower for the rest (Odisha 2.3%, Chhattisgarh 2.1%, Jharkhand 1.8%, and Bihar 1%). However, West Bengal grew by a CAGR of only 1% per annum during this period, while for Bihar and Chhattisgarh, the CAGR was 7 and 6% per annum, respectively. This period witnessed a downward trend in the growth of manufacturing output in India, whose main reasons are high-interest rates to tackle persistent inflation, a slowdown in investment, and a loss of business confidence (Economic Survey 2014–15). Also, growth rates in manufacturing were uniformly low worldwide because industrialized economies were experiencing slow growth, and emerging economies were finding it difficult to sustain growth as they were facing low demand in the global market as well as in their domestic economies (Economic Survey 2014–15).

As growth in the manufacturing sector is crucial for economic development, its poor performance during 2010–15 is a cause for concern. 'Make in India' is a new policy initiative for promoting the growth of the manufacturing sector. The 'Make

[4]The compound annual growth rate is calculated by taking the nth root of the total percentage growth rate, where n is the number of years in the period being considered. This can be written as CAGR = (End Value/Start Value)$^{(1/\text{No of years})}$−1.

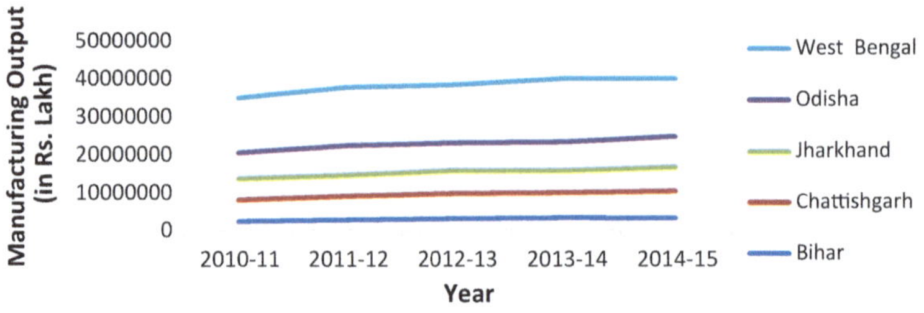

Fig. 1 State-wise production of manufacturing output (in Rs. Lakh). *Source* Author's calculation

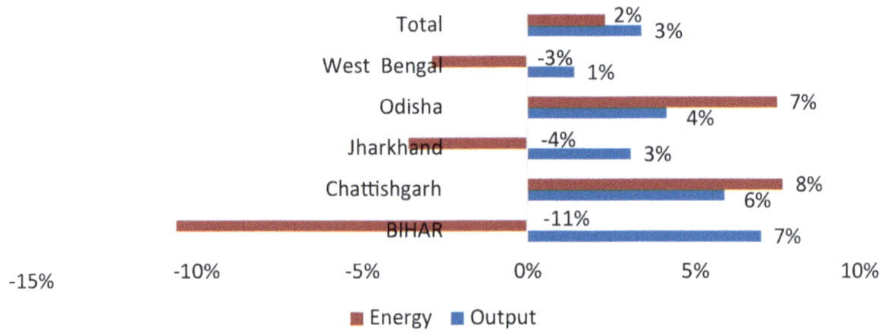

Fig. 2 Compound annual growth rate (CAGR) of production and energy consumption of five states of India, 2010–11 to 2014–15. *Source* Author's calculation

in India' program aims to facilitate investment, foster innovation, enhance skill development, protect intellectual property, and build a state-of-the art manufacturing infrastructure (Make In India 2017). As mentioned earlier, production in the manufacturing industries and energy consumption are interdependent. The CAGR of production of our five states is greater than the CAGR of energy consumption. Due to the low growth of output of the manufacturing sector as a whole, energy consumption was also decreasing (Fig. 2). This interdependence can also be captured by looking at energy intensity, as energy intensity is the ratio of energy consumption to production[5] (Paul and Bhattacharya 2004; EIA 2016). Figure 3 shows the overall trend of energy intensity for our study period.

It rose during 2010–11 and 2011–12, then declined during 2011–12 and 2012–13, and fell further until 2013–14. The decreasing trend was reversed after 2013–14. The trend line shows that there is a negative relationship between time and energy intensity, i.e., over our time period, energy intensity declined. Figure 4 provides industry-wise comparisons of energy intensity[6] while Fig. 5 provides source-wise

[5]According to conventional definition, a country's energy intensity is usually defined as energy consumption per unit of gross domestic product (GDP).

[6]The energy intensity has been calculated based on the data of average fuel consumed and value of output for the study period (2010–11 to 2014–15).

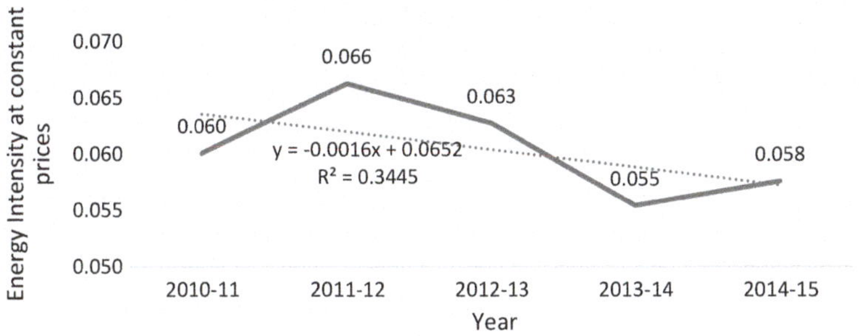

Fig. 3 Energy intensity of manufacturing industries of five states of India, 2010–11 to 2014–15. *Source* Author's calculation

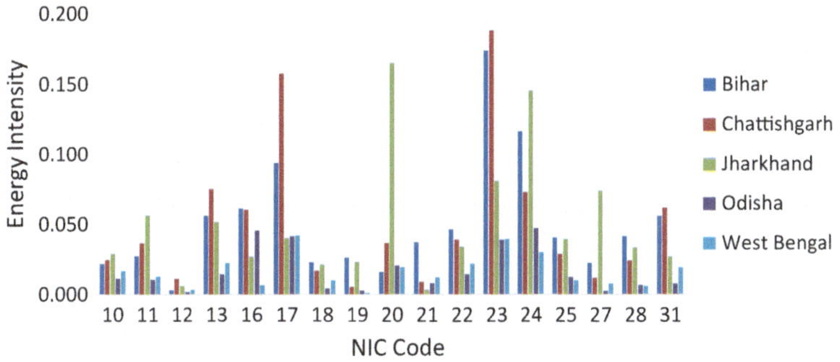

Fig. 4 Industry-wise energy intensity of manufacturing industries of five states of India, 2010–11 to 2014–15. *Source* Author's calculation

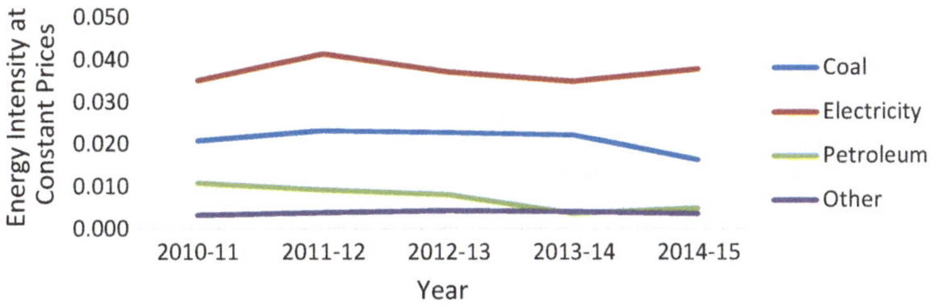

Fig. 5 Source-wise energy intensity of manufacturing industries of five states of India, 2010–11 to 2014–15. *Source* Author's calculation

comparisons of energy intensity. Figure 4 shows that the energy intensity of selected industries of Bihar, Chhattisgarh, and Jharkhand and is higher than for the other two states. These are textiles (13), wood and products of wood and cork except furniture (16), paper and paper products (17), chemicals and chemical industry (20), other nonmetallic mineral products (23), basic metals (24), and furniture (31).

Figure 5 shows the trends of energy intensity of coal, electricity, petroleum, and other fuels. During 2010–11 to 2011–12, the trend has been slightly upwards for coal and electricity, but it went down in the case of electricity and remained stable for coal in the period 2011–12 to 2013–14. After 2013–14, there was a slight increase for electricity and a slight decrease for coal. For petroleum, on the other hand, the trend was downwards all through 2010–11 to 2013–14, after which there was a reversal. For the other fuels, energy intensity remained largely stable for the entire study period.

We have now completed a brief overview of the pattern of energy consumption of the manufacturing industries in the five eastern Indian states which simultaneously indicates the necessity to identify the core factors behind the increase of energy (coal, electricity, petroleum, and other fuel) consumption in these states. We follow with the results of the decomposition analysis and what they imply.

4 Results of the Decomposition Analysis

The decomposition methodology using the log mean Divisia index I (LMDI I) method was applied to the manufacturing sector of the five eastern Indian states for the period 2010–11 to 2014–15, with 2010–11 as the base year. The following is a discussion of the results.

4.1 Coal

Among the conventional sources of energy, coal is the most significant source and it is mostly used in iron and steel production, sponge iron, cement production, textile industry, fertilizer industry, alumina refineries, paper manufacturers, brick industry, and the chemical and pharmaceutical industries (Indian Chamber of Commerce 2012; Qaisar and Ahmad 2014; CSO 2017). According to Energy Statistics (2017), industry-wise estimates of consumption of coal show that during 2014–15, steel and washery industries consumed 56.24 MT[7] of coal followed by cement industries (11.36 MT), paper industries (1.65 MT), and Textile (0.42 MT). During 2010–11 to 2014–15, the CAGR of coal consumption was only 1% per annum for the five states and India's CAGR of coal was approximately 9% per annum. Due to the slow growth rate of industry, coal consumption was falling in this period in our five states. Figure 6 shows the decomposition results of coal consumption of the manufacturing industries of the five states. It shows that intensity effect and activity effect are the factors that contribute most in increasing coal consumption of the five states during the study period.

[7]MT = Million Tons.

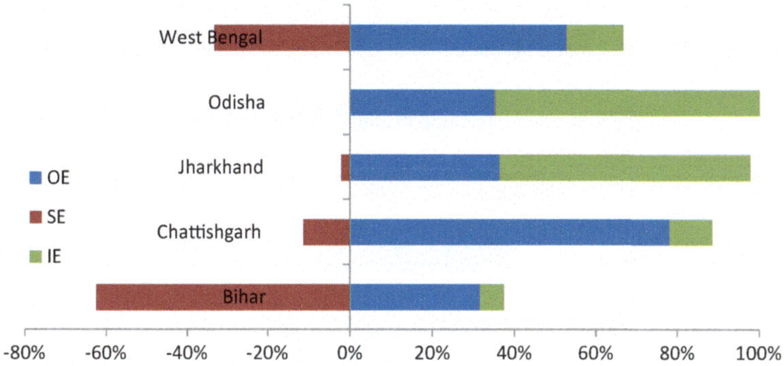

Fig. 6 Results of the decomposition analysis: coal consumption of the manufacturing industries of the five states of India. *Source* Author's calculation

Coal consumption in Jharkhand and Odisha mostly increased due to the intensity effect. The structural effect dominates over the two other effects in Bihar because, as Fig. 6 shows, it is negative. Therefore, total coal consumption is affected not only due to the intensity effect; activity is also a key factor in increasing the consumption of coal.

4.2 *Electricity*

Another crucial form of conventional energy demand is electricity. According to Energy Statistics (2017), estimated electricity consumption increased from 694.39 to 948.52 TWh during 2010–11 to 2014–15 with a CAGR of 8% per annum. Of the total consumption of electricity in 2014–15, the industrial sector accounted for the largest share (44%), followed by domestic (23%), agriculture (18%), and commerce (8%) (CSO 2017). The CAGR of electricity consumption by industries was 6% per annum in the five states during 2010–11 to 2014–15. Figure 7 shows the decomposition results that are necessary to identify the reason for the increase in electricity consumption.

Figure 7 shows that total electricity consumption of the manufacturing industries has increased due to a positive intensity effect during the study period in Bihar, Chhattisgarh, Jharkhand, and West Bengal, but not in Odisha (Fig. 7), and this suggests that the manufacturing industries of the first four states failed to achieve energy efficiency in electricity consumption. Only Odisha consumed electricity in a more efficient manner. It can be also seen from the above figure that the activity effect is another positively contributing factor for all states except Jharkhand for the entire study period.

As this study considered a broad data set of electricity consumption, it is not possible to identify the exact reason for the change in electricity consumption via the change of structure. The structural effect is negative for all states except Odisha.

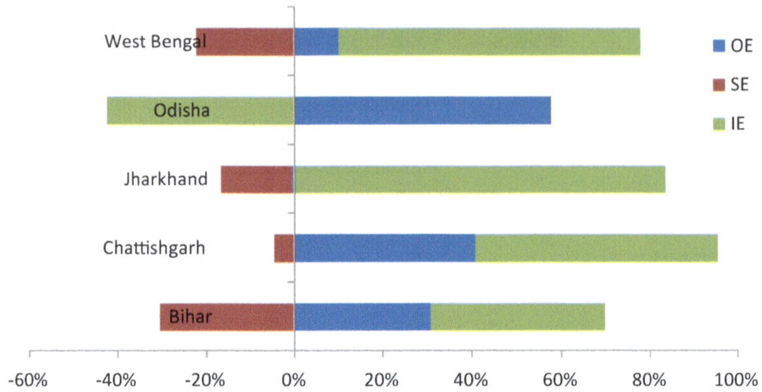

Fig. 7 Results of the decomposition analysis: electricity consumption of the manufacturing industries of the five states of India. *Source* Author's calculation

4.3 *Petroleum*

The petroleum industry is a major part of the Indian manufacturing sector. The main use of petroleum in manufacturing is for direct heat or steam generation. During 2010–11 to 2014–15, keeping pace with the trend in economic growth, the consumption of petroleum products in India has grown with a CAGR of 4% per annum. Figure 8 shows the results of the decomposition analysis; during this period, the intensity effect is negative except in the case of Odisha. Figure 8 suggests that the states are consuming petroleum in an efficient manner as the intensity effect of the states was negative for the entire period. The results show that the activity effect does contribute to the consumption of petroleum in these states, but to a much lesser extent. The structural effect is not a significant contributor in increasing the consumption of petroleum except in the case of Bihar.

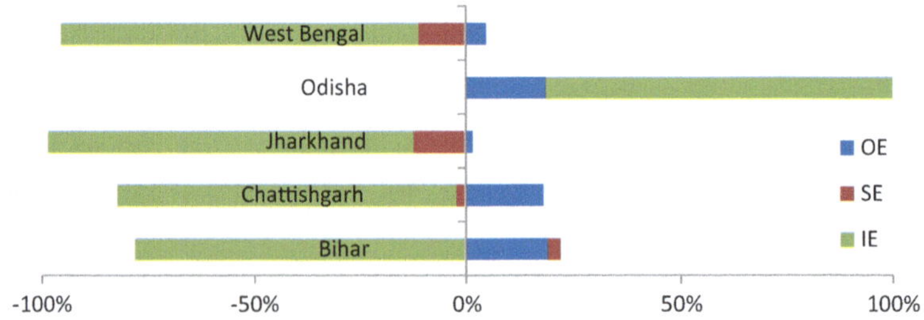

Fig. 8 Results of the decomposition analysis: petroleum consumption of the manufacturing industries of the five states of India. *Source* Author's calculation

4.4 Other Fuels

Besides coal, electricity, and petroleum, the manufacturing industries also depend on other fuels, for which the activity effect is one of the core contributing factors in increasing consumption in all five states (Fig. 9). These other fuels are basically industrial firewood or wood-based charcoal used for industrial purposes (Government of Bihar 2011). Bihar, Chhattisgarh, and Jharkhand consumed only 26% of other fuels during the study period, while 74% was consumed by Odisha and West Bengal (ASI 2010–11 to 2014–15).

The intensity effect is positive for Bihar, Jharkhand, Odisha, and West Bengal—this means that these states used the other fuels in an inefficient manner. Fuel consumption has decreased due to the negative intensity effect in Chhattisgarh, and the positive intensity effect is a minor contributor in Bihar. The structural effect is negative for Bihar, Chhattisgarh, and West Bengal and positive for Jharkhand and Odisha, but it is not a significant contributor in decreasing or increasing the consumption of other fuels in manufacturing. According to a report of the Indian Institute of Forest Management (2017), a few specific companies producing ferro-alloy and rubber in West Bengal as well as Tata Steel (specifically its high carbon ferro chrome plant) in Odisha used charcoal alternatives. This might be a reason for the negative structural effect in West Bengal. Our results also suggest that Odisha needs more sustainable practices to reduce its consumption of firewood or charcoal.

4.5 Summary of the Decomposition Results

This study used the log mean Divisia index (LMDI) to identify the core factor of increasing energy (coal, electricity, petroleum, and others) consumption in the

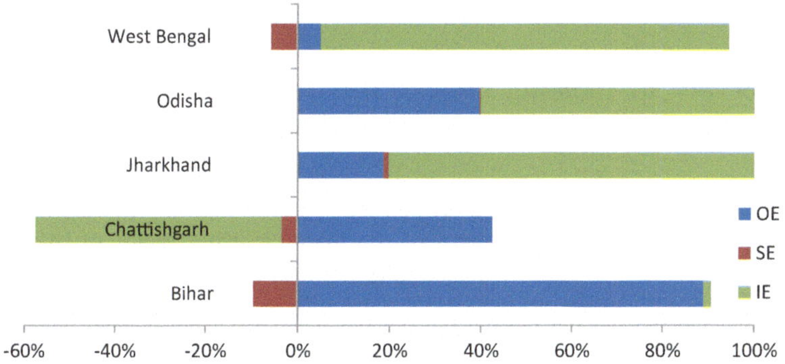

Fig. 9 Results of the decomposition analysis: consumption of other fuels in the manufacturing industries of the five states of India. *Source* Author's calculation

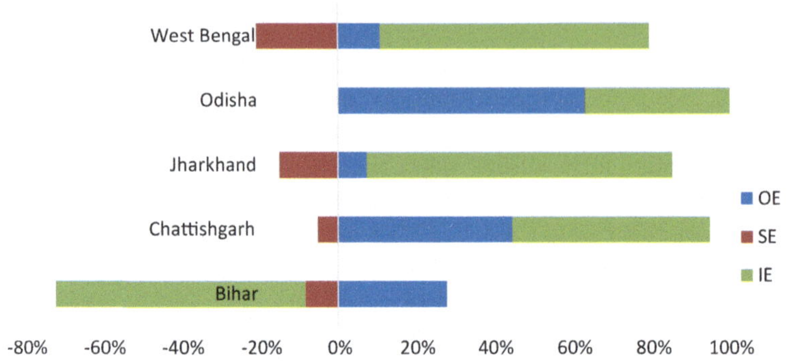

Fig. 10 Results of decomposition analysis: activity, structural and intensity effect of the manufacturing industries of the five states of India, 2010–11 to 2014–15. *Source* Author's calculation

manufacturing industries of the five eastern Indian states (Bihar, Chhattisgarh, Jharkhand, Odisha, and West Bengal), and the total change in energy consumption is decomposed into the activity effect, structural effect, and intensity effect. The results for the manufacturing sector of the five eastern states of India as obtained using data for 2010–11 to 2014–15 (with as base year 2010–11) show that the activity effect is the core contributing factor in increasing energy consumption in all the five states (Fig. 10).

Another contributing factor is the intensity effect, which is positive for Chhattisgarh, Jharkhand, Odisha, and West Bengal. But Bihar used energy in an efficient manner. The structural effect is a contributing factor in increasing energy consumption in Odisha because the effect is positive for the state. The contribution of the structural effect in case of Bihar and Chhattisgarh is not very significant for increasing energy consumption. Hence, the results show that the imposition of different policies related to energy efficiency did not reduce total energy consumption by the manufacturing industries of the five states.

5 Overview of Existing Policies for the Manufacturing Industries of the Five Eastern Indian States

This study identified the key forces behind increasing energy demand and helped us to understand the energy use pattern of the manufacturing sector of five eastern states of India. In this context, we need to review the existing policies of the sector in order to develop improved policies for the future. We know that India faces a dual challenge. As an emerging economy, India requires rapid economic growth, and at the same time, there is a pressing need to address climate change. In response, it has already implemented different policies to address the threats of climate change. Besides the Government of India's programs on climate responsible development goals, several state governments and local bodies have also introduced various incentive schemes to encourage industrial growth and promote environment-friendly options to produce output in an energy efficient manner. Here, we have mentioned some of the relevant policies which have been already implemented in the past few years.

Our concerned states (Bihar, Chhattisgarh, Jharkhand, Odisha, and West Bengal) have also promoted the renewable purchase obligation (RPO) for manufacturing industries; but between 2007 and 2010, Bihar and Odisha failed to achieve it (CII 2014; Pahuja et al. 2014). Then again, capacity building for industrial pollution management has been implemented by the state pollution control boards; this project scales up the cleanup and rehabilitation of polluted sites and facilitates the reduction of environmental and health risks through technical capacity building. With financial and technical assistance from the World Bank, West Bengal has begun implementing this initiative (CII 2014). The project became effective on October 13, 2010 and will last 5 years. It was envisioned to support the development of a policy, institutional, and methodological framework for the establishment of the National Program for Rehabilitation of Polluted Sites (CII 2014). Existing research (Heinrich Boll Foundation 2013; CII 2014; Pahuja et al. 2014) has identified the different programs which have already been put in place to control the consumption of energy in the eastern Indian states.

The Bihar Industrial Incentive Policy has been implemented in 2011. According to this policy, if firms produced energy through non-conventional sources, then 60% of the expenditure on plant and machinery will be subsidized and this facility will be available to existing units. However, there is no ceiling of expenditure has been fixed for availing this incentive. The Industrial Policy of Chhattisgarh (2009) includes mandatory measures for safeguarding the environment such as setting up of effluent treatment plants, hazardous waste management systems, solid waste disposal systems, and recycled water utilization. According to the Jharkhand Industrial Policy (2012), rainwater harvesting, recycling and reuse of waste water shall remain essential for industries. New plants will not be liable to pay 50% of electricity duty for a period of 10 years. Mega projects (with investment in fixed assets in excess of Rs. 100 crore) will be allowed to have captive power plants, to generate power from waste heat recovery. Such units will also enjoy 50% exemption from electricity duty for a period of 5 years. A Comprehensive Project Investment Subsidy (CPIS) to the tune of 20% will be given for investment made in pollution control equipment and environment-friendly alternative power generation equipment. The Odisha Industrial Policy (2015) says that to make the current industrialization process sustainable, maximum emphasis shall be laid on sound environment management practices. The West Bengal Industrial and Investment Policy (2013) incentives for industry are administered by the Industry Department under its benefit schemes. The state has announced a policy on co-generation and generation of electricity from renewable energy sources and has set a target of setting up 2706 MW capacity from the resources by the end of 13th plan. By 2017, 355 MW of power is targeted to be harnessed through co-generation facilities that are intended to be installed primarily in iron and steel, fertilizer, and chemical industries. Moreover, the policy envisages that iron and steel, fertilizer, and chemical industries having 2000 kVA and above as connected load should produce at least 5% of their requirement through captive power plants employing co-generation technology.

In spite of these various policies pursued by the eastern states, the report of the Heinrich Boll Foundation (2013) states that Jharkhand and West Bengal have highly polluted industrial units, while those of Bihar, Chhattisgarh, and Odisha are moderately and marginally polluted.

6 Conclusions

As economic development crucially depends on the manufacturing sector and this sector is also a major consumer of energy, policy for growth combined with the abatement of carbon emissions requires a detailed study of this sector. So, to design proper policies for the manufacturing industry of the eastern states of India, we need to examine the pattern of energy use in manufacturing sector and also to identify the key factors behind the increase in energy consumption. This identification is carried out with the help of the log mean Divisia index decomposition approach. This method decomposes the changes in energy consumption into the activity effect, structural effect, and intensity effect. The overall results show that the level of economic activity is the primary influencing factor, followed by energy intensity, in increasing the consumption of energy of the manufacturing units of the states of eastern India. On the other hand, the consumption of energy has not been significantly influenced by the structure of the manufacturing units.

A variety of industrial policies have been implemented to reduce energy consumption in this sector in the last few years, but the results suggest that the manufacturing sector of eastern India has not produced output in an energy efficient manner and the policy makers should not confine themselves to only increasing energy efficiency. Also, they have paid little attention to structural changes across sectors. As this analysis shows that level of activity is one of the key factors in increasing energy consumption, we can conclude that there is sufficient scope for refining and improving the technologies used in the manufacturing sector.

References

Ang BW (2004) Decomposition analysis for policymaking in energy: which is preferred method? Energy Policy 32(9):1131–1139

Ang BW, Choi KH (1997) Decomposition of aggregate energy and gas emission intensities for industry: a refined Divisia index method. Energy J 18(3):59–73

Ang BW, Zhang FQ (2000) A survey of index decomposition analysis in energy and environmental studies. Energy 25(12):1149–1176

Ang BW, Huang H, Mu AR (2009) Properties and linkages of some index decomposition analysis methods. Energy Policy 37(11):4624–4632

Ang BW, Mu AR, Zhou P (2010) Accounting frameworks for tracking energy efficiency trends. Energy Econ 32(5):1209–1219
Annual Survey of Industries (ASI) (2010–11 to 2014–15) Detailed report (factory sector), vol 1. Central Statistical Organisation (Industrial Statistics Wing), Department of Statistics, India
Bhattacharyya SC (2011) Energy economics concepts, issues, markets and governance. Springer, London. ISBN 9780857292674
Central Statistics Office (CSO) (2015) Energy statistics. Ministry of Statistics and Programme Implementation, New Delhi, Government of India
Central Statistics Office (CSO) (2016) Energy statistics. Ministry of Statistics and Programme Implementation, New Delhi, Government of India
Central Statistics Office (CSO) (2017) Energy statistics. Ministry of Statistics and Programme Implementation, New Delhi, Government of India
Choi KH, Ang BW, Ro K (1995) Decomposition of the energy-intensity index with application for the Korean manufacturing industry. Energy 20(9):835–842
Cian ED, Schymura M, Verdolini E, Voigt S (2013) Energy intensity developments in 40 major economies: structural change or technology improvement? ZEW discussion papers No. 13-052. Centre for European Economic Research, Germany
Confederation of Indian Industry (CII) (2014) The leap towards sustainable power in Eastern India (20 Aug 2014). India. (see https://www.pwc.in/en_IN/in/assets/pdfs/publications/2014/cii-the-leap-towards-sustainable-power-in-eastern-india.pdf). Accessed on 1 Sept 2015
Ghosh D, Dasgupta S, Ghosh A, Ghosh, G (2012) Manufacturing sector in Odisha: issues in energy consumption and sustainabilty transition. Odisha Environment Congress, Centre for Environment and Development, Bhubaneswar, pp 228–239
Ghosh D, Dasgupta S, Ghosh A, Ghosh G (2014) Energy consumption in the manufacturing sector in Odisha: complexities for sustainability transition due to size mix within the sector. J Ind Stat 3(2):199–223
Government of Bihar (2011) Bihar industrial incentive policy 2011. Department of Industries, Government of Bihar, India
Government of Chhattisgarh (2009) Industrial policy 2009–14. Department of Commerce and Industries, Government of Chhattisgarh, India
Government of India (2014–15) Economic survey 2014–15. Ministry of Finance, Government of India
Government of India (2015) India's intended nationally determined contribution is balanced and comprehensive: environment minister. Ministry of Environment and Forests, Government of India. http://pib.nic.in/newsite/PrintRelease.aspx?relid=128403. Accessed on 15 June 2017
Government of Jharkhand (2012) Jharkhand industrial and investment promotion policy 2012. Department of Industries, Mines and Geology, Government of Jharkhand, India
Government of Odisha (2015) Industrial Policy 2015. Industries Department, Government of Odisha, India
Government of West Bengal (2013) Investment and industrial policy of West Bengal 2013. Government of West Bengal, India
Heinrich Boll Foundation (2013) Compendium of state policies from the perspective of climate change mitigation. http://www.indiaenvironmentportal.org.in/content/389304/compendium-of-state-policies-from-the-perspective-of-climate-change-mitigation/. Accessed on 15 Sep 2015
Hoekstra R, Bergh JJ (2003) Comparing structural and index decomposition analysis. Energy Econ 25(1):39–64
IEA (2012) CO_2 emissions from fuel combustion highlights. International Energy Agency, Paris
IEA (2013) CO_2 emissions from fuel combustion highlights. International Energy Agency, Paris
IEA (2014) CO_2 emissions from fuel combustion highlights. International Energy Agency, Paris
IEA (2015a) CO_2 emissions from fuel combustion highlights. International Energy Agency, Paris
IEA (2015b) India energy outlook. International Energy Agency, Paris
IEA (2016) CO_2 emissions from fuel combustion highlights. International Energy Agency, Paris

Indian Chamber of Commerce (2012) The Indian coal sector: challenges and future outlook. https://www.pwc.in/assets/pdfs/industries/power-mining/icc-coal-report.pdf. Accessed on 13 June 2017

Indian Institute of Forest Management (2017) The feasibility of wood based charcoal in industrial processes. Indian Institute of Forest Management, Bhopal, India

IPCC (2013) Climate change 2013, physical science basis-summary for policymakers: working group I-fifth assessment report. Switzerland

Jenne J, Cattell R (1983) Structural change and energy efficiency in industry. Energy Economics 5 (2):114–123

Kaur H (2016) Manufacturing sector of India: challenging the 'Make in India' campaign. Int J Comme Bus Manage (IJCBM) 5(1):9–18

Liu XQ, Ang BW, Ong HL (1992) The application of the Divisia index to the decomposition of changes in industrial energy consumption. Energy J 13(4):161–177

Marlay R (1984) Trends in industrial use of energy. Science 226(4680):1277–1283

Mehodi AE, Aalami R (2011) Gasoline consumption analysis of the vehicles in developing countries by multiplicative LMDI approach: the case of Iran. In: International conference on applied economics, pp 397–402

Nasab EH, Aalami R, Dahr SF, Sadeghzadeh MA (2012) An analysis of energy consumption in transportation and industrial sectors—a multiplicative LMDI approach with application of Iran. Iran Econ Rev 16(22):1–17

Olivier JGJ, Janssens-Maenhout G, Muntean M, Peters JAHW (2016) Trends in global CO_2 emissions. The Hague: PBL Netherlands Environmental Assessment Agency; Ispra: European Commission, Joint Research Centre. Retrieved May, 2017, from http://www.pbl.nl/en/trends-in-global-co2-emissions

Pahuja N, Pandey N, Mandal K, Bandyopadhyay C (2014) GHG mitigation in India: an overview of the current policy landscape. Working paper. World Resources Institute, Washington, D.C. Available online at http://www.wri.org/publication/ghg-mitigation-ind-policy

Park S (1992) Decomposition of industrial energy consumption: an alternative method. Energy Econ 14(4):265–270

Paul S, Bhattachharya R (2004) Causality between energy consumption and economic growth in India: a note on conflicting results. Energy Econ 26(6):977–983

Planning Commission (2011) Low carbon strategies for inclusive growth—an interim report. New Delhi, India

Qaisar SH, Ahmad MA (2014) Production, consumption and future challenges of coal in India. Int J Curr Eng Technol 4(5):3437–3440

Ray BK, Reddy BS (2007) Decomposition of energy consumption and energy intensity in Indian manufacturing industries. Indira Gandhi Institute of Development Research, Mumbai, India

Sahu SK, Narayanan K (2010) Decomposition of industrial energy consumption in Indian manufacturing: the energy intensity approach. Indian Institute of Technology Bombay, India

Shahiduzzaman M, Alam K (2012) Changes in energy efficiency in Australia: a decomposition of aggregate energy intensity using logarithmic mean Divisia approach. Munich Personal Repec Archieve

Websites

EIA (2016) https://www.eia.gov/todayinenergy/detail.cfm?id=10191. Accessed on 5 Feb 2017

Make in India (2017) http://www.makeinindia.com/about. Accessed on 10 June 2017

The Statistics Portal (2016) http://www.statista.com/statistics. Accessed on 15 July 2017

UNFCCC (2015) http://www4.unfccc.int/submissions/INDC/Published%20Documents/India/1/INDIA%20INDC%20TO%20UNFCCC.pdf. Accessed on 16 Dec 2016

The manufacturer's authorised representative in the EU is Springer Nature Customer Service Centre GmbH, Europaplatz 3, 69115 Heidelberg, Germany. If you have any concerns regarding our products, please contact ProductSafety@springernature.com

Printed and bound by CPI Group (UK) Ltd, Croydon, CR0 4YY
30/09/2026
02225446-0008